Applications and Challenges of Geospatial Technology

Pavan Kumar • Meenu Rani
Prem Chandra Pandey • Haroon Sajjad
Bhagwan Singh Chaudhary
Editors

Applications and Challenges of Geospatial Technology

Potential and Future Trends

Springer

Editors
Pavan Kumar
Department of Geography
Jamia Millia Islamia
New Delhi, India

Meenu Rani
Department of Geography
DSB Campus, Kumaun University
Nainital, Uttarakhand, India

Prem Chandra Pandey
Institute of Environment
and Sustainable Development (IESD)
Banaras Hindu University
Varanasi, Uttar Pradesh, India

Haroon Sajjad
Department of Geography
Jamia Millia Islamia
New Delhi, India

Bhagwan Singh Chaudhary
Department of Geophysics
Kurukshetra University
Kurukshetra, Haryana, India

ISBN 978-3-319-99881-7 ISBN 978-3-319-99882-4 (eBook)
https://doi.org/10.1007/978-3-319-99882-4

Library of Congress Control Number: 2018962738

This Springer imprint is published by the registered company Springer Nature Switzerland AG
The registered company address is: Gewerbestrasse 11, 6330 Cham, Switzerland

Contents

Part I
General

Chapter 1
Introduction to Space Technology Challenges: Potential and Future Prospects

B. S. Chaudhary, Haroon Sajjad, Meenu Rani, P. C. Pandey, and Pavan Kumar

Abstract This book affords an outline of the future overview of the current position and short-term insights into the space technology and the issues in the fast-mounting geospatial technology. A prosperous marker in the space journey from the traditional to advance remote sensing technology varying in space has been portrayed within objectives and outcomes. The usefulness of spectral bands with dissimilar spectral signatures provides vast data acquisition for application and services. Urbanization, dynamic nature of agriculture, land use planning, ocean exploration, vegetation resource management, and other ecosystems are being effectively monitored by the satellite services from the space and have many future prospects. Space technology assumes greater significance for monitoring natural and human resources and analyzing judicious utilization of resources. The technology provides standardized solutions for assessing potential and planning process in different geographical regions. Thus, space technology with its different services like geographical information system (GIS) and global positioning system (GPS) can effectively be utilized for timely analysis and future planning of resources on the planet Earth. The book is divided into 5 sections spreading over 16 chapters. The first section discusses the usefulness of geospatial technology in various fields. Chapters 2, 3, 4 and 5 of Part II are devoted to water resource and its various aspects. Natural hazard risk was assessed through various models and presented in Chaps. 6, 7, 8, and 9 of Part III. Part IV deals with progress and perspective scenario of urban growth models and

B. S. Chaudhary
Department of Geophysics, Kurukshetra University, Kurukshetra, Haryana, India
e-mail: bschaudhary@kuk.ac.in

H. Sajjad · P. Kumar (✉)
Department of Geography, Jamia Millia Islamia, New Delhi, India

M. Rani
Department of Geography, DSB Campus, Kumaun University, Nainital, Uttarakhand, India

P. C. Pandey
Institute of Environment and Sustainable Development (IESD), Banaras Hindu University, Varanasi, Uttar Pradesh, India

P. Kumar et al. (eds.), *Applications and Challenges of Geospatial Technology*,
https://doi.org/10.1007/978-3-319-99882-4_1

covers Chaps. 10, 11, 12, and 13. Future challenges and prospects of geospatial technology have been examined in Chaps. 14, 15, and 16 of Part V.

Keywords Space technology · GIS · GPS · Potential and Future Prospects

Development of geospatial techniques has given a way for timely and accurate analysis of phenomenon of the surface of the earth. Use of varied spectral bands is widening the horizon of information. Intensive use of software system and sensor technology has proved effective for monitoring and assessing water resources, urbanization, ocean exploration, forest resource management, and agriculture practices. Geospatial technology assumes greater significance for examining potential and further exploration of natural resources. The technology is widely being used in multi- and interdisciplinary sciences for effective decision-making processes. Increasing population, growing demand for resources, and improper utilization of resources are major challenges in the era of climate change. Therefore, recent technology is essential for effective planning, management, and sustainable development of resources at various geographical scales. This technology can help not only in identifying the priority areas but also provide possible solutions. Integrated approach involving remote sensing (RS), geographical information system (GIS), and global positioning system (GPS) offers a range of possibilities for resource conservation, efficient utilization, and management.

Database is pertinent for any analysis and decision-making process. Geospatial database is employed for identifying, estimating, and analyzing varied phenomenon in spatiotemporal context. Accuracy and precision have improved considerably with the advancement of spatial and spectral resolutions. Development of the Defense Meteorological Satellite Program Operational Linescan Program (DMSP OLS) night-time light (NTL) data has enabled researchers to analyze gross domestic product (GDP), electricity consumption, population density, and land utilization index (LUI). Nowadays, applicability of geospatial techniques has become ubiquitous in terms of environmental impact assessment, resource management, and policy implementation. Digital elevation model (DEM), digital terrain model (DTM), and light detection and ranging (LIDAR) are providing edge for multidimensional visualization of spatial data. Remote sensing, GIS, and GPS are widely being used in land resource management; soil, water, and forest management; monitoring biodiversity; earthquake assessment; chlorophyll-a concentration; suspended sediment concentration; flood inundation; etc.

An attempt has been made by contributors to examine the effectiveness of geospatial techniques in various case studies. The present volume incorporates the papers on application on geospatial technology and instructive future prospects. The research papers have been grouped under five themes: General, Understanding of Sustainable Water Resource Management, Global Modelling of Natural Hazard Risk Assessment, Progress and Perspective Scenario of Urban Growth Models, and Future Challenges of Natural Resource Management and Their Potential over Space and Time.

Part I deals with usefulness of geospatial techniques in various domains. It discusses use of optical, microwave, as well as DMSP-OLS night-time light data. It covers wider applicability of geospatial technology in resource conservation and management, land use planning, forest resource management, biodiversity conservation, urbanization, ocean exploration, earthquake, and hazard risk assessment and agriculture.

Part II deals with Understanding of Sustainable Water Resource Management. It comprises of four chapters focusing on water resources and integrated watershed management. In Chap. 2, Rani et al. have made an attempt to examine the "Recharge Potential Mapping in Complex Hydrological System of Kosi Basin in Mid-Himalayan Region." In this study, a GIS-based weighted sum analysis approach was utilized to identify recharge potential sites in Upper Kosi basin of Indian Himalayan region GIS facilitated to provide precise and quick information on suitable recharge sites for rejuvenation of springs and hydrological sustainability of watershed. In Chap. 3, Vijendra Kumar Pandey has discussed about "Application of Geoinformatics for the Integrated Watershed Management and Development Planning, Bal Ganga Basin, Tehri Garhwal (Uttarakhand)." In this study, geoinformatics has been used as a tool for the resource prioritization, gap analysis, and development planning of the Bal Ganga watershed. In Chap. 4, Matin et al. addressed "A Semi-Analytical Approach to understand Remote Sensing based Backscattering Characteristics for Kerala Coast using In-Situ Observation." This paper has utilized a Quasi-Analytical Algorithm (QAA)-based distribution and variability of particulate backscattering coefficient (b_{bp}). In Chap. 5, Kaliraj et al. have attempted "Phytoplankton (Chl-A) Biomass Seasonal Variability in the Gulf of Mannar, South India: A Remote Sensing Perspective" through band combination analysis. Study entails remote sensing applications for assessing phytoplankton (chl-a) biomass variability and its major influencing factors (sea surface temperature, salinity, waves, and currents) in the Gulf of Mannar (GoM), southeast coast of India.

Part III deals with Global Modelling of Natural Hazard Risk Assessment. The section includes four chapters. In Chap. 6, Rajesh Kumar et al. have discussed "Flood Inundation and Hazard Mapping of 2017 Floods in the Rapti River Basin using Sentinel-1A Synthetic Aperture Radar Image." They analyzed the propagation of flood peaks and affected areas of Rapti River Basin using water level data and SAR images. In Chap. 7, Salaj et al. assessed "Application of Aster Remote Sensing for Lithological Mapping in the Udaipur District of Rajasthan, India." Visible-near infrared (VNIR) and short-wave infrared (SWIR) bands have been analyzed to discriminate lithology features in metasedimentary terrains of Aravalli Supergroup in Udaipur area of Rajasthan, India. In Chap. 8, Chaudhary et al. have made an attempt to build an "Interactive Approach for Earthquake Scenario Development and Hazards Resource Estimation." Remote sensing, GIS model builder, and syntax were used for analyzing earthquake scenarios development, planning, management, and resource estimation. In Chap. 9, Rani et al. in their paper "A Sediment Dynamic Modelling of Landsat OLI Image for Suspended Sediment Drift along the Southwest Coast of India" analyzed the spatiotemporal movement of suspended sediments in the shallow along the southwest coast of Thiruvananthapuram district, Kerala state in

India. This study revealed the effectiveness of geospatial technology to estimate sediment concentration and transportation in the shallow coastal water.

Part IV deals with Progress and Perspective Scenario of Urban Growth Models. It contains four chapters focusing on urban spatial patterns, urban expansion using optical, microwave, and NTL data. In Chap. 10, Elvidge et al. made an attempt to assess the "Inter-Calibration and Urban Light Index of DMSP-OLS Night-Time Data for Evaluating the Urbanization Process in Australia Capital Territory." This article employed scientific appraisal of the capabilities of NTL data set to identify and delineate the urban morphology, urban augmentation, and growth of suburban zones over the years in the Australian Capital Territory, using urban light index. In Chap. 11, Pandey et al. proposed statistical regression modeling-based framework to analyze the spatial patterns of urban growth with a focus on geographic determinants in their paper entitled "Modelling Spatial Patterns of Urban Growth in Pune Metropolitan Region, India." In Chap. 12 Rahaman et al. made an attempt to assess the "Analyzing Urban Sprawl and Spatial Expansion of Kolkata Urban Agglomeration using Geospatial Approach." The study discussed the spatiotemporal urban expansion of different municipalities and Municipal Corporation of Kolkata urban agglomeration of West Bengal, India, during 1990–2015. In Chap. 13, Suman Sinha has discussed about "Automated Extraction of Urban Impervious Area from Spectral-Based Digital Image Processing Techniques." Several spectral indices as vegetation index (soil-adjusted vegetation index (SAVI), normalized difference vegetation index (NDVI)), water index (modified normalized water index (MNDWI)), and urban indices (normalized difference built-up index (NDBI), built-up index (BUI), and index-based built-up index (IBI)) were implemented in the study.

Part V deals with Future Challenges of Natural Resource Management and Their Potential over Space and Time. This section comprises four chapters covering detection of waterlogged rice, forest structural diversity, and future challenges of remote sensing technology. In Chap. 14, Bambang H. Trisasongko has focused on "Hybrid Polarimetric Synthetic Aperture Radar for the Detection of Waterlogged Rice Fields." This chapter discussed the implementation of popular classification methods such as Random Forests (RFs) and Support Vector Machines (SVMs) and their variants to retrieve waterlogged condition of rice fields. In Chap. 15, Kumar et al. focused on the "Estimation of Forest Structural Diversity of Sariska Tiger Reserve, Rajasthan, India." A variety of different diversity indices including Margalef index (SR), Simpson's diversity (D), Shannon-Wiener index (H′), and Pielou's (J) have been used for community structure. In chapter 16: Sajjad et al. discussed the "Future Challenges and Perspectives of Remote Sensing Technology". In their paper the authors have enunciated the challenges of new technology and how these challenges can be accomplished.

Part II
Understanding of Sustainable Water Resource Management

Chapter 2
Recharge Potential Mapping in Complex Hydrological System of Kosi Basin in the Mid-Himalayan Region

Meenu Rani, Himanshu Joshi, Kireet Kumar, and Ashutosh Tiwari

Abstract Increasing water use and climatic variability threaten thousands of springs and spring-fed watersheds in the mid-Himalayan region. The decline in spring discharge resulted in shrinking cropland, out migration and is adversely affecting the economy of the region. Shallow aquifer and short retention time emphasize on need of disposition of site suitable artificial measures to recharge groundwater. Complex geological and tectonic formations and lithological and chronological variations on one hand and impact of undulated terrain and land use pattern on the other put obstruction in finding suitable recharge sites in Himalaya. In this study, a GIS-based weighted sum analysis approach was used to identify suitable sites for artificial recharge of groundwater in Upper Kosi basin of Indian Himalayan region. The tools of GIS facilitate study relief and structural aspect of basin, quantify the influence of one factor on the other and provide precise and quick information on suitable recharge sites for rejuvenation of springs and hydrological sustainability of watershed. The results indicated that 19.6% area lies under good to excellent while 46.9% area having fait to poor potential of groundwater recharge. Area under good to excellent recharge potential can be further considered for implementation of site suitable groundwater augmentation measures for sustained specific yield of an aquifer.

Keywords Spring rejuvenation · Recharges potential site · Mid-Himalayan region · Groundwater augmentation

---

M. Rani
Department of Geography, DSB Campus, Kumaun University, Nainital, Uttarakhand, India

H. Joshi (✉) · K. Kumar · A. Tiwari
GB Pant National Institute of Himalayan Environment and Sustainable Development (GBPNIHESD), Almora, Uttarakhand, India
e-mail: kireet@gbpihed.nic.in; ashutosh.gbpnihesd@gov.in

P. Kumar et al. (eds.), *Applications and Challenges of Geospatial Technology*,
https://doi.org/10.1007/978-3-319-99882-4_2

2.1 Introduction

Surface water and groundwater are complementary in the hydrological system and interact through various physiographic and diversified climatic landscapes. The dissimilarity of topographical characteristics with regard to soil, lithology, geomorphology, stratigraphy and variation in precipitation over time and space led to an unequal allocation of groundwater resource in different regions of the country (Jaiswal et al. 2003). Kumaun Himalaya, with respect to global climate change and rainfall variability, is among the world's most vulnerable regions in terms of water stress, specifically in the spring-fed river regime (Mishra 2014). The economy, demography, biodiversity, ecology and landscape of Kumaun are controlled by five watersheds of the region namely Ramganga, Kosi, Gaula, Saryu and Gagas. Kosi watershed which is a spring fed system, due to its dependence on precipitation and spring resurgence, is under critical water stress. Water availability is a function of runoff, groundwater flow and net storage. Drying up of perennial streams, evidence of died springs and diminishing discharge and rapid declining of summer flows in Kosi River are threatening not only human sustainability but also the biodiversity of this region. The major reasons suggested by researcher are rainfall variability (Negi and Joshi 2004) and complex hydrogeological behaviour of springs. Spring water in Kumaun is stored in the form of unconfined aquifers in most of the cases where the water comes out under the action of gravity, and therefore, rainfall variability and pattern affect spring water discharge. Secondly, the increasing water demand with increasing population and decreasing water discharge of spring widens gaps in water demand and supply. Rawat (2014) indicated that the perennial streams in Kosi watershed are disappearing inch by inch every year, and the total length of perennial streams has decreased from 225.89 km (a situation across 40 years back) to 41.57 km in 2014. Because of transformation process of perennial to seasonal streams in Kosi watershed, discharge of the Kosi River in summers has decreased very fast. The minimum summer discharge (i.e. the base flow) of the Kosi River was recorded 790 l/s in 1992 which drastically declined up to 80 l/s in 2013. Another research conducted by G.B. Pant National Institute of Himalayan Environment & Sustainable Development (GBPNIHESD) indicated that demand for water in Kosi watershed has increased from 8836 Cu m/day (2001) to 10,910 cu.m/day (2011) for human and 6110 cu.m/day (2007) to 7393 cu.m/day (2014) for livestock and predicted rise in water demand from 45 to 85% in the next 18 years for different socio-economic scenarios. With climatic changes and declining precipitation, drying springs are unable to feed future water demand of increasing population. Peculiar hydrological characteristics of a watershed, geology, slope, aspect and climate (Jaturon et al. 2014; Kumar and Shankar 2014) and is reflected in terms of land use practice, occupational structure and social, cultural, floral and faunal biodiversity. Therefore, the study of a watershed needs to be conducted in systematic approach considering one and each component affecting or affected by the different parameters acting in watershed.

Researches on discharge of Kosi River indicated that future of Kosi River is depressing. The diminishing discharge of springs in Himalaya continuously reported by various researchers (Valdiya and Bartarya 1989, 1991; Negi and Joshi 2004; Joshi and Kothyari 2003) indicated that these traditional sources of water have become unsustainable to fulfil future water demand. In view of perturbed hydraulic situations and increasing water demand, the need of the hour is to formulate mechanical and biological treatment for rejuvenation of spring-fed rivers. Geomorphologic features combined with structures such as joints/fractures and lithology controls not only the flow and occurrence but also quality of groundwater. Generally, the conventional methods of investigation like field-based hydrogeological and geophysical resistivity survey are costly (Singh and Prakash 2003) and do not always consider the varied factors that control the groundwater movement and occurrence in aquifer (Oh et al. 2011). Results are therefore not as consistent as they may perhaps be in case of complex terrain using these traditional techniques of groundwater exploration (Murthy 2000). In view of above constraints of conventional techniques, groundwater potential investigation required a cumulative approach which can count each factor responsible for peculiar hydrological characteristics of a watershed. Recent advancement in techniques of remote sensing has proved indispensable for environmental monitoring, geographical and geomorphologic mapping, climatic condition, hazard mapping, resource estimation and management, urban planning and many more applications (Chowdhury et al. 2009). Hence, in search for groundwater potential mapping, it offers the current spatial character of general information on landforms, geology, soils, LULC, drainage and slope very quickly and reliably, even with less expenditure and labour than traditional techniques (Gumma and Pavelic 2013). Various researches have been carried out to delineate groundwater recharge potential throughout the world by utilizing remote sensing and GIS-based methods. Researchers established that GIS provide spatial autocorrelations between governing factors in complex hydrogeological system by incorporating spatial data with database of water resources and present more realistic and extensive view of complete watershed (Fortes et al. 2005; Chenini et al. 2010). Various studies have been conducted in various regions of the globe in order to identify and delineate groundwater recharge potential zones using advanced remote sensing and GIS-based techniques (Krishnamurthy et al. 2000; Shaban et al. 2006; Solomon and Quiel 2006; Tweed et al. 2007; Riad et al. 2011). During the past few decades, researchers found multi-criteria decision analysis (MCDA) as an effective method that provides a framework for water resource management and planning (Pietersen 2006; Jha et al. 2010).

Widening gap between increasing water demand due to increasing population and decreasing supply due to diminishing discharge of springs and streams in Kosi basin needs implementation of an appropriate water management system that has capacity to cope with the situation. Groundwater recharge is the only remaining solution with planners and resource managers to preserve existing water resources and restraint depleting groundwater levels. Identification of site suitable groundwater recharging structures is a prerequisite in this effort. Integrating thematic layers of controlling factors such as LULC, lithology, geomorphology, structure, slope, drainage density, and soil with expert knowledge and knowledge-driven factor analysis provides a way to look into complexities that arises due to different factors and assessing the overall recharge capacity. In view of depleting water sources and to augment groundwater, the present study is an effort to delineate groundwater recharge potentiality in the complex hydrogeological mountainous watershed for identification of sites for groundwater augmentation as the necessity to revitalize dying spring and streams of Kosi basin.

2.2 Materials and Method

2.2.1 Characteristics of Study Area

The Upper Kosi watershed lies in Kumaun region of Kumaun Himalaya; fed by numerous tiny springs and rainfall, river Kosi is said to be the backbone of economy, basis of biodiversity and habitat sustainability. Hydrology of Kosi watershed is controlled by different landscapes (tectonically and fluvial), soil and geology and modified by various landforms (dissected valley, terraces, ridge), microclimate and demography (Fig. 2.1).

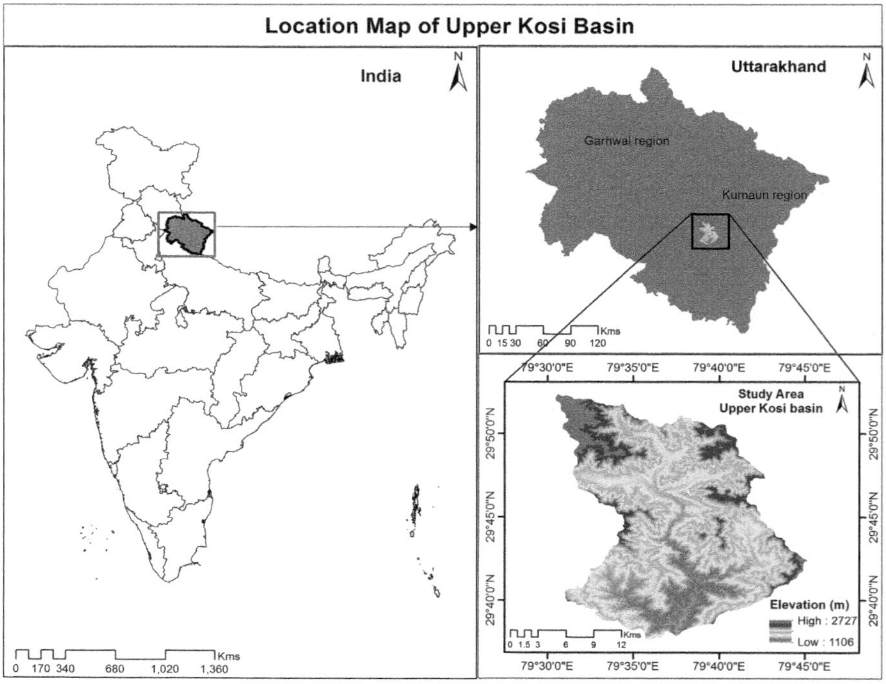

Fig. 2.1 Location of study area

The Kosi River originates from its northernmost point at Pinath (near Kausani, Almora district), which flows downward and joins Ramganga in Ramnagar as its major tributary. Geographically, the catchment of Upper Kosi River (the northernmost sub-watershed of Kosi River basin) has its spatial extent between 29° 37′ 30″ N to 29° 52′ 20″ N and 79° 31′ 00″ E to 79° 45′ 00″ E which covers an area of about 462 sq.km. The absolute relief of the catchment ranges between 1106 m and 2758 m above mean sea level.

The climate of the watershed is temperate with summer, rainy and winter as three distinct seasons. As climate is regarded a controlling factor for land use, complexities of the landscape promote variations in microclimatic phenomena and influence the soil and vegetation. The watershed represents diverse agricultural land (irrigated to rain-fed, less fertile terraces to highly productive valley, food grains (wheat, paddy), vegetables, etc.) but highly dependent upon water supply from Kosi River. The wide and open valley and bench terraces containing thick layer of alluvium and suitable climate condition were prime factors that encouraged the settlements of human population in the region. The ultimate storehouse of water in Kumaun region is the mountain groundwater in the form of spring water (locally known as 'naula', 'dhara' and 'gadhera') at up/middle/down slopes. Most of the people residing in rural (64,202, Census 2011) places heavily depend on these resources for drinking and all other uses. The undulating topography, diverse microclimatic conditions, etc. also pose some difficulty in the efficient distribution of natural water resources. Springs in the Kosi watershed are now under stress due to climatic variation, increasing population/water demand and poor management of water resources. Water stress and drying spring are raising question on the sustainability of Kosi watershed. Water conservation and proper management are the only solution left for hydrological sustainability of the watershed.

2.2.2 Data Collection and Preprocessing

Watershed boundary was delineated from integrated use of Survey of India (SOI) topographical map and CartoSat-2 Digital Elevation Model (DEM) acquired from NRSC-Bhuvan. Slope of the area was generated with contour interval of 10 m using DEM. SOI maps were first scanned and converted into digital format and geo-referenced (projection-UTM, spheroid and datum-WGS 84, Zone 44 North). Geo-rectified satellite data of IRS Resourcesat-2 (LISS-IV sensor) was utilized for preparing land use/land cover map (1:10000 scales) after performing radiometric enhancement for better analysis and identification of features. For assimilating information on soil and geology, soil map was procured from National Bureau of Soil Survey and Land Utilisation Planning (NBSS&LUP), Nagpur, and geology map procured from Geological Survey of India (GSI) on 1:50,000 scale. The secondary information acquired on lineaments, geomorphology and lithology was collected and co-registered using high-resolution LISS-IV sensor (5 m). All these maps were rectified using ERDAS Imagine 2013 and put into geo-database for further processing in ArcGIS-10.3 software.

2.2.3 Multi-criteria Decision and Weightage Sum Analysis

The major challenge faced by hydrogeologist is the quantification of controlling factors and proportional influence of one factor for controlling GW recharge and discharge over the others. Multi-criteria decision analysis (MCDA) evaluates multiple conflicting criteria in decision-making where conflicting criteria are typical in valuing options. GIS-based multi-criteria decision analysis provides good functionality for mapping potential of groundwater revival (Chenini et al. 2010). In case of complex hydrogeology and formations, multiple factors influence the recharge potential of a region but in different proportion. In multi-criteria evaluation technique, experts have liberty and judgment on relative weights of controlling factors for assessment of recharge potential (Kaliraj et al. 2014). The judgment lies in the assigning appropriate weightage to these factors. Principal geomorphologic and hydrogeological controlling factors on groundwater flow systems in groundwater–surface water interactions have been identified, and weightage has been assigned based on their relative impact (Magesh et al. 2012). Integration of these controlling factors and their potential weights has been calculated using weighted sum analysis method in ArcGIS.

2.3 Results and Discussion

2.3.1 Relief Aspect and Recharge Potentiality of Basin

Upper Kosi River is a seventh-order stream, elevation between 1150 m and 2700 m asl, following dendritic to sub-dendritic pattern at major scale as major tributaries join main streams at angle less than 90°. This type of pattern develops in a region underlain by homogeneous material. It indicated that the subsurface geology has a similar resistance to weathering throughout the catchment, and there is no apparent control over the direction of river. But at minor scale, the first- and second-order streams join higher-order stream at sharp angle at many places showing trellis pattern. Trellis pattern at local level indicated the development of folded topography. Rectangular pattern at some place indicated the impact of joint and fractures at local scale. Drainage density is dominant factor for identification of potential recharge sites. Drainage density is termed as the closeness of spacing the drainage channels and computed by dividing the total length of the stream by total basin area (Singh et al. 2014).The occurrence of a natural drainage system is an indirect sign of high porosity and permeability due to its direct relationship with surface runoff (Krishnamurthy et al. 2000). A region with low drainage density causes more

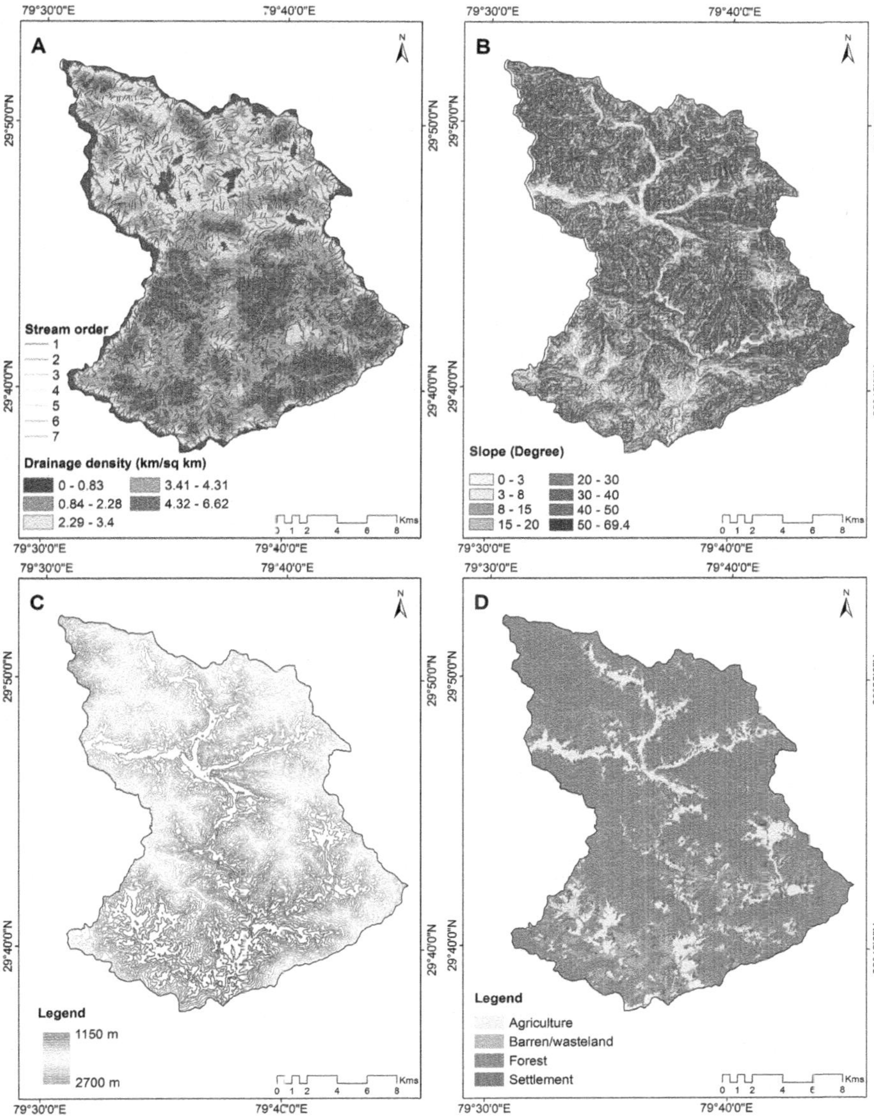

Fig. 2.2 Relief aspects of basin controlling hydrological behaviours of the region. (**a**) Stream order and density, (**b**) slope angel, (**c**) contours and (**d**) land use and land cover

infiltration and lowered runoff. It means that areas having high drainage density have lower potential of groundwater recharge and not suitable for implementation of groundwater augmentation measures (Dinesh Kumar et al. 2007). Drainage density of the basin lies between 0 and 6.6 km/km^2 indicating various degrees for recharging groundwater (Fig. 2.2a).

Degree of slope also influences the surface water infiltration directly. Steep slope enhances velocity of surface runoff, therefore reducing percolation (i.e. infiltration is inversely proportional to the slope) thus adversely affecting the process of groundwater recharge (Adiat et al. 2012). Steep slopes have less potential of recharging as it allows water to flow downwards providing inadequate time to permeate. Alternatively, flat terrain escalates the process of groundwater recharge by storing rainwater and by providing restrained evaporation environment. Out of total catchment area of watershed, 3.6% is under <3° and considered as flat surface with comparatively high recharging capacity, but at the same time, 90% of this area is under river channel which is again not good for recharging groundwater due to saturated soil and thus low infiltration capacity. Therefore, this area has been masked while preparing potential recharge zone map. 52.6% area is having >20° slope and considered higher runoff zone with lower groundwater recharge (Fig. 2.2b, c). Dense forest cover obstructs reduces the velocity of surface runoff thus increases infiltration, while build-up (settlement, roads) allows flow of water due to higher degree of relative imperviousness therfore contributes to very low infiltration. In Upper Kosi watershed, 64% area is covered under forest and provides fair amount of infiltration (Fig. 2.2d). But at the same time, 40.8% (153.4 km^2) area lies under reserved forest where intervention is not possible without collaboration with forest department. Relief aspects of basin such as slope, drainage, stream order and density affect the recharge potential by controlling runoff and, therefore, need to be studied in identifying ideal site for groundwater augmentation.

2.3.2 Structural Aspect and Recharge Potentiality of Basin

Soil permeability coefficient among the leading factors in recharging groundwater (Eid et al. 2006). Major soils in the watershed are 'Typic Udorthents associated with Dystric Eutrochrepts' which are a moderately deep, coarse loamy soil with moderate erosion tendency (61.4%). Further, Dystric Eutrochrepts associated with Typic Udorthents are also found at upslope and hilltop which are characterized as deep, fine loamy soils, with slight erosion (18.4%), and Typic Udorthents as moderately deep, coarse loamy soil (20.2%) found at high slopes which is moderately eroded soil (Fig. 2.3a). Ayazi et al. (2010) stated that lithology influences the porosity and permeability of aquifer rocks. Basin lithology is dominant by quartzite of *Berinag formation*, schist and gneiss of *Saryu_Gumalikhet_Munsiyari Formation*, slate of *Ratgaura* formation, dolomite of *Deoban (Gangolihat) Formation* and gneiss of Augen Gneiss (after Valdiya 1978, Fig. 2.3b). Two elongated tectonic belts of sedimentary/meta-sedimentary rocks are separated by ESE-WNW trending crystalline zone, namely, Almora-Dudhatoli. The outer sedimentary belt, i.e. Krol Belt, lied to the southern side of the crystalline mass, while the inner sedimentary belt, i.e. Deoban-Tejam zone, lied to the north (Gansser 1964; Valdiya 1978). The impact of variable rock group reflects spatial variability in recharging capacity in the study area. Lineaments reflect a general surface appearance of fractures lied underground

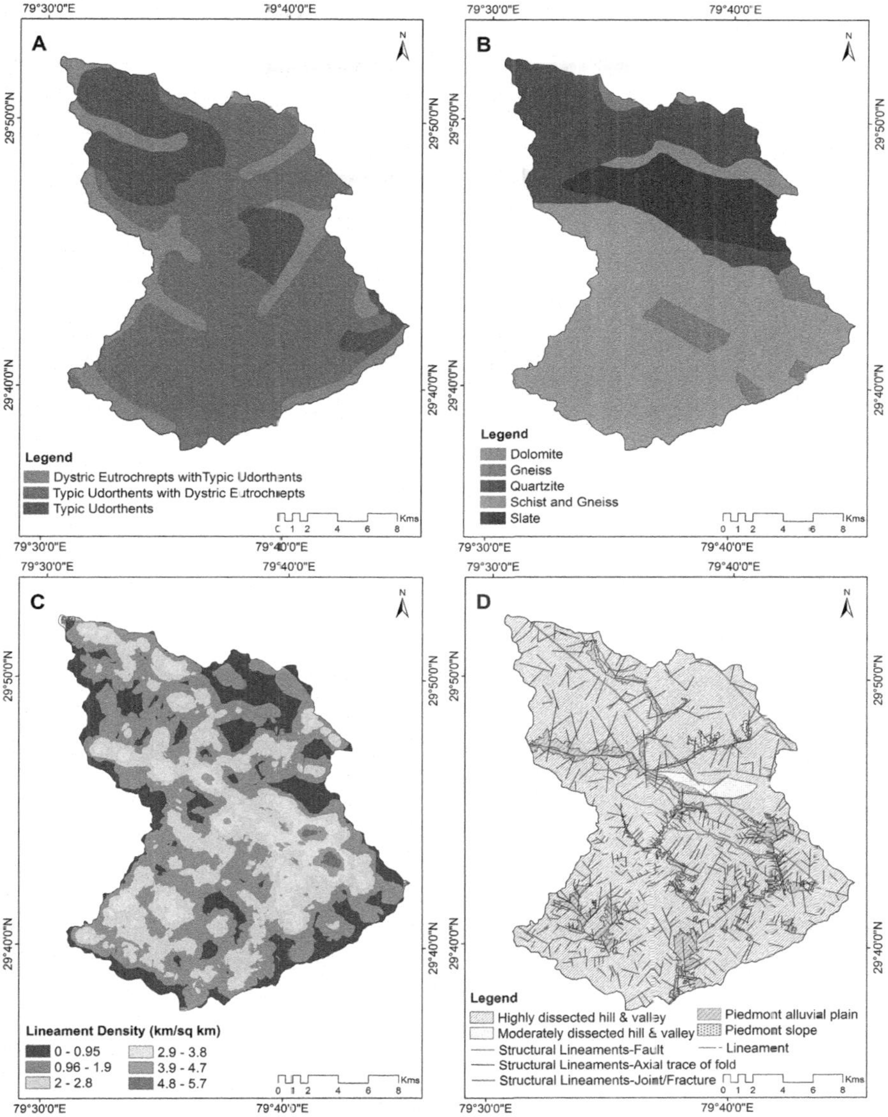

Fig. 2.3 Structural aspects of basin controlling hydrological behaviours of the region. (**a**) Soils, (**b**) lithology, (**c**) lineament density and (**d**) geomorphology and lineaments type

(Pradhan 2009). Lineaments in the form of faults and fracture/joints represent geological discontinuities and act as pathway for groundwater percolation and storage. The contribution of lineaments on runoff, infiltration and groundwater recharge are well documented by Subba Rao et al. (2001) and Chenini et al.

(2010). Quantification of lineaments was done either on the basis of density (Dinesh Kumar et al. 2007) or presence and absence of lineament (Babus and Sashikumar 2010). The Upper Kosi basin is characterized by structural lineament (Fault, fracture/joint) and fluvial lineament (drainage parallel) as indicated in Fig. 2.3c. For quantification of the relative influence of lineament, the density of lineament (km/km^2) was computed.

Evolution of landforms provides helpful information to understand the incidence of permeable and porous zones; and therefore, for recharging groundwater the study of geomorphology of a region is considered as an essential component. Four major geomorphic units have been identified in the study area, viz. highly dissected hills and valley, moderately dissected hills and valley, piedmont alluvial plain and piedmont slope (Fig. 2.3d). Major geomorphology of the area is highly dissected hills and valley (89%) which has moderated capacity of infiltration. Krishnamurthy et al. (2000) suggested that less compact zone with higher degree of fracturing and weathering assist runoff infiltration and therefore are comparatively suitable in hard rock terrain for recharging groundwater. This way, the structural aspect of basin has been thoroughly studied and assigned comparative weightage based upon expert knowledge and published literature for assigning impact of these factors in evaluation of recharge potential.

2.3.3 *Recharge Potential Evaluation and Mapping*

With the aim to evaluate potential of recharge in the watershed, relief aspect and structural aspect of watershed have been studied. Simple approach of assigning weightage has been adopted for quantification of complex hydrogeology. Different relief and structural components as well as LULC of watershed have been classified into different categories by assigning rank on 1–5 scale. Rank has been assigned based upon comparative recharge capacity of different classes as a controlling factor. Rank 1 has been assigned to class in a layer having higher recharge potential than other categories. Similarly, higher rank has been assigned to class with lower potential for recharge. Different layers and their ranks have been shown in Table 2.1. After assigning rank, weighted sum analysis method was used for calculation of a multi-criteria analysis between thematic layers of controlling factors.

The weightage assigned to each factor was equally distributed in different classes of corresponding factor for computing overall weightage. Thereafter, sum of weightage in each class was computed and divided into five equal categories indicating excellent to poor recharge potential. Recharge potential zones are identified though weighted sum indicated that 5.1%, 14.5% and 33.4% area in the basin lies under excellent, good and moderate recharge potential, respectively, while 31.7% and 15.2% area under fair and poor recharge potential. Recharge potential area identified in Fig. 2.4 can be used as a base map for identification of ideal sites for implementing suitable groundwater augmentation measures and conservation of watershed.

Table 2.1 Hydrogeological controlling factors, their categorization and rank assigned

Controlling factors and their weight	Categories	Rank on scale (1–5)
Slope (19.2)	< 3 = excellent	1
	3–8 = good	2
	8–15 = moderate	3
	15–20 = poor	4
	>20 = nil	5
Lithology (15.4)	Dolomite = excellent	1
	Schist = good	2
	Slate = moderate	3
	Quartzite = poor	4
	Gneiss = nil	5
Soil (11.5)	Typic Udorthents associated with Dystric Eutrochrepts = excellent	1
	Typic Udorthents = good	2
	Dystric Eutrochrepts associated with Typic Udorthents = moderate	3
Geomorphology (11.5)	Piedmont alluvial plain = excellent	1
	Piedmont slope = good	2
	Moderately dissected hill and valley = moderate	3
	Highly dissected hill and valley = poor	4
Land use /land cover (15.4)	Agriculture = excellent	1
	Forest = good	2
	Wasteland/ barren = moderate	3
	Build-up/settlement = poor	4
Stream density (15.4)	<0.83 = excellent	1
	0.84–2.28 = good	2
	2.29–3.40 = moderate	3
	3.41–4.31 = poor	4
	>4.31 = nil	5
Lineament density (11.5)	<0.95 = negligible	5
	0.96–1.90 = poor	4
	1.91–2.80 = moderate	3
	2.81–3.80 = good	2
	> 3.80 = excellent	1

Groundwater augmentation measures in the form of chal-khal, percolation tank, contour trenches, bioenineering structures like live check dams, rip-rap drains etc. store water in recharge zone and facilitate infiltration which helps in increasing the aquifer storage and sustenance of drying spring. Recharge potential in current study has been predicted using remote sensing-based data and GIS tools and found that these data and technique provide excellent platform for quantification of spatial data for assessing groundwater recharge potential. But the limitation of identifying recharge potentiality is that the error or accuracy in methods of proportionate weight

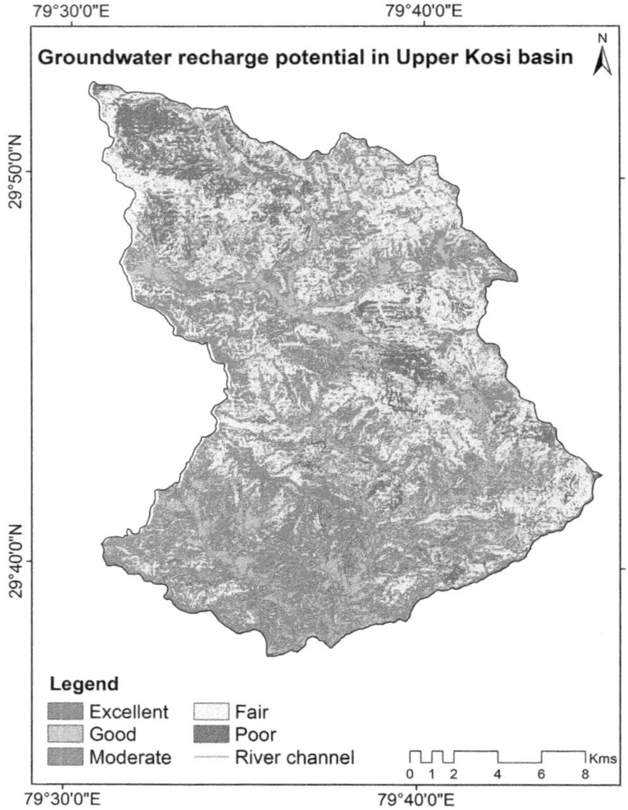

Fig. 2.4 Groundwater recharge potential in Upper Kosi basin, Almora, Uttarakhand

assignment is not assessed. On the other hand, some factors work independently in recharging groundwater. In case of lineament, the control of other factors becomes negligible because the water flows through joints or cracks independent of other factors (slope, lithology, LULC, etc.). Similarly, in case of drainage, the area around channel is saturated with very poor or nil recharge potentiality, while other factors in same area may indicate good or positive sign of recharge potential (low slope, agricultural land, alluvial deposit, etc.). In such cases, the water body buffer has been masked and assigned poor potential. The technique of assigning weightage and ranking is good for comparing and identifying in case of remaining factors, but alternative techniques can be developed through discussion with expert for quantification of these two factors, viz. lineament and drainage.

2.4 Conclusion

Thematic data availability, level of accuracy, hydrogeological conditions and government policies play a vital role in groundwater management activities. The relationship of hydrogeological factors is extremely useful for recharge estimation and groundwater resources evaluation in any kind of topographical region. Remote sensing data with integration of GIS techniques provide facility for utilizing various hydrogeological components like slope, lithology, structure, rainfall, soil, LULC and drainage on single platform and provide quantification techniques of these components in a judgmental way for identification of groundwater recharge potential zones. The study has established the ability of remote sensing data together with GIS technique for delineation of groundwater projection, especially in the complex hydrogeological terrain. Identifying the groundwater recharge potential zones in Upper Kosi basin of Almora district using remote sensing and GIS methods is competent to curtail the time, manpower and money. Although the proficiency in assigning weightage and quantification of accuracy level (ground truth data accuracy) is a limitation, it provides spatial view of recharge potential area with various degrees of variability and provides flexibility of decision-making by modifying the weightage. The recharge potential map prepared in this study can provide a guideline for construction of recharge structures and adopting conservation measures to preserve hydrological sustainability of the watershed. The recommended methodology will guide researcher and water departments for future groundwater exploration, development, and management in mountainous catchment. Future scope of the new emerging space technology-based modelling techniques lied in (1) understanding and quantification of independent and interdependent factors that control the spatial distribution hydrological factors, (2) challenges of modelling cumulative impact of these factors with altering land use and changing climate, (3) identification of suitable sites for augmentation work for recharging aquifer for sustainability of groundwater resource.

Acknowledgements The financial support provided by the National Mission on Himalayan Studies (NMHS), Ministry of Environment, Forest and Climate Change (MoEF&CC), New Delhi, for conducting this research is duly acknowledged. The authors also wish to acknowledge the Principal Investigator of the project Er. Soukhin Tarafdar, Scientist D, G.B. Pant National Institute of Himalayan Environment & Sustainable Development (GBPNIHESD), Garhwal Unit, Srinagar, Garhwal, Uttarakhand. The efforts and hard work of Junior Project Fellow Mr. Darshan K Bhatt, Field Assistants Mr. Vikram Negi, Mr. Pramod Kandpal and Mr. Mohit Chandra are appreciated for collecting water samples and conducting field surveys. Authors are grateful to the Director, GBPNIHESD, Kosi-Katarmal, Almora, Uttarakhand, for providing lab facility and field instruments.

References

Adiat KAN, Nawawi MNM, Abdullah K (2012) Assessing the accuracy of GIS-based elementary multi criteria decision analysis as a spatial prediction tool—a case of predicting potential zones of sustainable groundwater resources. J Hydrol 440:75–89

Ayazi MH, Pirasteh S, Arvin AKP, Pradhan B, Nikouravan B, Mansor S (2010) Disasters and risk reduction in groundwater: Zagros mountain Southwest Iran using geo-informatics techniques. Adv Disaster 3:51–57

Babus OG, Sashikumar MC (2010) Delineation of groundwater recharge potential zones for Tirrupur block using GIS. J Soil Water Conserv 9:154–158

Chenini I, Mammou AB, May MY (2010) Groundwater recharge zone mapping using GIS-based multi-criteria analysis: a case study in Central Tunisia (maknassy basin). Water Resour Manag 24:921–939

Chowdhury A, Jha MK, Chowdary VM, Mal BCP (2009) Integrated remote sensing and GIS-based approach for accessing groundwater potential in West Medinipur District, West Bengal. India Int J Remote Sens 30:231–250

Dinesh Kumar PK, Gopinath G, Seralathan P (2007) Application of remote sensing and GIS for the demarcation of groundwater potential zones of a river basin in Kerala, southwest cost of India. Int J Remote Sens 28:5583–5601

Eid MM, Abdel Rahman MT, Zaghloul E, Elbeih S (2006) Integrated remote sensing and GIS for proposing groundwater recharge locations: case study at west El-Nubariya Canal, Egypt. Egypt J Remote Sens Space Sci 9:113–134

Fortes PS, Platonov AE, Pereira LS (2005) GISAREG- a GIS based irrigation scheduling simulation modelling to support improved water use. Agric Water Manag 77:159–179

Gansser A (1964) Geology of the Himalayas. Wiley InterScience, New York, p 289

Gumma MK, Pavelic P (2013) Mapping of groundwater potential zones across Ghana using remote sensing, geographic information systems and spatial modeling. Environ Monit Assess 185:3561–3579

Jaiswal RK, Mukharjee S, Krishnamurthy J, Saxena R (2003) Role of remote sensing and GIS techniques for generation of groundwater prospect zones towards rural development-an approach. Int J Remote Sens 24:993–1008

Jaturon K, Wiewwiwun R, Srilert C (2014) Hydrogeologic characteristics and groundwater potentiality mapping using potential surface analysis in the Huay Sai area, Phetchaburi province. Thailand Geosci J 18:89–103

Jha MK, Chowdary VM, Chowdhury A (2010) Groundwater assessment in Salboni block, West Bengal (India) using remote sensing, geographical information system and multi-criteria decision analysis techniques. Hydrogeol J 18:1713–1728

Joshi BK, Kothyari BP (2003) Chemistry of perennial springs of Bhetagad watershed: a case study from Central Himalayas, India. Environ Geol 44(5):572–578

Kaliraj S, Chandrasekar N, Magesh NS (2014) Identification of potential groundwater recharge zones in Vaigai upper basin, Tamil Nadu, using GIS-based analytical hierarchical process (AHP) technique. Arab J Geosci 7:1385–1401

Krishnamurthy J, Mani A, Jayaraman V, Manivel M (2000) Groundwater resources development in hard rock terrain-an approach using remote sensing and GIS techniques. Int J Appl Earth Obs Geoinf 2:204–215

Kumar SGR, Shankar K (2014) Assessment of groundwater potential zones using GIS. Geosci Front 2:1–10

Magesh NS, Chandrasekar N, Soundranayagam JP (2012) Delineation of groundwater potential zones in Theni district, Tamil Nadu, using remote sensing, GIS and MIF techniques. Geosci Front 3:189–196

Mishra A (2014) Changing climate of Uttarakhand, India. J Geol Geophys 3(2014):1–5

Murthy KSR (2000) Groundwater potential in a semi-arid region of Andhra Pradesh-a geographical information system approach. Int J Remote Sens 21:1867–1884

Negi GCS, Joshi V (2004) Rainfall and spring discharge patterns in two small drainage catchments in the western Himalayan Mountains. India Environ 24:19–28
Oh H-J, Kim Y-S, Choi J-K, Park E, Lee S (2011) GIS mapping of regional probabilistic groundwater potential in the area of Pohang City. Korea J Hydrol 399:158–172
Pietersen K (2006) Multiple criteria decision analysis (MCDA): a tool to support sustainable management of groundwater resources in South Africa. Water 32:119–128
Pradhan B (2009) Groundwater potential zonation for basaltic watersheds using satellite remote sensing data and GIS techniques. Cent Eur J Geosci 1:120–129
Rawat JS (2014) Impact of climate change in the non-glacial fed Himalayan River system, a case study from the Kosi River in district Almora, Uttarakhand state (India). In: 3rd international conference on hydrology and meteorology, Hyderabad, India
Riad PH, Billib M, Hassan AA, Salam MA, El Din MN (2011) Application of the overlay weighted model and Boolean logic to determine the best locations for artificial recharge of groundwater. J Urban Environ Eng 5:57–66
Shaban A, Khawlie M, Abdallah C (2006) Use of remote sensing and GIS to determine recharge potential zone: the case of occidental Lebanon. Hydrogeol J 14:433–443
Singh AK, Prakash SR (2003) An integrated approach of remote sensing, geophysics and GIS to evaluation of groundwater potentiality of Ojhala sub-watershed, Mirjapur District, U.P., India. In: Asian conference on GIS, GPS, aerial photography and remote sensing, Bangkok
Singh P, Gupta A, Singh M (2014) Hydrological inferences from watershed analysis for water resource management using remote sensing and GIS techniques. Egypt J Remote Sens Space Sci 17:111–121
Solomon S, Quiel F (2006) Groundwater study using remote sensing and geographic information systems (GIS) in the central highlands of Eritrea. Hydrogeol J 14:1029–1041
Subba Rao N, Chakradhar GKJ, Srinivas V (2001) Identification of groundwater potential zones using remote sensing techniques in and around Gunur town, Andhra Pradesh. India J Indian Soc Remote Sens 29:69–78
Tweed SO, Leblanc M, Webb JA, Lubczynski MW (2007) Remote sensing and GIS for mapping groundwater recharge and discharge areas in salinity prone catchments, South-Eastern Australia. Hydrogeol J 15:75–96
Valdiya KS (1978) Extension and analogues of the Chail nappe in the Kumaun Himalaya. Indian J Earth Sci 5:1–19
Valdiya KS, Bartarya SK (1989) Diminishing discharge of mountain springs in a part of Kumaun Himalaya. Curr Sci 58:417–426
Valdiya KS, Bartarya S (1991) Hydrogeological studies of springs in the catchment of the Gaula River, Kumaun lesser Himalaya, India. Mt Res Dev 11(3):239–258

Chapter 3
Application of Geoinformatics for the Integrated Watershed Management and Development Planning, Bal Ganga Basin, Tehri Garhwal (Uttarakhand)

Vijendra Kumar Pandey

Abstract The primary issue of the twenty-first century is optimum utilization of natural resources and sustainable development planning in order to accelerate the pace of socioeconomic development. The key issue of the developing countries is how to address the development of backward region and remote hilly areas. They need specific attention of the decision-makers to cater the rapidly, growing population that has created pressure on shrinking natural resources. It is need of the time to prepare development plans considering the optimal use of resources and its sustainability. Despite remarkable advances in agriculture technology, there is prevalence use of conventional methods of cultivation in the various regions. Either it was because of inadequate resources, undulating terrain, or irascible area. The degradation of natural resources was mainly due to the lack of technological and appropriate methodologies to address these issues. Soil degradation and lack of irrigation facility are the primary aspects liable for poor agricultural growth and productivity. Integrated watershed management approach can effectively resolve the issues of natural resource management, livelihood, food security, and environmental issues. In this study, geoinformatics has been used as tools for the resource prioritization, gap analysis, and development planning of the Bal Ganga watershed. This study is an attempt to check the applicability of geoinformatics techniques in the development planning process and resource optimization. Survey of India topographic sheets has been used to digitalize contours and prepare digital elevation model that was used for the physiographic analysis in the ArcGIS software. The watershed is assumed as a hydrologic unit which controls the ecological processes, such as soil and water resources. Bal Ganga Basin has the potential for higher agriculture growth with the optimal use of resources. The watershed is situated in the Lesser Himalayas; there is eminence effect of physiographic control on the land use practices. Thematic layers such as slope condition, aspect, types of soil, and soil depth are the essential

V. K. Pandey (✉)
Center for the Study of Regional Development, School of Social Sciences, Jawaharlal Nehru University, New Delhi, India

P. Kumar et al. (eds.), *Applications and Challenges of Geospatial Technology*,
https://doi.org/10.1007/978-3-319-99882-4_3

parameters that affect crops growth were prepared in the GIS environment and prioritized for the development planning. These independent variables are prepared using ArcGIS 9.3, and land use capability was delineated for the suitable area of cultivation. Demographic data of the villages and socioeconomic variables has been attached, and level of development has been analyzed. As per the analysis of potential resource region, cultivation zone has been identified for comprehensive land use planning.

Keywords Geoinformatics · Integrated watershed development · Land degradation · Resource region · Uttarakhand

3.1 Introduction

The watershed management approach has become an increasingly significant question in the developing countries (Bulkley 1977). This is stimulated by government departments as well as nongovernmental organizations to find out suitable management methodologies for the aggregate productivity of natural resources (Ratna Reddy et al. 2017). The main approaches which are related to watershed management have experienced a paradigm shift during the past few decades. Thoguh, there is no universal methodology that has been developed for attaining watershed management (Nath Roy 2005; Chowdary et al. 2009). The significant advance in the watershed management approach is community participation (Wani and Garg 2008; Thapa 2000; Saravanan 2002; Pirani and Mousavi 2016; Johnson et al. 2001; Batabyal 2002; Arya 2005). Watershed management has gained eminence appreciation in the developing countries for the integrated resource management which will be focused on the livelihood security of the poor. The frequent crops failure due to climatic variability made the situation worse and become difficult to address the situation of large-scale river valley projects forced a paradigm shift from watershed to micro-watershed. It become major concern for the state and community to scripts a new ground for the government towards as a "learning organization" which has thrust to address the changing social order and environmental condition (Eswaran and Samra 1997).

The watershed projects in India have long history implemented during the Vedic period (Bhan 2013). The recent development in the watershed programs in India can be dated back with the implementation of Command Area Development and Water Management (CADWM) program, which initially started during 1974 (NITI Ayog 2015). Further, Drought-Prone Areas Programme (1973–1974) and the Desert Development Programme (1977–1978) were brought into the watershed mode in 1987 (Department of Land Resource, Ministry of Rural Development 2001). The Integrated Wasteland Development Programme (IWDP) was launched in 1989 under the aegis of the National Wasteland Development Board and also aimed the development of wasteland (Department of Land Resource, Ministry of Rural

Development 2003). These three major programs were implemented with watershed development guidelines from 1995 with the National Watershed Development Project in Rainfed Areas (NWDPRA) and the Watershed Development in Shifting Cultivation Areas (WDSCA) (Department of Land Resource, Ministry of Rural Development 2001). The objective of these schemes are to improve the irrigation potential, soil conservation, and utilization and optimize the agricultural production and productivity and promotion of agroforestry through integrated and coordinated approach of efficient watershed management (Department of Land Resource, Ministry of Rural Development 2001; Bhan 2013; Bulkley 1977). Due to the partial success, a new comprehensive scheme "Haryali" was launched during 2003 which has encompassed rainwater harvesting, erosion control, irrigation, and microenterprise as a part of watershed programs (Department of Land Resource, Ministry of Rural Development 2003).

Integrated watershed management (IWM) is considered as optimum approach for improving agricultural productivity in the rain fed or drought-prone regions (Naqvi et al. 2015; Pandey et al. 2007; Pirani and Mousavi 2016; Prasannakumar et al. 2012). It was estimated that India has planned and implemented micro-watershed development programs at large scale (Farrington et al. 1999). The GoI has further implemented Integrated Watershed Management Programme (IWMP) during 2008–2009 (Department of Land Resource, Ministry of Rural Development 2008). The scheme has focused on the selection of micro-watershed and application of *Geoinformatics* for the preparation, implementation, and monitoring of IWMP projects. The objectives of the IWMP are overall development of the micro-watershed and mainly focused on the involvement of community and stakeholders in each stage of the project.

The integrated watershed management programs were strengthened since the 1980s after the successful implementation of Sukhmojari Projects in Haryana (Agarwal and Narain 2010; Arya 2005). It was a an innovative approache to create new functionaries that meant to intensification of community involvement, sustainability of programs, and its impacts on rural livelihood. Many government departments as well as nongovernment organizations have promoted watershed development schemes to address the basic issues with the objectives of soil conservation, soil erosion, water harvesting, afforestation, and encouraging suitable technologies for the resourceful and sustainable use of natural resources. However, many of these watershed projects have not achieved the desired objectives, due to the poor community participation (Johnson et al. 2001; Wani and Garg 2008; Thapa 2000; Village et al. 1990; Wani et al. 2014).

The development of watershed management approach in India has significantly evolved in the last two decades. During the first phase, it was primarily based on the physical parameters and conservation of natural resources. In the late 1990s, there was a significant paradigm shift observed in the watershed management approach. The watershed approaches not only were included in the conservation of natural resources but also focused on the increasing productivity of natural resources and optimum utilization with active people participation (Wani et al. 2014 Tennyson and White 2005). It was also adopted that modern tools and techniques for sustainable utilization and better coordination were found between the government agencies. The development of remote sensing technology and geographic information system (GIS) has added a new dimension to the watershed development planning (Tim and Mallavaram 2003; Pandey et al. 2007; Murty 1998; Khan et al. 2001; Hiese et al. 2011; Gosain and Rao 2004; Abdelrahman et al. 2016). The geoinformatics tools have proved to be a catalyst for the analysis of geophysical parameters such as estimation of soil erosion, water conservation, forest loss, and land use changes in the watershed (Al-Nasrawi et al. 2016; Bhaduri et al. 2000; Gosain and Rao 2004; Naqvi et al. 2015; Prasannakumar et al. 2012; Pandey et al. 2007). The availability of global digital elevation model (GDEMs) and Google Earth images has promoted the use of technology to address the issues and progression toward the development of Civil GIS (Al-Nasrawi et al. 2016; Hiese et al. 2011).

The main issues of the twenty-first century are food security, drinking water, and environmental attributes (Schmidhuber and Tubiello 2007). The technological advances which were made in the agricultural apparatuses, many of the countries around the world are unable to keep up with the fast paced food production and increasing population. Stagnation or deterioration in agricultural production can be due to excess use of fertilizer, degradation of soil and irrigation facility, and lack of knowledge about soil fertility restoration methods. It was required to provide appropriate training and demonstration to the farmers about new farming methods to address the basic issue of resource management and to increase agricultural productivity. The water scarcity is increased in the hilly areas where sufficient precipitation occurs in the form of snowfall and rain. This is because of the lack of water harvesting approach and deterioration of water quality. This also has negative impacts on soils and land resources and potential risks of the enhanced deterioration of the environment (Kukal and Bawa 2013; Khan et al. 2001; Dhruvanarayana and Babu 1983; Nath Roy 2005). The study aims to explore the application areas of geoinformatics technique for the watershed management, prioritization of the issues, and needs and gap analysis for sustainability in the hilly terrain of Uttarakhand.

3.1.1 Why Watershed

Watershed is a well-defined area or catchment, which drains through a common outlet. It is geographically delineated from the interfluve or water divide. Watershed includes detailed characteristics of soils, landforms, vegetation, and land use. Hydrologic processes within the watershed which are infiltration, runoff, subsurface flow, and evapotranspiration are interlinked and can be appropriately assessed. It is a basic hydrologic unit to study biophysical parameters of the region to prepare development plans (Pirani and Mousavi 2016). Each watershed has a unique physical setting which requires specific attention to address the problems. These could be soil erosion processes and transport of nutrients and pollutants by surface and subsurface flow of water within the watershed. The surface runoff is directly related to topography, slope length, and gradient as well as the direction which influences geochemical process (Reddy et al. 2004). The fluvial processes show watershed physical and vegetative characteristics controlling erosion and minimizing land degradation, vegetative loss, and water quality. These issues require a holistic approach to understand the hydrologic processes at watershed level in order to check the process and approaches to manage it. Soil degradation is a physical process that has positive impacts on anthropogenic activity such as land use practices. Deforestation and water scarcity need to be addressed with land degradation to improve the geophysical environment within a watershed. Therefore, the watershed is a basic unit for the sustainable development planning (Pirani and Mousavi 2016; Reddy et al. 2004; Saravanan 2002; Schmidhuber and Tubiello 2007).

3.1.2 Natural Resource Degradation

The degradation of soil and water resources has been very crucial and considered as a global risk (Oldeman 1994). Due to the degradation, lands lost productive capacity. The sources which degraded lands are wind and water. These two account for the 82% (250 out of 305 Mha) of the total degraded area (Oldeman 1994). In order to check the soil erosion, drainage, and slope gradient, relationship-based strategy is required in a catchment area. The watershed management approach in this regard is very useful to effectively address the erosion processes and also restoration of degraded land. A watershed is geographic unit which can be used as an ideal unit to study the degradation process and decreasing productivity of natural resources. The lack of adequate management techniques for the natural resources, which are fragile, can cause deterioration of environments (Saravanan 2002). The quality of life and livelihood of people are strongly interlinked with the natural resources quality and function of the watershed. If the soil resources are good, its productivity is higher which sustains livelihood pattern of people residing in the particular watershed. In India, average soil loss was estimated at 16.5 ton/ha/year (Dhruvanarayana and Babu 1983). Sedimentation rate of reservoirs of Indian subcontinent has been examined to be 2–20 times more than that these were predicted during the design stage (Galay and Evans 1989). The average annual soil loss is calculated ranging from 21 to 555 ton/ha/year for the Asian rivers (Holeman 1968).

3.2 Study Area

The Bal Ganga is a right bank tributary of Bhilangna River which itself is an important tributary of the Bhagirathi River. It originates near a peak (4846 mts) opposite of Shastru Tal in the district of Tehri Garhwal. The geographical location of the watershed is 78° 32′ 37″ E longitude 30° 25’ 42″ N latitude to 78° 49′ 5″ E longitude to 30° 43’ 16″ N latitude (Fig. 3.1). The watershed covers part of Ghansali and Pratap Nagar Tehsil. Bal Ganga is remarkably known for its flat summit surface near Chamiyala which is a town and market of the area. The climate of the area is varied to subtropical in the valley to cold temperate on the upper reaches. January is

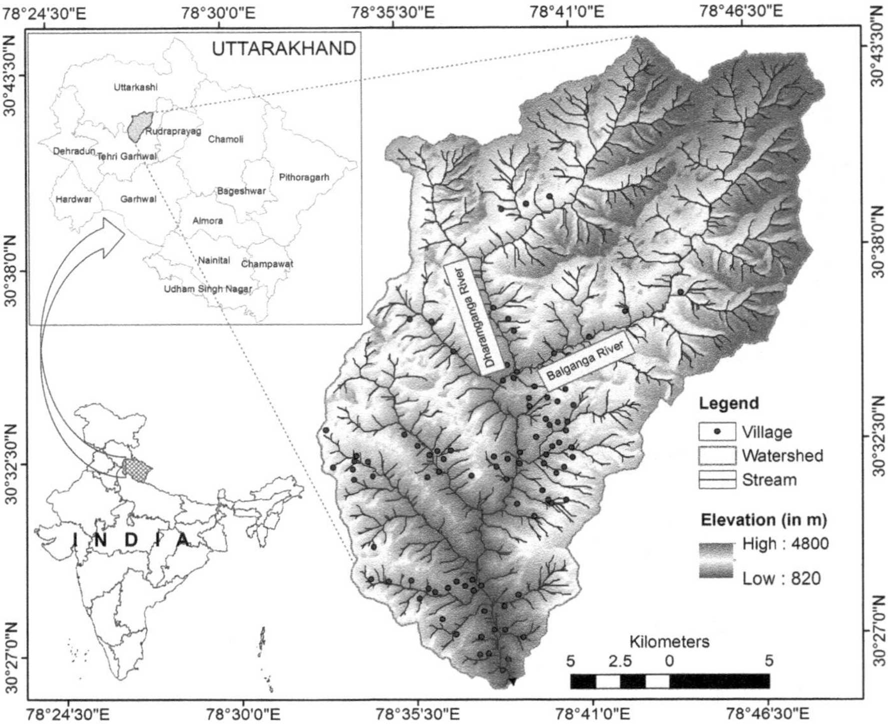

Fig. 3.1 Location map of the Balganga Basin, District Tehri Garhwal (Uttarakhand)

the coldest and June is the warmest month in the area. Mean temperature of the warmest months is recorded 32 °C and coldest months 19.6 °C. Relative humidity touches to 85.6% in the monsoon season. Approximately 85% rainfall is received in monsoon season (July–September) in the study area. Winter rainfall is also associated with the passage of the western disturbances and is in the form of snowfall on higher elevations. The average annual rainfall in the district is 1395 mm which occur in 87 rainy days (Indian Meteorological Department).

In the present study, 96 villages were studied of Pratapnagar Tehsil of Tehri Garhwal district. Total population of the study area is 34,081 persons with a population density of 45 persons/km^2. The number of household resides in these 96 villages is 6250.

3.3 Materials and Method

To achieve the objectives of the present study, it is necessary to choose an appropriate methodology. There are a number of digital elevation models (DEMs) and satellite images that were available in the open-source domain (Table 3.1). In order

Table 3.1 Database and software available for geoinformatics applications

Type of database	Data source	Spatial resolution	Applications	Web links
DEMs	SRTM	90 m	Topographic analysis	https://earthexplorer.usgs.gov
	ASTER	30 m		https://earthexplorer.usgs.gov
	CartoDEM	30 m		http://bhuvan.nrsc.gov.in/data
	GEOTOPO30	30 m		https://earthexplorer.usgs.gov
Satellite image	Landsat 8	30 m (*15* m panchromatic)	Land use/land cover, infrastructures and utilities, etc.	*glovis.*usgs.gov
	Sentinel 2	10 m		https://earthexplorer.usgs.gov
	IRS AWiFS	56 m		http://bhuvan.nrsc.gov.in/data
	IRS LISS-3	23.5 m		http //bhuvan.nrsc.gov.in/data

to prepare physiographic analysis survey of India topographic sheets and SRTM, 1-arc v.3 DEMs (digital elevation models) were used. ArcGIS 10.3 software was used for the DEM processing, and satellite imageries were analyzed using Erdas Imagine 14.0 software. Rainfall and temperature data and other ancillary data have been procured from the various sources of the study area. The base map was prepared on RF 1:12,500 scales. Survey of India topographic sheets was used to digitize contours, drainage, and other infrastructures for the development planning. DEM (digital elevation model) of the area is prepared from contour map and also used for the preparation of slope, aspect, and relief maps. Land use/land cover of the area was extracted using the satellite images. Soil map was prepared from the National Bureau of Soil Survey and Land Use Planning (NBSS&LUP). Google Earth images were used to geo-rectify cadaster map on RF 1:4000 for the IWMP and resource management.

3.3.1 Physiography Variables

Bal Ganga basin lies in the Lesser Himalayan range having rugged topography. Minimum elevation of the watershed is 820 m at the confluence with Bhilangana River, and the highest elevation is 4780 m at the peak summit which separates the watershed (Fig. 3.2a). It has unique topography as high undulating peak, wide flat surface in the middle of the watershed, and narrow outlet. In the middle part of the basin, there is wide and flat river terrace deposited by fluvial processes.

Slope gradient is an important factor which controls human interface to the environment. According to the s Slope inclination, controls the human activity and reaching a particular point, it seeded. Average slope inclination of the area is varying from 10 to 30° which are categorized at moderate slope category. Low slope angle covers around 3% area (Fig. 3.2b). The total area of low and moderately inclined slope accounts 63% area of the watershed. Particularly, this slope category is used for plantation and growing crops. Subsequently, moderate-high slope covers 31% area and high slope angle which is 40–60° inclination covers around 5% area of watershed (Table 3.2).

Aspect is a very important physical factor which regulates geo-environmental process in the hilly area (Fig. 3.2c). It is slope direction toward the sun. This gives details about the average sunshine hours to the slope and also one of the important agents that control physical weathering process. This is also important for the growing crops in the hill tract.

Geomorphology is confined to the process that is operating on the surface of the area. It could be fluvial, glacial, and aeolian in the hilly area. Most of the terrain of the watershed is moderately dissected followed by very highly dissected according to the coverage of the area. Highly dissected slope shows high land degradation and covered by exposed rocky surface often sometime by sparse vegetation. River terrace is best used for growing crops in the area. Rice cultivation is done on the floodplain and terraces.

3.3.2 Soil

The soil is basic element for the development of a watershed. The soil of Bal Ganga watershed has been classified as per multiple criteria such as soil depth; soil texture; sand, silt, and clay composition; and soil structures (Fig. 3.2d). This has also included soil fertility. The soil of valley floor is more fertile and has good depth. The soil of upper slope has low soil depth and poor textures.

3.3.3 Land Use/Land Cover

Land use/land cover of the watershed has been changed significantly over a period of time. The land use map prepared using topographic map being prepared by Survey of India, surveyed during 1962–1963, has shown some contrasting result in comparison to the satellite imagery dated 16 April 2006 IRS LISS III, path 046, row 097. Table 3.2 shows that cultivated land has been decreased around 2.5%, and further decreases have been noticed in the forest area near about 4% (Table 3.3). These changes have been directly related to land degradation as the area under barren land

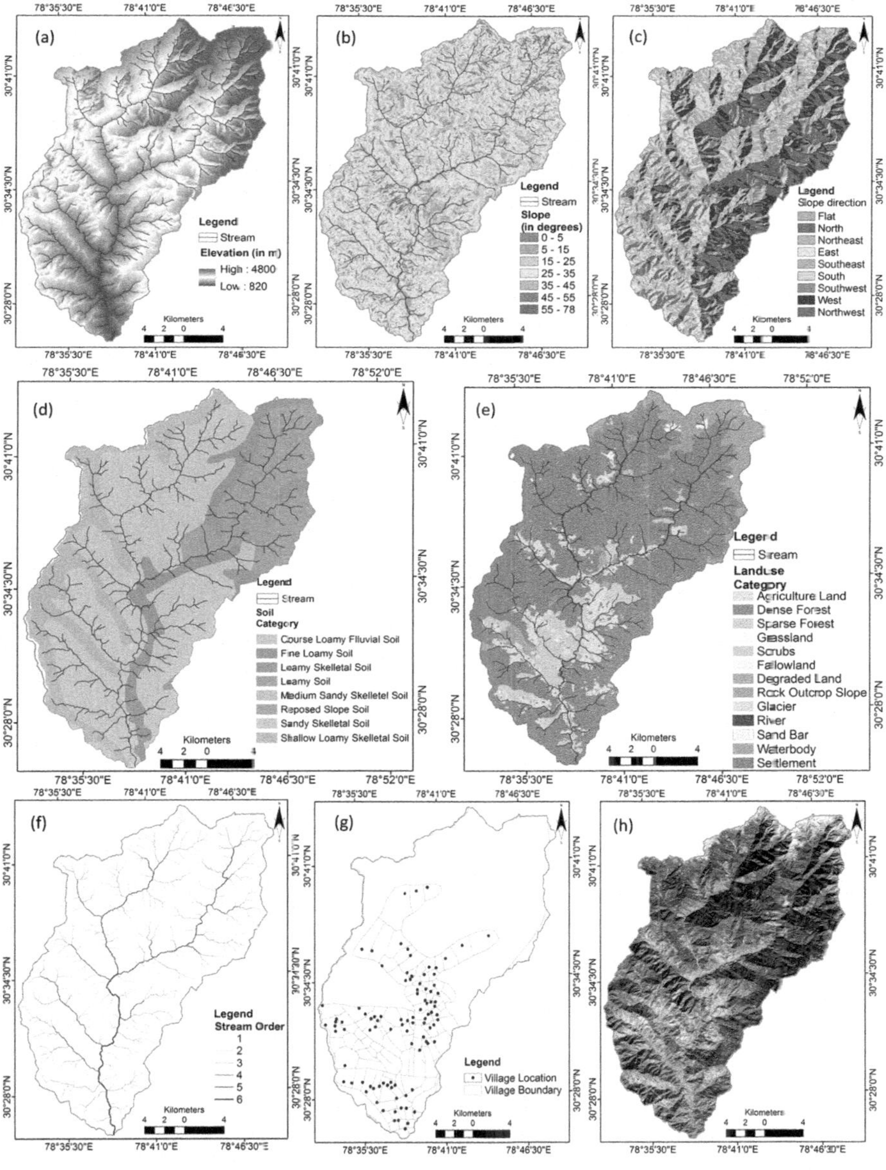

Fig. 3.2 Physiographic variables of the study area: (**a**) physiography; (**b**) slope; (**c**) aspect; (**d**) soil; (**e**) land use and land cover; (**f**) drainage network; (**g**) village boundary and location of settlement; and (**h**) false color composite (FCC) satellite imagery showing vegetation cover

Table 3.2 Slope distribution of the Bal Ganga watershed

S. No.	Slope category	Description	Area (km^2)	% Area
1	< 10	Gentle	19.15	3.93
2	10–20	Moderate	101.82	20.88
3	20–30		190.88	39.14
4	30–40	Moderately high	150.53	30.87
5	40–50	High	23.93	4.91
6	50–60		1.24	0.26
7	> 60	Very high	0.11	0.02
Total			487.66	100.00

Table 3.3 Land use changes in Bal Ganga watershed

Land use based on topographic sheet (1962–1963)			Land use based on imagery (2006)		
Class name	Area (km^2)	% area	Class name	Area (km^2)	% area
River	1.63	0.33	Waterbodies	0.82	0.17
Inland water	0.04	0.01			
Open area	54.83	11.27	Open area	40.34	8.27
Quasi open	17.95	3.69			
Alpine grasses	2.22	0.46	Scrubs	28.57	5.86
Forest	314.94	64.72	Forest	335.14	68.72
Dense forest	17.87	3.67			
Cultivated land	62.98	12.94	Cultivated land	76.91	15.77
Settlement	1.38	0.28	Settlement	4.74	0.97
Landslides	3.27	0.67	Rocky slope	1.16	0.24
Glaciers/snow covered	10.42	2.14			
Total	487.53	100.00	Total	487.67	100.00

(open area) has increased 8.27–11.27% (Fig. 3.2e). Those areas have been classified as scrubs in 1962–1963 presently classified as separating alpine grasses and scrubs. Some new category has also been introduced to improve the classification processes.

3.3.4 Drainage Network

Dendritic drainage pattern is observed in the watershed (Fig. 3.2f). It shows uniform lithology and bedding plane. The major right bank tributary is Dharamganga, Medh Gad, Gangadd Gad, and Argad Gad. Left bank streams are Jandria and Ghatu Gad which pours water into the Bal Ganga.

3.3.5 *Villages*

In the Bal Ganga watershed, villages are located in two major clusters. First, villages are situated along the river valley, terraces, and moderately sloping surfaces which are suitable for cultivation and construction of houses and road networks. Approximately, 97 villages are located in the watershed, which is found in a clusters (Fig. 3.2g). The first is located between Chakrushera village and Jyoncana village, and the second cluster is located between Gajwan Gaon village and Koti village. The average village population size is 350 people.

3.3.6 *Road Network*

Villages of the watershed are connected with pucca road categorized as district and link road. These roads run along the Bal Ganga River starting from Ghansali town to Budha Kedar village (confluence at Bal Ganga and Dharamganga River). Other villages have recently connected with road constructed under Pradhan Mantri Gram Sadak Yojana (PMGSY). These roads are constructed by local PWD divisions. The major roads run through the valley can be seen in the satellite images (Fig. 3.2h).

3.4 Integrated Resource Planning

The approaches of integrated watershed studies have gained momentum because it evolves multiple problem solution environments within a catchment. It was observed that natural processes are interlinked together and need a comprehensive approach to address the issues. It was noted in the study area that in upper catchment area due to the steep slope and denudation, land degradation is prevalent. Mass movements induced by the torrential streams devastate forest, soil, and economic resources. In the river valley, scarcity of water for irrigation has persisted. Land degradation is directly related to soil erosion, high-velocity surface runoff, and vegetation damage, particularly on the upper reaches. On the other side, these are related to slope inclination, forest species composition, and institutional frameworks regulating use of the natural resources. The barren slope which is exposed to the natural process of denudation is most vulnerable to land degradation rather than the vegetated surface. The main forces in mountain area of the land surface transformation processes are physiography and climate. These forces influence works differently and sometimes individually or in combinations to the biophysical environment of the watershed. In this regard, a combined approach has been required to solve the problems (Fig. 3.3).

The effective implementation of watershed plans and management strategies depends on the efficiency of institutions and community participation. Comprehensive development plans have been prepared for the study area using problem solution

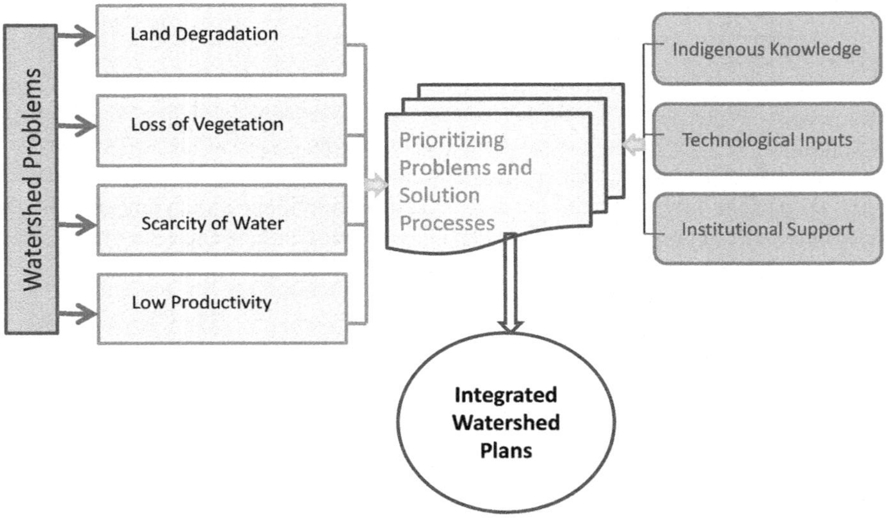

Fig. 3.3 Flowchart for the integrated watershed management planning and resource prioritization

matrix. Land capability classes have been prepared for the area to identify a suitable area of cultivation. It was also analyzed to prepare development according to the feasibility and level of economic development. Community participation was ensured to study the local issues, their priorities, and indigenous knowledge to tackle the issues with the sustainability of the project (Fig. 3.4). In this study, physiographic variables such as physiography, slope, aspect, soil, and drainage were analyzed to prepare land capability classification of the area using geoinformatics techniques. Land use/land cover map was prepared to understand the present scenario of primary activities in the area, influencing the environment. GIS techniques prove a panacea for the integrated watershed management. Topographic analysis, cadaster map digitalization, and preparation of development to implementation of land management as well as monitoring using the satellite imagery can be effectively done. All the biophysical layers and resource prioritization were overlaid in the GIS environment, and optimum regions were identified for the development planning.

The Bal Ganga watershed was divided into three principle zones based on the resource prioritization and identification of the main problems and suggests effective measures (Fig. 3.5). During the field visits, it was assessed that due to the topographic constraints, upper reaches were barren land and fallow land. Mostly, community grasslands were found on the upper reaches slope that were used for the fodder. In the discussions with the community, it was noted that these areas were earlier used for agriculture, but due to drying up of springs and streams, scarcity of water for irrigation facility turned these areas into fallow lands. If the irrigation facility is available, these areas can be used for cropping. Forest on the lower slope covered the first zone, and upper reaches are barren land (Fig. 3.6). These areas have vast potential for off-season vegetables and herbals and aromatic plants. Villagers

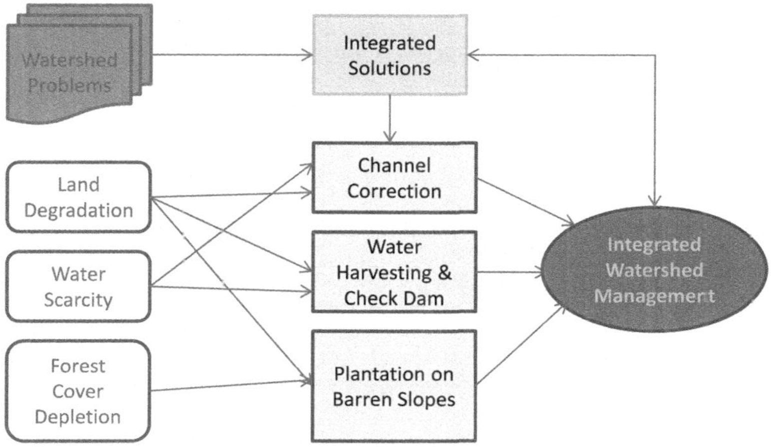

Fig. 3.4 Watershed issues and challenges and approaches to address the problems for the sustainable development planning

Fig. 3.5 (**a**) Broad Bal Ganga valley used for intensive cultivation; (**b**) debris fall damaging forests; (**c**) high-tech nursery promoting community as well as departmental afforestation program; and (**d**) Boorha Kedar temple situated at the confluence of Bal Ganga and Dharamganga river

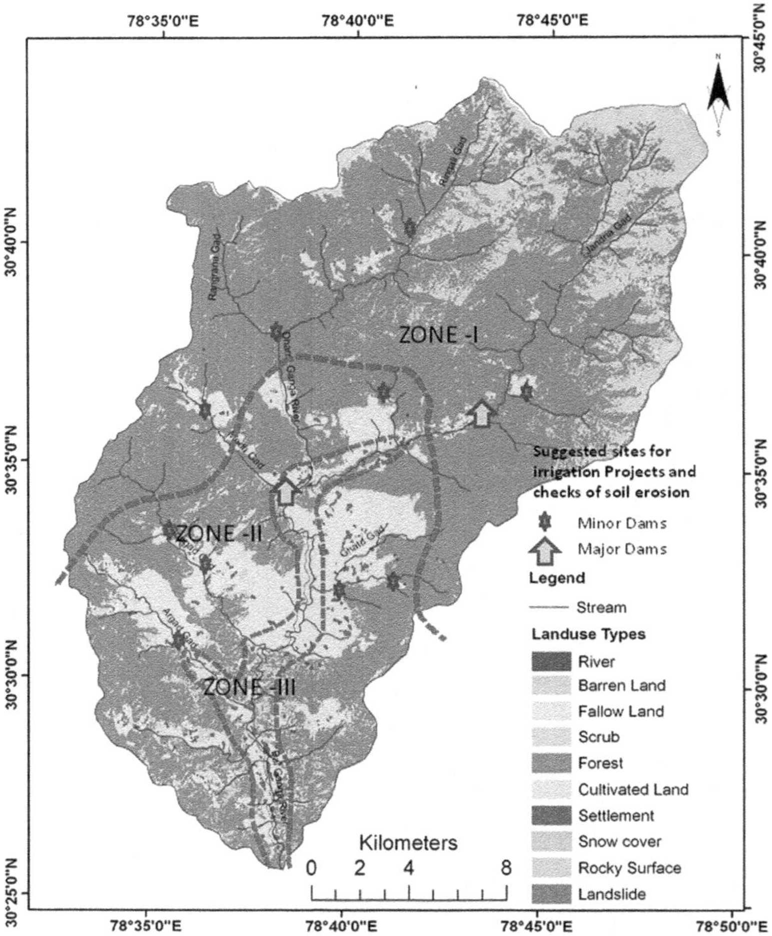

Fig. 3.6 Resource planning and development zones prepared for the IWMP

were very keen if the government agency will help them for the water resource conservation using check dams and small ponds. This water can be used for irrigation of crops and drinking water. Some villagers have told that during pre-monsoon months, they face acute shortage of drinking water. These barren lands are the source of land degradation during the monsoon season. This causes havoc on the lower slopes and in the river valleys. Slope failures, soil erosion, and vegetation loss are the main issues in this zone. In order to minimize the effect of channel correction and plantation, check dams have been suggested.

Second zones have major problems of soil erosion, land degradation, forest loss, etc. These have economic values and very important for the prosperity of the watershed. It requires a scientific approach to agriculture practices, community plantation, and awareness among individuals about social forestry on the barren and fallow land. Third zones witness severe flash floods during monsoon season due to sudden catastrophic rainfall. The area also witness landsliding that dispalce slope materials and causes debris deposits is other issues which are being a concern as it helps to floods and erodes fertile soil of the valleys.

3.5 Conclusions

This paper is an attempt to raise the issues, challenges, and potentialities of the watershed development planning which could be studied and managed in an efficient way using the geoinformatics technique. Land degradation, soil erosion, and scarcity of water and forest damages are interlinked in such a way that a holistic approach is needed to solve these issues. Land degradation is not only deteriorating the soil and decreasing the farm production but also overall natural resource productivity which could lead a region toward prosperity. How the topographic variable influences the approaches to implement integrated watershed management within a watershed can be prioritized using the GIS techniques. Physiography, slope, soil, forest, and drainages can be integrated to prepare development plans for the watershed level. In order to implement any project successfully, community participations and change of approaches in both the community and institution are required working together in coordination to sustain ecology and natural resources. Community participation and institutional cooperation has been highlighted to effectively manage the natural resources. Considering the step-by-step approaches to study different stages of watershed planning and implementation, geoinformatics techniques can play a crucial role, particularly in the hilly areas. The paper explains key challenges which are needed to be addressed and methods to be applied for the community participation and institutional integration.

Acknowledgment I would like to thank the District Administration, Tehri Garhwal, for their kind support during the field visits and providing the required data. The author also acknowledges the USGS for the SRTM DEM and Landsat L8 satellite images were downloaded from https://earthexplorer.usgs.gov/.

References

Abdelrahman MAE, Natarajan A, Hegde R (2016) Assessment of land suitability and capability by integrating remote sensing and GIS for agriculture in Chamarajanagar district, Karnataka, India. Egypt J Remote Sens Space Sci 19(1):125–141. https://doi.org/10.1016/j.ejrs.2016.02.001

Agarwal A, Narain S (2010) Equitable sharing of benefits in Sukhomajri. GIST, pp 1–4

Al-Nasrawi AKM et al (2016) Civil-GIS incorporated approach for water resource management in a developed catchment for urban-geomorphic sustainability: Tallowa dam, southeastern

Australia. Int Soil Water Conserv Res 4(4):304–313. https://doi.org/10.1016/j.iswcr.2016.11.001
Arya SL (2005) Participatory integrated watershed management a success story of "Sukhomajri Project" (Haryana), India
Batabyal AA (2002) Participatory watershed development: challenges for the twenty-first century by JohnFarrington; Cathryn Turton; A. J James J Asian Stud 61(4):1407–1408 http://www.jstor.org/stable/3096502
Bhaduri B et al (2000) Assessing watershed-scale, long-term hydrologic impacts of land-use change using a GIS-NPS model. Environ Manag 26(6):643–658
Bhan S (2013) Land degradation and integrated watershed management in India. Int Soil Water Conserv Res 1(1):49–57. https://doi.org/10.1016/S2095-6339(15)30049-6
Bulkley JW (1977). Integrated watershed management: past, present, and future, p 18
Chowdary VM et al (2009) Integrated water resource development plan for sustainable management of mayurakshi watershed, India using remote sensing and GIS. Water Resour Manag 23 (8):1581–1602
Department of Land Resource, Ministry of Rural Development, Govt. of India (2001) Guidelines for Watershed Development (Revised-2001). Available at: http://dolr.nic.in/dolr/guidewd.asp
Department of Land Resource, Ministry of Rural Development, Govt. of India (2003) Guidelines for Haryali, 2003. Available at: http://dolr.nic.in/HariyaliGuidelines.htm
Department of Land Resource, Ministry of Rural Development, Govt. of India (2008) Common guidelines for watershed development projects
Dhruvanarayana VV, Babu R (1983) Estimation of soil Erosion in India. J Irrig Drain Eng 109 (4):419–434 Available at: http://ascelibrary.org/doi/10.1061/%28ASCE%290733-9437%281983%29109%3A4%28419%29
Eswaran H, Samra JS (1997) Challenges in ecosystem management in a management in a watershed context in Asia. In: Lal R (ed) Integrated watershed management in the global ecosystem. CPC Press, Boca Raton, pp 19–32
Farrington et al (1999) Participatory watershed management. Oxford University Press, Oxford
Galay VJ, Evans R (1989) Sediment transport modeling. In: Proceedings of the international symposium, New Orleans, American Society of Civil Engineers, New York, pp 1–12
Gosain AK, Rao S (2004) GIS-based technologies for watershed management. Curr Sci 87 (7):948–952
Hiese N et al (2011) Application of geoinformatics in land resource management at micro-watershed level for sustainable development of Sanis, Wokha District, Nagaland. In: Dimensions and Directions of Geospatial Industry, Hyderabad, pp 1–10
Holeman JN (1968) The sediment yield of major rivers of the world. Water Resour Res 4 (4):737–747
Johnson N et al (2001) User participation in watershed management and research. Water Policy 3:507–520. Available at: http://www.sciencedirect.com/science/article/pii/S1366701702000144
Khan MA, Gupta VP, Moharana PC (2001) Watershed prioritization using remote sensing and geographical information system: a case study from Guhiya. India J Arid Environ 49 (3):465–475
Kukal SS, Bawa SS (2013) Temporal variations in runoff and soil loss in relation to soil conservation practices in catchments in Shiwaliks of lower Himalayas. Int Soil Water Conserv Res 1 (2):19–25. https://doi.org/10.1016/S2095-6339(15)30036-8
Murty JVS (1998) Watershed management. New Age International, New Delhi
Naqvi HR et al (2015) Soil erosion planning using sediment yield index method in the Nun Nadi watershed. India Int Soil Water Conserv Res 3(2):86–96 Available at: http://linkinghub.elsevier.com/retrieve/pii/S2095633915300836
Nath Roy U (2005) Sustainability of watershed projects: a case of Punjab Shivalik, India Available at: http://citeseerx.ist.psu.edu/viewdoc/download?doi=10.1.1.738.5673&rep=rep1&type=pdf
NITI Ayog (2015) Selected best practices in water management, New Delhi

Oldeman L (1994) Global extent of soil degradation. Soil resilience and sustainable land use, pp 19–36. Available at: http://library.wur.nl/isric/fulltext/isricu_i26803_001.pdf
Pandey A, Chowdary VM, Mal BC (2007) Identification of critical erosion prone areas in the small agricultural watershed using USLE, GIS and remote sensing. Water Resour Manag 21 (4):729–746
Pirani FJ, Mousavi SA (2016) Integrating socio-economic and biophysical data to enhance watershed management and planning. J Hydrol 540:727–735. https://doi.org/10.1016/j.jhydrol.2016.05.072
Prasannakumar V et al (2012) Estimation of soil erosion risk within a small mountainous sub-watershed in Kerala, India, using revised universal soil loss equation (RUSLE) and geo-information technology. Geosci Front 3(2):209–215. https://doi.org/10.1016/j.gsf.2011.11.003
Ratna Reddy V, Saharawat YS, George B (2017) Watershed management in South Asia: a synoptic review. J Hydrol 551:4–13. Available at. https://doi.org/10.1016/j.jhydrol.2017.05.043
Reddy VR et al (2004) Participatory watershed development in India: can it sustain rural livelihood? Dev Change 35(2):297–326. Available at: http://eds.b.ebscohost.com/eds/pdfviewer/pdfviewer?vid=5&sid=6e200a5d-df28-4002-b9e0-b1d8112672a6@sessionmgr112&hid=121
Saravanan VS (2002) Institutionalising community-based watershed management in India: elements of institutional sustainability. Water Sci Technol 45(11):113–124
Schmidhuber J, Tubiello FN (2007). Global food security under climate change. Proc Natl Acad Sci USA 104(50):19703–19708. Available at: http://www.pnas.org/content/104/50/19703 short
Tennyson L, White R (2005) Preparing for the next generation of watershed management programmes and projects. In: Proceedings of the Asian regional workshop, p 163
Thapa GB (2000) Integrated watershed management: basic concepts and issues. Hum Settlement Dev II:12–23
Tim US, Mallavaram S (2003) Application of GIS technology in watershed-based management and decision making. Watershed Update 1(5):1–6
Village S, Shiwalik H, Soil C (1990) Forest and watershed protection in India. In: World pp 1–4
Wani SP, Garg KK (2008) Community watersheds for improved livelihoods through consortium approach in drought prone rain-fed areas. J Hydrol Res Dev 23:55–77
Wani SP, Chander G, Sahrawat KL (2014) Science-led interventions in integrated watersheds to improve smallholders' livelihoods. NJAS – Wageningen J Life Sci 70–71:71–77. https://doi.org/10.1016/j.njas.2014.07.001

Chapter 4
A Semi-analytical Approach to Understand Remote Sensing-based Backscattering Characteristics for Kerala Coast Using In Situ Observation

Shafique Matin, Sisir Kumar Dash, and Tune Usha

Abstract A quasi-analytical algorithm (QAA)-based distribution and variability of particulate backscattering coefficient (b_{bp}) was studied for Kerala coast, India. A total of 28 observations were made in the coastal stretch of about 410 km from Kasaragod to Ernakulam for up to 50 m depth. Optical data were collected using a hyperspectral underwater radiometer to evaluate the b_{bp}, water-leaving radiance (L_w) and chlorophyll-a (Chl-a) concentration. We aimed to achieve three objectives, i.e. (1) QAA-based b_{bp} calculation using underwater radiometer and its sensitivity to downwelling irradiance (*Ed*) and surface radiance (*Es*), (2) validation of the relationship between b_{bp} and Chl-a concentration for inshore and offshore coastal waters and (3) the relationship of L_w with QAA-based b_{bp} and in situ Chl-a. We observed that the range of b_{bp} values varied between 0.07 and 0.002 m^{-1}, with a maximum b_{bp} value between 1200 and 1400 h for inshore waters. *Ed* and *Es* are independent variables and were placed at the denominator to calculate b_{bp}, where *Ed* is found relatively more sensitive than *Es*. The correlation between b_{bp} and Chl-a found growing with depth (< 20 m R^2: 0.067, > 20 m R^2: 0.487), due to the increasing complexity of coastal waters (Case II). While relating the Chl-a and b_{bp} with *Lw*, showed a poor corleation with a low R^2 value of 0.229 and 0.203, respectively, signifying the maximum scattering due to other suspended matters with less contribution from Chl-a pigment in highly turbid coastal waters.

Keywords Quasi-analytical algorithm · Hyperspectral radiometer · Remote sensing reflectance · Backscattering coefficient · Chlorophyll-a

---

S. Matin (✉)
Teagasc Food Research Centre, Ashtown, Dublin, Ireland

S. K. Dash · T. Usha
National Centre for Coastal Research (NCCR), Pallikaranai, Chennai, India
e-mail: skdash@nccr.gov.in; usha@nccr.gov.in

P. Kumar et al. (eds.), *Applications and Challenges of Geospatial Technology*,
https://doi.org/10.1007/978-3-319-99882-4_4

4.1 Introduction

The spectral dependency of particulate backscattering is essential to understand the dynamics of ocean optics. Backscattering carries information with its shape, slope and spectral curve and provides information about the particle composition (Boss et al. 2004; Smith and Baker 1981). The particulate backscattering ratio and its spectral dependency have been extensively studied in the ocean colour inversion, inferring particle properties and light field modelling. However, information on actual in situ spectral variability of the particulate backscattering and its properties is limited and only available for global scale (Whitmire et al. 2007). A difference in backscattering coefficient due to particle size and ocean depth is the key to understand the coastal ecology, its biogeochemistry, which also includes dynamism of particle (Loisel et al. 2007). Apart from its empirical relationship with Chl-a and other organic constituents in open sea, information about particle size has tremendous importance and can also be related to know about pelagic ecosystems (Kostadinov et al. 2010).

Many researchers have utilised and reported several applications of spectral particulate backscattering ratio in interpreting ocean colour data (Whitmire et al. 2007). Such interpretations are based on the interlinking of optical information acquired from a hyperspectral sensor. For example, ocean reflectance '*R*' and the sensor irradiance (ratio of upwelling and downwelling radiance) are directly proportional to the ratio of absorption and backscattering, respectively (Gordon and Morel 1983). The outcome proportionality here depends upon the quantum of dissolved and particulate materials present in the sea-water. The absorption dynamics of dissolved and particulate materials present in the sea-water has been adequately explored using laboratory experiments, but there is still a gap in understanding of the backscattering properties especially for coastal turbid waters (Stramski et al. 2004). As the reason, many existing algorithms to estimate ocean dynamics use the particulate backscattering coefficient in place of backscattering ratio and scattering coefficient.

Sea waters have been categorised into two broad categories (i.e. Case I and Case II) based on several factors including its turbidity and particulate density. The global distribution of Case I water is ambiguous, although the deep-sea waters (offshore) are frequently referred as Case I water (Lee and Hu 2006). By definition, a Case I water is defined as the water for which phytoplankton and their associated materials like bacteria, heterotrophic creatures and, dissolved organic matter (DOM) are majorly reasonable for backscattering. On the other hand, mapping chlorophyll concentration in Case I water is comparatively easier because only plankton influences the optical properties, and scattering and absorption by debris are correlated with chlorophyll concentrations. The Case II waters are fed by the coastal rivers or suspended sediments in addition to phytoplankton. Optical properties are typically controlled by three independent components: (1) phytoplankton and their associated debris, (2) colour DOM also known variously as yellow material, and (3) inorganic traces and suspended remains. Concentrations of these three must be retrieved simultaneously; and is a complex problem that is not yet fully resolved, although progress is being made, taking benefit from the Medium Resolution Imaging Spectrometer (MERIS) sensor from the NIR and red range of the spectrum. The standard MERIS product uses a neural net algorithm, which concurrently retrieves yellow substance and suspended sediment. The Case I waters are typical of the open ocean, away from coasts and river mouths. Along the Kerala coast, the Chl-a concentration ranges from very low in the oligotrophic waters to very high (up to 21 mg m^{-3}) in coastal areas. Observation of chlorophyll concentration from space extents over three grades of magnitude and the likelihood of a point having chlorophyll concentration are contrariwise proportional to its concentration. For this reason, chlorophyll concentrations are generally presented on a logarithmic scale for the blooming periods. Thus phytoplankton blooms frequently occurring in the upwelling areas are a particular problem for ocean colour processing due to the presence of very high chlorophyll concentration.

Ideally, Case I waters are controlled by phytoplankton and other supplementary material, but enormous number of phytoplankton cells in the water means that the water-leaving radiance (L_w) may be high enough to trigger the Case II classification flag, and some blooms may have sufficient reflectance above 700 nm to cause problems for aerosol retrieval and thus for the atmospheric correction. The Case I waters may have a range of backscattering likelihood between 0.2% and 2% dependent majorly on the chlorophyll concentration and other constituents (Gordon et al. 1988) that has been estimated with very few in situ data and almost none from the Indian Ocean (Morel 1988).

Watercolour and turbidity are two major factors that affect extensively to the energy levels that a camera or underwater radiometer detect. Globally, a major part of the Case I water falls in the tropical and subtropical areas. More than 50% of the Case I waters are in the higher latitudes particularly in the North Hemisphere and fall in the category of open ocean with very few river feds. Unlike the high latitude oceans, the rivers and backwaters along the Kerala coast have sharp headwater gradients, that feed high sediment loads, especially in the monsoon season. Dynamic deltas at the river mouth are another effect of high sediment load and heavy rainfall which carries rich biodiversity and healthy ecosystems through flooding. However, river catchments are rapidly changing due to deforestation, anthropogenic interference and sea-level rise.

So as to use the geoinformatics technique to quantify watercolour or quality, it is significant to comprehend the complexities of water-light interface especially for the coastal waters due to its high uncertainties related with space and time (Nechad et al. 2016). On the other hand, in situ measurements of primary production and biomass are particularly critical, although few novel promising techniques have been recently proposed (Graff et al. 2015; Riser and Johnson 2008). To reduce the space-time coverage sampling limitations, bio-optical oceanography has implemented optical sensors on a variety of in situ platforms. Platforms comprise from research vessels and moorings to gliders and profiling floats, each with specific complementary space-time observation scales (Claustre et al. 2010; Dickey 2003). Such platforms enabled to monitor bio-optical properties that serve as proxies for major biogeochemical variables.

The difference in watercolour and quality can be observed by multispectral instruments, but minute information may not be warranted due to its broader bandwidth. Satellite-based hyperspectral approach to cover the inherent optical properties of water has the limitation with respect to cloud cover and other atmospheric attenuations. In practice, however, the fundamental problem of this approach is the difficulty in finding the true remote sensing reflectance for a given location and time, from which reflectance errors of individual pixels and uncertainties of all pixels can be referenced (Hu et al. 2013). Satellites like Aqua MODIS (Moderate Resolution Imaging Spectroradiometer) are found to be good for such estimations but gives under-estimated values, and the matchup between satellite and in situ values is also partly consistent (Deidun et al. 2011). A critical area of research in aquatic optics is to extract tangible information about water column optical properties, which becomes more difficult in optically complex coastal waters. The capability to acquire information about optical properties in coastal waters can be facilitated with the application of high-resolution underwater spectral sensors. Such enhanced sensor provides the opportunity to develop and evaluate algorithms that distinguish subtle features in the radiance spectrum which may be significant to discrimination of optically important materials. In many cases, such algorithms are based on theoretical and semi-analytical relationships relating remote sensing reflectance to properties of absorption and backscatter. Narrow spectral bandwidth sensors (e.g. hyperspectral) are capable of acquiring information about the optical properties in coastal waters. High spectral resolution sensor provides the opportunity to distinguish subtle features in the radiance spectrum using object-specific algorithms, which may be significant to discriminate different optically active substances. Such algorithms are explicitly based on the theoretical and semi-analytical relationships linking remote sensing reflectance to properties of absorption and backscattering.

With the changing climate and related dynamics of river discharge, it is complex to understand the coastal variabilities. In absence of any regional model and other causative factors, river discharge supposed to have a major influence on the water quality of coastal regions for Kerala coast. Researchers like Seo et al. (2009) relied on a regional coupled model to understand the effect of surface run-off into the Indian Ocean because of rainfall. The model treated river discharge by salinity relaxation to climatology. They observed that the precipitation affected considerably well-enough due to river discharge through its effect on barrier layer thickness and therefore on inherent optical properties of coastal water.

Algorithms, derived from the radiative transfer equation like quasi-analytical algorithm (QAA; Lee et al. 2002), are based on the relationship between R_{rs} and inherent optical properties of water. The algorithm uses the multiband below-surface remote sensing reflectance (r_{rs}) as the input information to calculate absorption and backscattering coefficients analytically. Such algorithms can be applied to sensors having narrow bandwidth hyperspectral instruments and can also be used for subsurface irradiance reflectance values as input data for the analysis. The performance of these algorithms relies on accurate spectral models for discrete water constituents as the absorption coefficients such as chlorophyll pigments, colour dissolved organic matter (CDOM) or suspended sediments. Such regression was developed using specific water type and may give better correlation if the data being analysed were collected from the same source because of the nature of the regression (Lee et al. 2002). The quasi-analytical algorithm found to be estimating an under-value of $a_{nw}(\lambda)$ (non-water absorption coefficient) for all the waters in different seasons if the slope of the regression is at the lower end, more specifically at a wavelength of 400 nm. Conversely, a wavelength longer than the reference wavelength may give rise to uncertainties in deriving $a_{nw}(\lambda)$ and may also result in a very high RMS error (Root Mean Square) and absolute percentage difference if taken at 440 nm (Li et al. 2016).

In this study, we utilised the above-surface remote sensing reflectance (R_{rs}) values collected using the hyperspectral underwater radiometer instrument. The collected in situ data were used to calculate the QAA-based b_{bp} and also the sensitivity of b_{bp} to *Ed* and *Es*. The relationship between b_{bp} and Chl-a concentration was also compared for inshore and offshore coastal waters. Finally, the association of L_w with QAA-based b_{bp} and in situ Chl-a was analysed.

4.2 Materials and Method

The study was carried out in a coastal stretch of about 410 km between Kasaragod to Ernakulam district of Kerala (Fig. 4.1). A hyperspectral underwater radiometer (Satlantic HyperOCR II) was used to collect optical data samples. The radiometer instrument can be used in different modes based on the data requirement and analysis. It can be either used for a free-fall profiling to collect data at a different depth or in floating mode using a detachable float cover to acquire data for near-surface measurements. The device can also be used on a lowering frame base that has been designed to use different sensors. This is an interchangeable platform that allows optical sensors like the Satlantic multispectral OCR-500 instrument as well as Satlantic hyperspectral HyperOCR OCR-3000 sensors. The instrument also has few optional features like a conductivity sensor and connection ports. It contains an option to integrate two external sensors into the instrument. This makes the instrument versatile and the best available option for measuring ocean dynamics.

The radiometer instrument was specifically arranged to minimise the issues of self-shadowing and cruise-induced instabilities. We collected data for seven transects from 10 to 50 m depth, on-board CRV Sagar Purvi between 28 January to 4 February 2013 (Table 4.1). In each transect, four depths (10, 20, 30 and 50 m) were considered for sampling with river mouths as a landmark. Rivers considered for sampling are Tejaswini (Kasargod), Valapattanam (Azhikkal), Chaliyar (Beypore), Bharathapuzha (Ponnani), Periyar (Munambam), Kochi (Cochin) and Valiazheekal. The hyperspectral underwater radiometer is capable of providing 136 channels of optical data in the spectra ranging from 350 to 800 nm wavelength with 10 nm spectral resolution. The optical data were collected at different time intervals in the clear sky between 0900 and 1600 h to get the maximum light penetration. Apart from the deployment mode we also collected radiometer data in buoy (floating) mode to understand the changes at different depth intervals. Major emphasis has been given

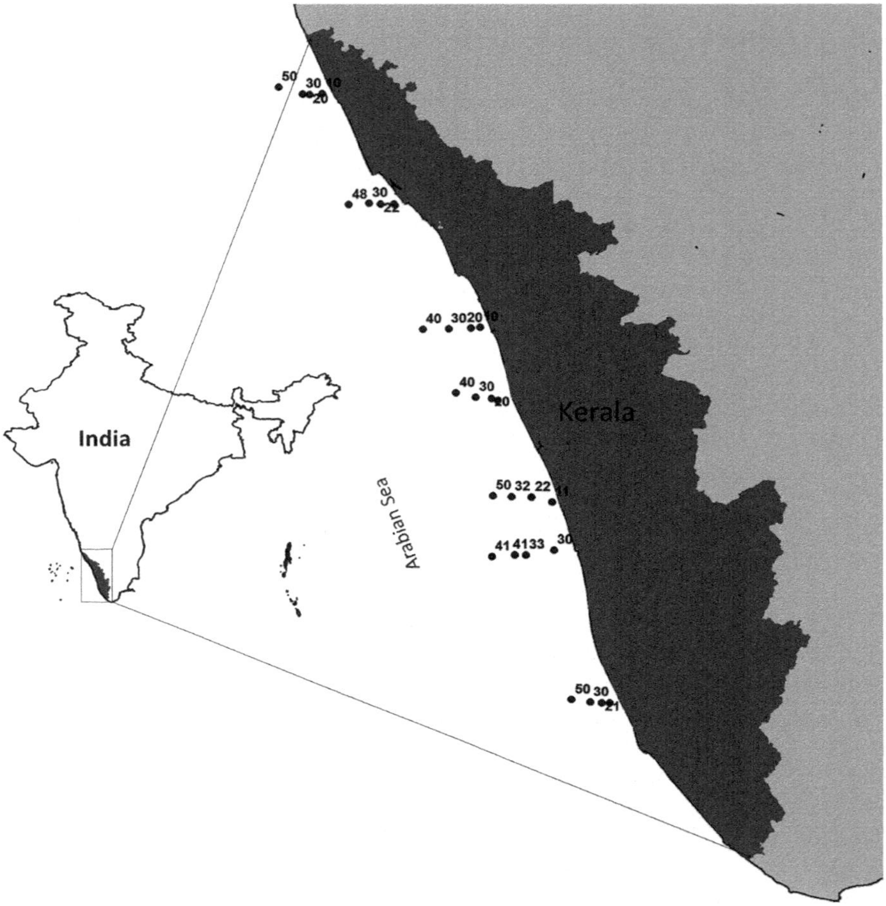

Fig. 4.1 Showing sampling location in seven transects along south-west Kerala coast at different sea depth (10–50 m), selected at the mouth of the major rivers in the state

to data collected at around 1300 h to avail better lighting conditions and also to synchronised the satellite pass.

To calculate the particulate backscattering coefficient (b_{bp}) by means of the multiband QAA (Lee et al. 2002), initially, the raw R_{rs} data acquired from the hyperspectral underwater radiometer were converted into r_{rs} using the following formula provided by Lee et al. (2002):

$$r_{rs}(\lambda) = \frac{R_{rs}(\lambda)}{0.52 + 1.7\, R_{rs}(\lambda)} \tag{4.1}$$

Table 4.1 Showing details of sampling plots. Sampling locations were selected considering major river mouths as landmark along the Kerala coast

Plot no.	Location; date	Time	Location	Depth (m)
Cast-1	Off Ponnani; 29-01-2013	10:02	Lat, 10°47.0 N; long, 75°50.0E	11
Cast-2	Off Ponnani; 29-01-2013	11:27	Lat, 10°47.5 N; long, 75°48.0E	20
Cast-3	Off Ponnani; 29-01-2013	13:46	Lat, 10°48.1 N; long, 75°43.2E	30
Cast-4	Off Ponnani; 29-01-2013	15:56	Lat, 10°49.5 N; long, 75°37.2E	40
Cast-1	Off Azhikal; 30-01-2013	09:42	Lat, 11°51.0 N; long, 75°18.6E	12
Cast-2	Off Azhikal; 30-01-2013	11:07	Lat, 11°51.0 N; long, 75°14.6E	22
Cast-3	Off Azhikal; 30-01-2013	13:25	Lat, 11°51.3 N; long, 75°11.1E	30
Cast-4	Off Azhikal; 30-01-2013	15:29	Lat, 11°50.8 N; long, 75°04.9E	48
Cast-1	Off Kasaragod; 31-01-2013	09:20	Lat, 12°26.4 N; long, 74°56.9E	10
Cast-2	Off Kasaragod; 31-01-2013	11:14	Lat, 12°26.2 N; long, 74°53.8E	20
Cast-3	Off Kasaragod; 31-01-2013	12:58	Lat, 12°26.3 N; long, 74°51.0E	30
Cast-4	Off Kasaragod; 31-01-2013	15:16	Lat, 12°28.5 N; long, 74°43.7E	50
Cast-1	Off Beypore; 01-02-2013	09:01	Lat, 11°10.3 N; long, 75°27.2E	40
Cast-2	Off Beypore; 01-02-2013	11:28	Lat, 11°10.4 N; long, 75°35.1E	30
Cast-3	Off Beypore; 01-02-2013	14:08	Lat, 11°10.7 N; long, 75°41.7E	20
Cast-4	Off Beypore; 01-02-2013	15:19	Lat, 11°10.9 N; long, 75°44.5E	10
Cast-1	Off Munambam; 02-02-2013	09:13	Lat, 10°14.2 N; long, 76°06.4E	11
Cast-2	Off Munambam; 02-02-2013	10:50	Lat, 10°15.7 N; long, 76°00.2E	22
Cast-3	Off Munambam; 02-02-2013	13:29.	Lat, 10°15.8 N; long, 75°54.2E	32
Cast-4	Off Munambam; 02-02-2013	15:39	Lat, 10°16.1 N; long, 75°48.4E	50
Cast-1	Off Valiazheekal: 03-02-2013	09:12	Lat, 09°08.9 N; long, 76°23.8E	13
Cast-2	Off Valiazheekal: 03-02-2013	10:43	Lat, 09°08.9 N; long, 76°21.4E	21
Cast-3	Off Valiazheekal: 03-02-2013	12:59	Lat, 09°09.1 N; long, 76°17.9E	30
Cast-4	Off Valiazheekal: 03-02-2013	14:52	Lat, 09°10.0 N; long, 76°12.2E	50
Cast-1	Off Cochin: 04-02-2013	09:26	Lat, 09°57.4 N; long, 75°53.8E	41
Cast-2	Off Cochin: 04-02-2013	11:48	Lat, 09°58.2 N; long, 76°01.7E	30
Cast-3	Off Cochin: 04-02-2013	13:54	Lat, 09°58.1 N; long, 76°05.7E	20
Cast-4	Off Cochin: 04-02-2013	15:40	Lat, 09°58.1 N; long, 76°08.7E	10

where R_{rs} is above-surface remote reflectance and T is equal to 0.52 and Q is 1.7, considered from hydro light to deep waters and at a nadir-viewing sensor angle.

$$R_{rs}(\lambda) = \frac{\mathrm{Lw}(\lambda)}{\mathrm{Ed}(\lambda, 0^{+})} \tag{4.2}$$

In the QAA there are two empirical aspects in the derivation of the backscattering coefficients and total absorption. One relays to the assessment of the total absorption coefficient at a reference wavelength of 555 nm, while the other one to the assessment of the spectral slope Y of the available particles backscattering coefficient. In the equation the reference wavelength was considered by keeping in mind that where $r_{rs}(\lambda_0)$ the elastic scattering and the absorption $a(\lambda_0)$ can be well-measured with maximum accuracy. So, the reference wavelength considered here ($r_{rs}(\lambda_0)$) from elastic scattering can be measured without any error and well-used to calculate absorption $a(\lambda_0)$ considering the same wavelength (λ).

$$b_{bp}(555) = \frac{u(555)\ a(555)}{1 - u\ (555)} - b_{bw}(555) \tag{4.3}$$

Here, *u555* is the ratio of backscattering coefficient with absorption and backscattering coefficient combined and can be obtained using the formula provided by Lee et al. (2002). $u(\lambda)$ is the ratio of backscattering coefficient with absorption and backscattering coefficient combined and can be obtained using the following formula:

$$u(\lambda) = \frac{-g_0 + \left[(g_0)^2 + 4\ g_1 r_{rs}(\lambda)\right]^{1/2}}{2g_1} \tag{4.4}$$

Here $g_0 = 0.0895$ and $g_1 = 0.1247$ are the averaged values taken from the Gordon et al. (1988) and Lee et al. (1999) calculated for open and coastal waters. The backscattering coefficient of the pure sea water (b_{bw}) was referenced from Smith and Baker (1981). The value for λ_o can be replaced with different wavelengths, such as 640 nm or 670 nm for high-absorbing clear water for a better measurement of $r_{rs}(\lambda_o)$ and accurate estimate of $a(\lambda_o)$.

$$a(555) = 0.0596 + 0.2\left[a(440)_i - 0.01\right] \tag{4.5}$$

$$a\ (440) = \exp(-2.0 - 1.4 + 0.2\rho\hat{}(2)),\quad \text{where}, \rho = \ln\ \left[\frac{r_{rs}(440)}{r_{rs}(555)}\right] \tag{4.6}$$

The b_{bw} is the backscattering coefficient of pure sea water that was taken from Smith and Baker (1981).

$$\mathrm{Y} = 2.2\left\{1 - 1.2\exp\left[-0.9\frac{r_{rs}(440)}{r_{rs}(555)}\right]\right\} \tag{4.7}$$

Here *Y*, which is the wavelength-dependent factor, was taken from the empirical relation derived by Lee et al. (1996). Lee et al. (1996) used this at the wavelengths of 440 nm and 555 nm as an exponent value in the backscattering. To calculate the backscattering coefficient and other wavelengths than 555 nm, we used the b_{bp} at 555 nm as reference value using the following formula:

$$b_{bp}(\lambda) = b_{bp}(555) * (555/\lambda)^{Y} \tag{4.8}$$

We also measured Chl-a concentration data at each point location simultaneously with the R_{rs} data. The Chl-a concentration data were collected using the fluorescence sensor mounted on the same hyperspectral underwater radiometer. The phytoplankton-based Chl-a concentration estimation (Eq. 4.9) was carried out at 440 nm wavelength where Chl-a absorbs maximally (Loisel et al. 2007).

$$\mathrm{Chl-a} = \left(\frac{K(\lambda)-K_w(\lambda)}{\mathrm{Xc}^{(\lambda)}}\right)^{1/e(\lambda)} \tag{4.9}$$

where X_c, K_w and $e^{(\lambda)}$ were taken from the Morel and Maritorena (2001) model. *K(λ)* coefficients can be expressed as functions of the *C* under the general power law which includes a coefficient X_c and exponent e, both function of λ. The L_w is the radiance that exits the ocean to the sensor mostly from the surface and subsurface waters which give better information about Chl-a concentration; hence we analysed the relationship between L_w and Chl-a and later validated this with observed QAA values (Morel 1988).

$$\mathrm{K}(\lambda) = \mathrm{Kw}(\lambda) + \mathrm{Xc}(\lambda)\,\mathrm{Ce}(\lambda) \tag{4.10}$$

The water-leaving radiance (L_w), which is basically the reflected radiance that leaves the ocean water and reaches to the sensor, is mostly from the surface and subsurface waters. These water-leaving radiance (L_w) may give better information about Chl-a concentration on the surface; hence we analysed the relationship between L_w and Chl-a and also validated with observed QAA values. We used the calculated values of L_w using the given formula:

$$\mathrm{L}_w(0^+,\lambda) = \mathrm{L}_u(0^-,\lambda)\left(\frac{1-\rho\,(\lambda,\theta)}{\eta_w 2\,(\lambda)}\right) \tag{4.11}$$

Here, $\rho(\lambda,\theta)$ taken is the Fresnel reflectance index (FRI) of seawater. Note that the default value for ocean water is 0.021, and $\eta w(\lambda)$ is the Fresnel refractive index of ocean water and the default value for is 1.345.

Precisely, to achieve the first objective, i.e. QAA-based b_{bp} calculation, we used the existing algorithm proposed by Lee et al. (2002), and the sensitivity of the *Ed* and *Es* data was carried using two different correlation analyses. To validate the relationship of b_{bp} with Chl-a as the second objective, we used the b_{bp} values calculated in objective one and plotted against the Chl-a value derived using Eq. (4.9). The third objective to understand the association of L_w with QAA-based b_{bp} and in situ Chl-a was carried out by plotting an individual regression analysis for L_w with both b_{bp} and Chl-a.

4.3 Results and Discussion

Samples collected at different locations and times resulted into a wide range of values which helped to understand the variability of particle availability and response. The b_{bp} was observed maximum between 1200 and 1400 h and minimum at the 0900 h and ranged from 0.0035 m^{-1} to 0.039 m^{-1} (Fig. 4.2). For offshore coastal waters, the b_{bp} variation was relatively lesser (0.0085–0.0215 m^{-1} and 0.00947–0.039 m^{-1}, respectively, for 30 and 50 m depths) than the inshore water.

Chl-a concentrations were found very high (up to 13 ug/l) in all the inshore coastal water (<20 m depth) samples and low in offshore (> 30 m depth) coastal waters (0.03 ug/l). The high Chl-a concentration found in most of the samples may be attributed to the presence of high organic feeding by rivers. The relationship between b_{bp} and Chl-a was observed becoming stronger with the increasing distance from the shoreline (Fig. 4.3). In the inshore coastal waters, the correlation (R^2) showed a minimum of 0.067, whereas in offshore waters, the value increased up to 0.487.

It is always intricate to understand the multifaceted nature of the coastal waters due to its optical inconsistency with space and time. Many unknown factors in the measurement procedure and limitations in the algorithm development may

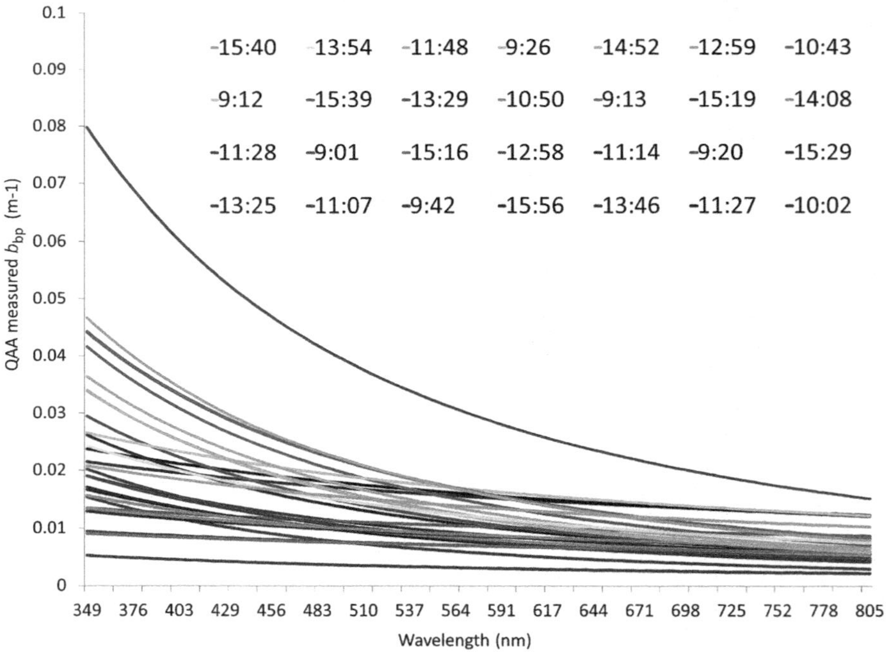

Fig. 4.2 QAA-based particulate backscattering coefficient (b_{bp}) for all the 28 plots at different depth and time. Plotting timing is mentioned for each day. Depth information is available in Table 4.1

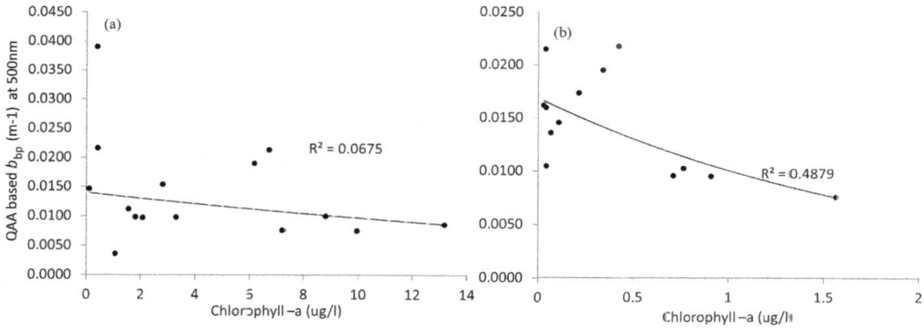

Fig. 4.3 QAA-based particulate backscattering coefficient (b_{bp}) plotted against chlorophyll-a observed using underwater radiometer at 500 nm wavelength at (**a**) inshore 10 m to 20 m and (**b**) offshore 30 m to 50 m sea depth

contribute to various levels of uncertainties in optical properties calculation including b_{bp}. It is always recommended to use above-surface remote sensing reflectance (R_{rs}) instead of surface irradiance (*Es*) to calculate the b_{bp} of coastal turbid waters (Lee et al. 2002), whereas few studies (Latha et al. 2013) have used *Es* sensors for b_{bp} calculation in absence of *Ed* sensor. Here we have calculated b_{bp} considering both *Ed* and *Es* and observed that *Ed*-based b_{bp} is having a better relationship with *Es*-based b_{bp} for Chl-a concentration (Fig. 4.4), although the values obtained from *Ed*- and *Es*-based b_{bp} is following a similar trend. To validate the efficiency of both the results, we calculated b_{bp} by considering both *Ed* and *Es* values. It was clearly evident that *Ed*-based b_{bp} calculation gives a better relationship instead of *Es* (Fig. 4.4).

We observed that water-leaving radiance (L_w) is generally high among all the samples near 500 nm wavelength (Fig. 4.5). At the same wavelength, the hyperspectral underwater radiometer also measures the Chl-a concentration to achieve the maximum penetration of coastal turbid water. The relationship of water-leaving radiance (L_w) with calculated b_{bp} and Chl-a was also evaluated to validate b_{bp} as a function of absorption by particle (not reflectance). Insignificant relationships were observed for L_w and Chl-a (R^2: 0.229) and L_w and b_{bp} (Fig. 4.6; R^2 : 0.203).

4.4 Conclusions

In the current study, we evaluated the hypothesis of using a semi-analytical approach like QAA to understand remote sensing-based backscattering characteristics for Case I and Case II waters and determine that such observations are only good if the water is less productive. The investigation also reveals the statement that QAA-based b_{bp} algorithm should not be used for the b_{bp}–Chl-a relationship for turbid coastal waters as it may contain many particulate materials other than

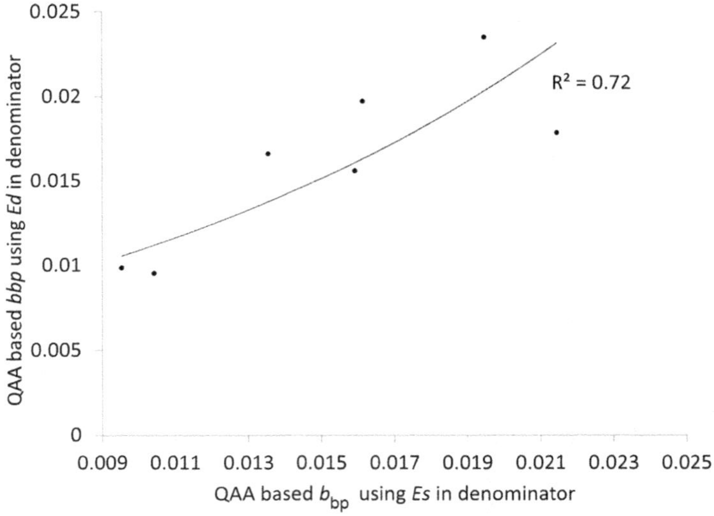

Fig. 4.4 Correlation between backscattering coefficient (b_{bp}) calculated using downwelling irradiance (Ed) and surface irradiance (Es) at 500 nm wavelength

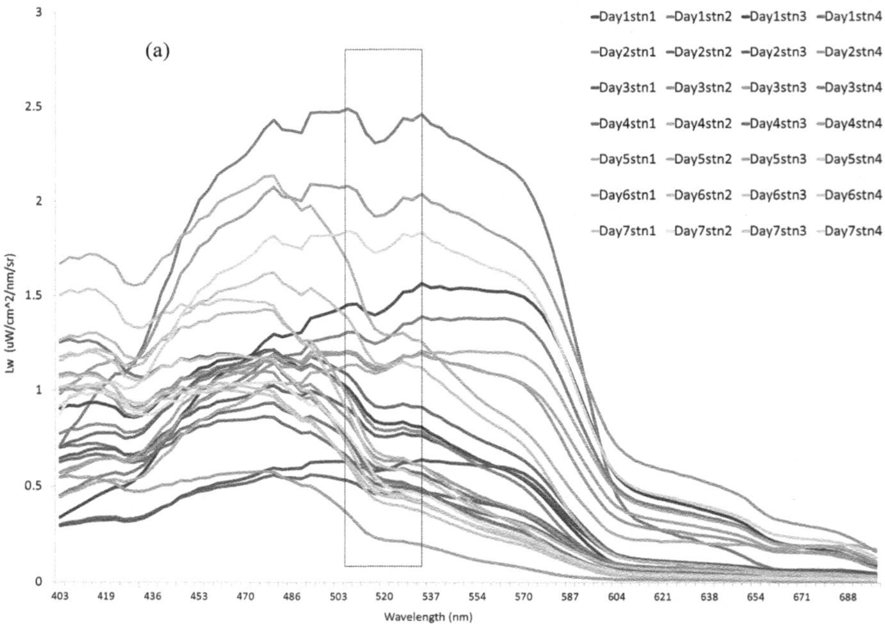

Fig. 4.5 Showing water-leaving radiance (L_w) for all the 28 plots between 400 and 700 nm. Dotted box represents wavelength band with highest L_w value

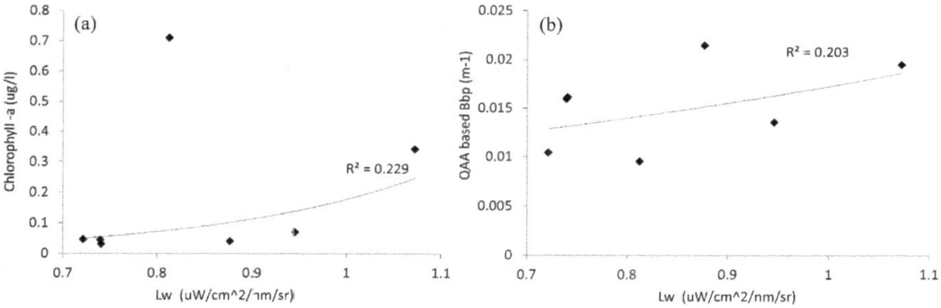

Fig. 4.6 Water-leaving radiance plotted against (**a**) chlorophyll-a and (**b**) QAA-based particulate backscattering coefficient (b_{bp}) for all seven plots in 50 m depth for 500 nm wavelength

dissolved or suspended materials. The algorithm may also not give better relationship if water is pigment-rich (like in ponds or lakes) or in a blooming condition. We argued on the correlation of L_w with Chl-a and b_{bp} for turbid coastal waters and identified that maximum scattering caused due to the other suspended matters not by Chl-a. We conclude that due to a broad spectrum of optical properties of waters found at different latitudes with varied environmental, a single empirical function cannot be applied to all, but may be for the Case I quality waters, where optical properties of water vary only with concentrations of chlorophyll present. Further research is required by considering other contributing constituents for backscattering in coastal waters with more number of observations in offshore waters to better discriminate the scattering coming from many other molecules.

Acknowledgement The authors are thankful to Project Director of ICMAM in facilitating the work.

References

Boss E, Stramski D, Bergmann T, Pegau WS, Lewis M (2004) Why should we measure the optical backscattering coefficient? Oceanography 17:44–49

Claustre H, Bishop J, Boss E, Bernard S, Berthon JF, Coatanoan C, Johnson K, Lotiker A, Ulloa A, Perry MJ, D'Ortenzio F, D'andon OHF, Uitz J (2010) Bio-optical profiling floats as new observational tools for biogeochemical and ecosystem studies- Proceedings of oceanObs'09: sustained ocean observations and information for society (Vol. 2), Venice, Italy

Deidun A, Drago A, Gauci A, Galea A, Azzopardi J, Mélin F (2011) A first attempt at testing correlation between MODIS Ocean colour data and in situ chlorophyll-a measurements within Maltese coastal waters. Proc of SPIE 8175:1–8

Dickey TD (2003) Emerging ocean observations for interdisciplinary data assimilation systems. J Mar Syst, 40– 41C, 5–48

Gordon HR, Morel A (1983) Remote assessment of ocean color for interpretation of satellite visible imagery: a review (eds: Barber RT, Mooers CNK, Bowman MJ, Zeitzschel B). Springer, New York

Gordon HR, Brown OB, Evans RH, Brown JW, Smith RC, Baker KS, Clarks DK (1988) A semianalytic radiance model of ocean color. J Geophys Res 93:10909–10924
Graff JR, Westberry TK, Milligan AJ, Brown MB, Dall'Olmo G, van Dongen-Vogels V, Reifel KM, Behrenfeld MJ (2015) Analytical phytoplankton carbon measurements spanning diverse ecosystems. Deep-Sea Res Pt I 102:16–25
Hu C, Feng L, Lee ZP (2013) Uncertainties of SeaWiFS and MODIS remote sensing reflectance: implications from clear water measurements. Remote Sens Environ 133:168–182
Kostadinov TS, Siegel DA, Maritorena S (2010) Global variability of phytoplankton functional types from space: assessment via the particle size distribution. Biogeosciences 7:3239–3257
Latha TP, Nagamani PV, Rao BS, Amarendra K, Rao KH, Choudhury SB, Dash SK (2013) Particle backscattering variability in the coastal waters of Bay of Bengal: a case study along off Kakinada and Yanam regions. Geosci Remote Sens Lett 10(6):1517–1521
Lee ZP, Hu C (2006) Global distribution of case-1 waters: an analysis from SeaWiFS measurements. Remote Sens Environ 101:270–276
Lee ZP, Carder KL, Peacock TG, Davis CO, Mueller JL (1996) Method to derive ocean absorption coefficients from remote-sensing reflectance. Appl Opt 35(3):453–462
Lee ZP, Carder KL, Arnone R (2002) Deriving inherent optical properties from water color: a multiband quasi-analytical algorithm for optically deep waters. Appl Opt 41:5755–5772
Lee ZP, Carder KL, Mobley CD, Steward RG, Patch JS (1999) Hyperspectral remote sensing for shallow waters: 2. Deriving bottom depths and water properties by optimization. Appl Opt 38:3831–3843
Li S, Song K, Mu G, Zhao Y, Ma J, Ren J (2016) Evaluation of the quasi-analytical algorithm (QAA) for estimating Total absorption coefficient of turbid inland waters in Northeast China. Appl Earth Observ Remote Sens 9(9):4022–4036
Loisel H, Meriare X, Berthan, Poteau A (2007) Investigation of the optical back scattering ratio of marine particles in relation to their biogeochemical composition in the eastern English Channel and southern North Sea. Limnol Oceanogr 52(2):739–752
Morel A (1988) Optical modeling of the upper ocean in relation to its biogenous matter content (case I waters). J Geophys Res 93:10749–10768
Morel A, Maritorena S (2001) Bio-optical properties of oceanic waters: a reappraisal. J Geophys Res 106:7163–7180
Nechad, B., Dogliotti, A., Ruddick, K., Doxaran, D. (2016) Particulate backscattering and suspended matter concentration retrieval from remote-sensed turbidity in various coastal and riverine turbid waters. In: Proceedings of ESA living planet symposium, Prague, 9–13 May 2016, ESA-SP 740
Riser SC, Johnson KS (2008) Net production of oxygen in the subtropical ocean. Nature 451 (7176):323–325
Seo H, Xie SP, Murtugudde R, Jochum M, Miller AJ (2009) Seasonal effects of Indian Ocean freshwater forcing in a regional coupled model. J Clim 22:6577–6596
Smith RC, Baker KS (1981) Optical properties of the clearest natural waters. Appl Opt 20:177–184
Stramski D, Boss E, Bogucki D, Voss KJ (2004) The role of seawater constituents in light backscattering in the ocean. Prog Oceanogr 61:27–56
Whitmire AL, Boss E, Cowles TJ, Pegau WS (2007) Spectral variability of the particulate backscattering ratio. Opt Express 15(11):7022

Chapter 5
Phytoplankton (chl-a) Biomass Seasonal Variability in the Gulf of Mannar, South India: A Remote Sensing Perspective

S. Kaliraj, N. Chandrasekar, and K. K. Ramachandran

Abstract The phytoplankton is being a primary producer that lies at the base of food web for aquatic flora and fauna of marine and coastal ecosystems. The study describes remote sensing applications for assessment of phytoplankton (chl-a) biomass variability and its major influencing factors (sea surface temperature, salinity, waves and currents) in the Gulf of Mannar (GoM), southeast coast of India. Multi-temporal Landsat ETM+ images acquired on 2016 and 2017 are used for mapping assemblage of phytoplankton (chl-a) and its concentration at site-specific level. Band combination analysis along with mathematically derived coefficient of determination values has been used to extract phytoplankton (chl-a) concentration using spectral reflectance properties (0.4–10 μm). Multispectral images acquired on different times are used for the mapping of phytoplankton concentration with its seasonal variability. Highest assemblage of phytoplankton (chl-a) is measured at concentration of 48.8 μg/L, and it reflects 0.25% at wavelength of 0.55 μm (green) and 1.0 μm (SWIR), respectively. Coastal water comprises higher chl-a concentration which can observe majority wavelength in blue (0.45 μm) and red (0.65 μm) that distinguish phytoplankton from coastal water. The result reveals that chl-a concentration has significantly decreased to 36.20 μg/L, and this reflects 0.16% of wavelength at 0.55 μm (green). The chl-a concentration is further decreased to 33.33 μg/L, and it reflects 0.17% at the wavelength of 0.55 μm. Seasonal assessment shows higher chl-a concentration during pre-monsoon. Coastal water in shallow depth (<5.0 m) area has estimated higher chl-a concentration within a distance of 1.0 km from the shore during post-monsoon. However, the chl-a concentration has significantly decreased in monsoon due to highly fluctuating hydrodynamic conditions

S. Kaliraj (✉) · K. K. Ramachandran
Central Geomatics Laboratory (CGL), ESSO-National Centre for Earth Science Studies (NCESS), Ministry of Earth Sciences, Government of India, Thiruvananthapuram, India
e-mail: s.kaliraj@ncess.gov.in; raman.kk@ncess.gov.in

N. Chandrasekar
Centre for GeoTechnology, Manonmaniam Sundaranar University, Tirunelveli, Tamil Nadu, India

Francis Xavier Engineering College, Tirunelveli, Tamil Nadu, India

P. Kumar et al. (eds.), *Applications and Challenges of Geospatial Technology*,
https://doi.org/10.1007/978-3-319-99882-4_5

that reduce availability of nutrients. Spatial variability of chl-a assemblage is mainly regulated by changing salinity and sea surface temperature. The coastal waters with a salinity level of 33.67 psu (practical salinity unit) at the temperature of 26.44 °C is found favour higher phytoplankton concentration (48.88 mg/m^3) in the post-monsoon, whereas considerable reduction of primary production to 17.10–33.33 mg/m^3 at the salinity level of 33.0–34.5 psu and this has been observed during monsoon. The phytoplankton concentration increases to 26.85–36.20 mg/m^3 at the salinity of 34.0–35.5 psu with the optimum temperature range of 28.0–30.5 °C during pre-monsoon. Sea surface temperature (SST) involves growth and productivity of phytoplankton (chl-a) in coastal waters. The phytoplankton productivity increases up to 33.33–48.48 mg/m^3 in the coastal water, and SST ranges from 26.0 to 28.0 °C; however, it is decreased to 22.5–24.5 mg/m^3 at the SST level of 28.5–30.0 °C. Growth and productivity of phytoplankton have increased in various parts during post-monsoon than monsoon and pre-monsoon because of the occurrence of optimum SST and salinity in coastal water and prevailing favourable hydrodynamic forces and climatic conditions. It is observed that phytoplankton (chl-a) concentration is gradually decreased with an increase of depth and distance in the Gulf of Mannar region.

Keywords Phytoplankton biomass · Seawater salinity · Sea surface temperature · Landsat ETM+ image · Remote sensing · Gulf of Mannar

5.1 Introduction

The coastal waters are characterized by diversity having the largest numbers of flora and fauna such as phytoplankton, zooplankton, sea grasses, coral reefs, etc. Phytoplankton biomass is a primary food source and is highly productive ecosystem of fish communities. (Chauhan et al. 2002). Coastal and marine ecosystems are dependent on phytoplankton biomass (chl-a) and its concentrations with high temporal and spatial variability due to availability of nutrients, which is often too difficult to characterize using in situ survey and with limited equipments available onboard (George and Heaney 1978; Sathe and Jadhav 2001; Selvavinayagam et al. 2003; Laws 2013). The phytoplankton communities in the coastal waters are diverse in nature and include approximately tens of thousands of phytoplankton species (Solanki et al. 2001). Seawater at each 100 ml contains thousands of flora and fauna especially phytoplankton species within every 100 m^2 of the area (Desortiva 1981; Bosart and Sprigg 1998). Phytoplankton productivity is highly sensitive to physical and chemical properties of coastal water that are frequently influenced by environmental and climatic factor with noteworthy shifting blooms or phenology of marine species, i.e., changing of salinity, temperature, and nutrient resulting from natural and anthropogenic impacts (Dey and Singh 2003). Traditionally bio-optical models have been used to categorize phytoplankton in the coastal waters (Beaver and Crisman 1991). Multispectral remote sensing images have contributed significantly to oceanographic studies because of higher spectral, spatial and radiometric resolutions, as evidenced by many scientific studies. The advancement of image processing techniques can retrieve phytoplankton assemblage from coastal water and other suspended matters (Barale 2010). Mapping of phytoplankton using spectral reflectance properties of image provides insight for understanding phenology of marine ecosystems in the coastal water (Nagamani et al. 2011). Satellite image-derived phytoplankton (chl-a) biomass in coastal water estimated at spatial scale is highly correlated to in situ sampling measurement and used to identify its spatio-temporal variability. Multispectral remote sensing images are being most suitable for oceanographic studies such as phytoplankton (chl-a) concentration, sea surface temperature, sea surface salinity, movement of suspended nutrients, etc. (Roy El Hourany et al. 2017). Spectral reflectance properties of coastal water are a primary indicator for assessment of phytoplankton concentration in the coastal waters that can use multispectral images for the purposes (Singh and Chaturvedi 2010). Assemblage of phytoplankton communities in coastal water and their dispersal have noticed spatial and temporal variability, so that the chl-a estimation may provide potential result by using remotely sensed images like Landsat, OCM, MODIS and Sentinel, etc. depends on spatial, spectral and temporal resolutions. Phytoplanktons occur in coastal water that reflects a variety of spectral reflectances at visible and infrared wavelengths depending on their phenology and concentration. Spatial diversity of phytoplankton communities may reflect noticeable amount of energy (spectral radiance), which quiet differently from leaving radiances by clear water and other suspended particles. Depth variation of coastal water determines vertical distribution of phytoplankton that affects primary productivity as well as energy transfer from surface to deep tropic levels. Spectral reflectivity of phytoplankton distinguishes optical properties from clear seawater, allowing its relatively straightforward retrieval from remote sensing images. Seasonal variability of phytoplankton biomass using multi-temporal satellite images may provide considerable output due to variability of optical reflectance in coastal waters during different seasons (Sarangi 2011; Chen and Quan 2013). Satellite remote sensing can be used as crucial tool for discriminating spectral and spatial properties of suspended matters and clear water, and this technique may help to differentiate phytoplankton assemblage (Behrenfeld and Boss 2006), and variability in spectral reflectance is increasingly applied to retrieve phytoplankton (chl-a) biomass in coastal water (Gelder et al. 1997; Ahn et al. 2008).

Quantitative measurement of phytoplankton (chl-a) concentration in coastal water has noticed high correlation with measured radiances from the images like Landsat TM and ETM+, Ocean Colour Monitor (OCM), MODIS, SeaWiFS, Sentinel OLCI, etc. (Vaillancourt et al. 2004). Monitoring seasonal variability of phytoplankton (chl-a) in coastal water can be achieved successfully using multi-temporal images that may be an alternating technology to in situ measurement. Mapping the spatial assemblage of phytoplankton (chl-a) using multispectral images may provide insights of spatio-temporal dynamics, bloom initiation, peak and declining stages (Siegel et al. 2013). Growth and productivity of phytoplankton (chl-a) are mainly dependent on SST and salinity which can be assessed using satellite images along with their spatial dispersal (Reynolds and Smith 1994; Racault et al. 2015). Physico-chemical properties like sea surface temperature (SST) and salinity (Burrage and Wesson 2008) can be extracted from the satellite images like Landsat ETM+, MODIS, OCM and SeaWiFS (Balzano et al. 2015). Landsat ETM+ images provide significant results in chl-a mapping compared to other images, since the sensor onboard has superior radiometric sensitivity to the reflected spectra from the suspended objects in the seawater. Many studies are executed to extract phytoplankton from the seawater using various band combination analyses and multivariate algorithms that segregate phytoplankton concentration exclusively based on the spectral radiance values from satellite images (Burrage and Wesson 2008; Wallhead et al. 2014).

In the Gulf of Mannar located along the southeast coast of India, seawater shows spectral variability due to dispersal of organic and inorganic materials, whereas phytoplankton and suspended sediments can distinctly extract from seawater (Sarangi 2011). Band combination analysis calculates phytoplankton by differentiating spectral reflectance at different wavelengths (Chauhan et al. 2002; Sarangi et al. 2008). Spectral variability observed from VNIR and SWIR bands of Landsat ETM+ images shows seasonal changes of phytoplankton (chl-a) concentration in coastal water at pixel scale (Nagamani et al. 2008; Palmer et al. 2015). The Gulf of Mannar experiences seasonal variability of phytoplankton (chl-a) concentration due to changes in SST and salinity. The present investigates spatial variability of phytoplankton and its influencing factors at seasonal scale using multispectral Landsat ETM+ images.

5.2 Gulf of Mannar: Geographical Profile

The Gulf of Mannar is the world's second largest marine bioreserve that lies in the Bay of Bengal sea, southeast coast of India, which is located between Tuticorin and Rameswaram Island. The GoM covers the longitude of 78° 5′ 00′–79° 30′ 00′E and latitude of 8° 47′ 00′N–9° 15′ 00′N with the geographical area extent of 140 km (see Fig. 5.1). The GoM consists of 21 small islands located in various parts with the distance of 8 Km from the coastline. The islands of GoM endow a variety of marine ecosystems that comprise dunes, beaches, mangroves, sea grass, seaweeds and coral reefs, which provide enough nutrients for phytoplankton growth and productivity (Selvavinayagam et al. 2003). The Gulf of Mannar extends with the length of 140 km towards SW-NE direction and the total area of 10,500 km^2. The Gulf of Mannar consists of Quaternary to recent sediment formations, which exhibit various types of coral reefs such as fringing, patch and coral pinnacles. The GoM prevails sub-tropical climatic condition with optimum temperature of 21–24 °C and average rainfall of 10–16 cm. The GoM experiences lower waves and current flow, transporting sediments with enriched nutrients from various parts of the area; however, sea surface is often choppy during late spring and through summer due

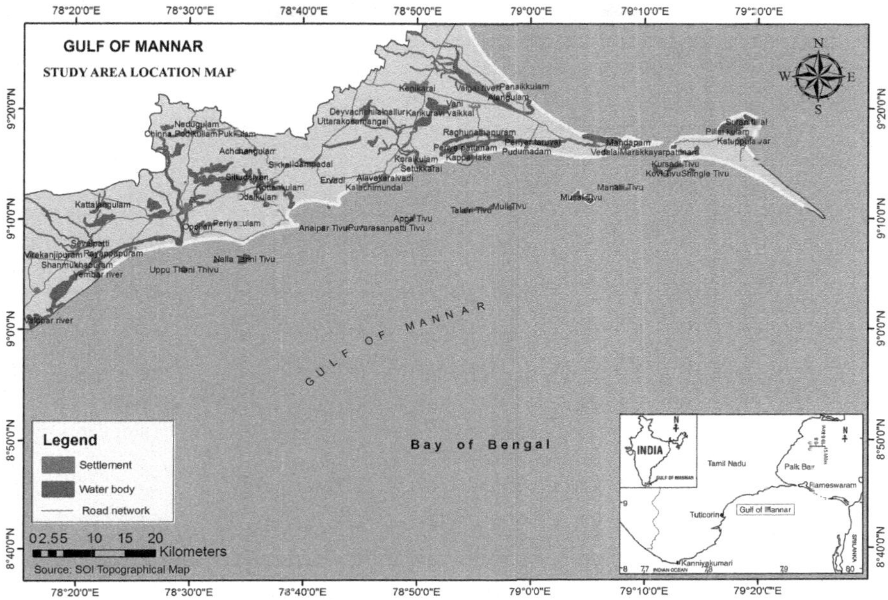

Fig. 5.1 Geographical location of the Gulf of Mannar

to tropical cyclones and storm surge. The productivity of phytoplankton influences the life cycle of marine ecosystems, and changes in phenology may decrease growth and productivity of zooplankton and fish communities (Sarangi et al. 2005; Singh and Chaturvedi 2010). In the GoM region, the growth and productivity of phytoplankton are frequently influenced by physical and chemical properties, i.e. sea surface temperature and salinity. Seasonal variability of phytoplankton biomass is important for understanding the flora and fauna communities in the GoM marine reserve.

5.3 Materials and Method

Landsat ETM+ image shows spectral reflectance variability that differentiates phytoplankton (chl-a) and clear seawater in the Gulf of Mannar. Landsat ETM+ (Enhanced Thematic Mapper Plus) visible, NIR, and mid-IR at spatial resolution of 30 m have been widely used for estimation of phytoplankton concentration in the seawater (see Fig. 5.2). Phytoplankton (chl-a) shows highest reflection in red (band 6) and absorption in near-infrared band (band 5). Spectral reflectance variability of these bands involves the extraction of distinct assemblage of chl-a from seawater and other inorganic matters suspended in the near offshore area (Barale 2010; Sarangi 2011).

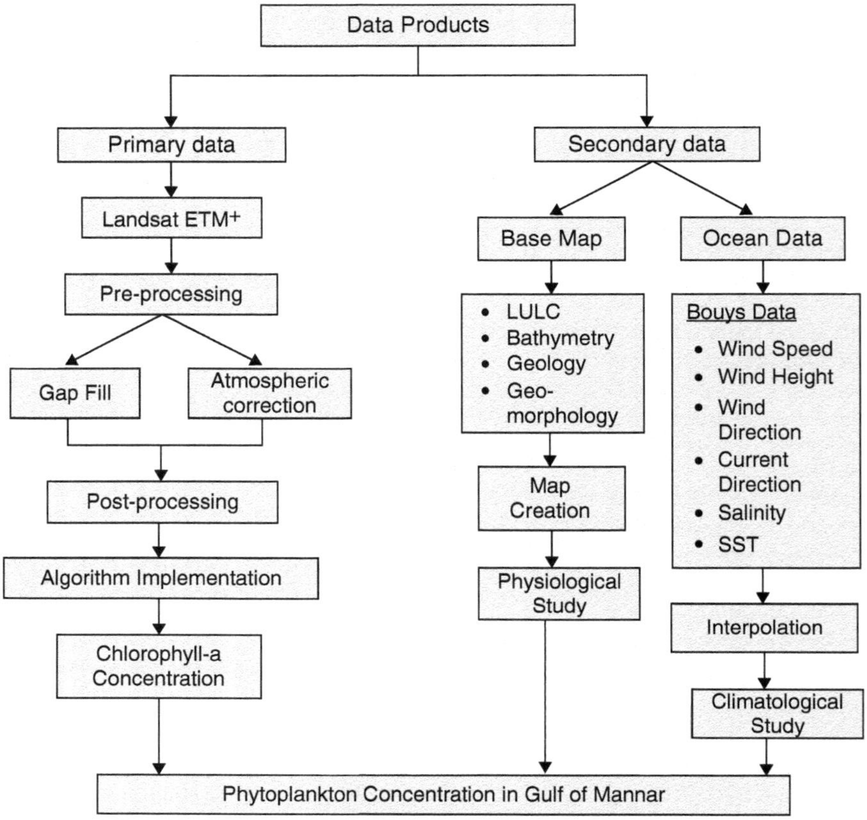

Fig. 5.2 Methodological flow chart of the present study

Landsat ETM+ images acquired on April and October 2016 and February 2017 have been used for estimation of seasonal variability of phytoplankton concentration. The images are systematically processed using advanced image processing techniques to obtain their actual spectral reflectance to extract the occurrence of phytoplankton concentration in the coastal water. To obtain the actual reflectance of phytoplankton, suspended matter and clear water, all the VNIR and SWIR bands of Landsat ETM+ image are atmospherically corrected using FLAASH (Fast Line-of-sight Atmospheric Analysis of Spectral Hypercubes) model in ENVI 4.7 software. Subsequently, the images have been applied for systematic band combination analysis using mathematical algorithm to demarcate the phytoplankton assemblage from seawater and other suspended matters.

5.3.1 *Extraction of Phytoplankton (chl-a) Concentration*

Landsat ETM+ images of VNIR and SWIR bands are analysed using band combination technique by applying coefficient values using GIS software for the assessment of phytoplankton (chl-a) concentration. Landsat ETM+ includes seven bands with 30 m spatial resolution except band 6 (TIR with 60 m) such as visible bands, namely, ETM_1 (blue, 0.45–0.51 μm), ETM_2 (green, 0.525–0.605 μm) and ETM_3 (red, 0.63–0.690 μm). NIR band of ETM_4 (0.75–0.90 μm), SWIR bands of ETM_5 (1.55–1.75 μm) and ETM_7 (2.09–2.35 μm). The regression analysis of ETM_6 (TIR) is used to measure SST at the wavelength of 10.40–12.50 μm. The band combination analysis of Landsat ETM+ bands of VNIR and SWIR is used to extract phytoplankton (chl-a), in which regression analysis of band 2 and band 4 using coefficient of determination ($R^2 = 0.677$) and error of dependent parameter is estimated as (root mean square error) RMSE = 0.958 gμ/l). Phytoplankton (chl-a) and suspended matter have been differentiated based on actual spectral radiance using linear equation applied on ETM_1, ETM_2, and ETM_3 by applying coefficient of determination value ($R^2 = 0.542$), and error of dependent parameter is estimated as (RMSE = 1.465 mg/l). The band ratio analysis of Landsat ETM+ image is used to establish logarithmically the transformation of spectral reflectance of all objects to demarcate phytoplankton (chl-a) concentration in the Gulf of Mannar. Estimation of phytoplankton (chl-a) concentration using standard mathematical equation with laboratory-derived coefficient of determination values is expressed as follows:

$$\begin{aligned}\mathrm{Chl(a)} = {} & 15.1329 - 0.1573(\mathrm{ETM}_1) + 0.0394(\mathrm{ETM}_2) + 0.2671(\mathrm{ETM}_3) \\ & - 0.5383(\mathrm{ETM}_4) + 0.1214(\mathrm{ETM}_5) - 0.0606(\mathrm{ETM}_6) \\ & + 0.1007(\mathrm{ETM}_7)\end{aligned} \quad (5.1)$$

5.3.2 *Sea Surface Temperature Estimation*

Sea surface temperature (SST) is estimated using Landsat ETM+ TIR (band 6) based on object emissivity properties. The regression algorithm is applied on ETM+ band 6 (10.40–12.50 μm) for converting spectral radiance into surface temperature (Reynolds and Smith 1994; Kilpatrick et al. 2001; Solanki et al. 2001). Landsat ETM+ image's band 6 (TIR) is used to estimate sea surface temperature (SST) using regression algorithm, and the equation is represented as follows:

$$\mathrm{SST} = 36.3409 - 0.2613\ (\mathrm{ETM}\ 6_1) + 0.0010\ (\mathrm{ETM}\ 6_2)^2 \quad (5.2)$$

Where, coefficient of determination is $R^2 = 0.157$ and RMSE = 0.814.

5.3.3 Seawater Salinity Estimation

Seawater salinity is mainly dependent on chemical properties (Na and Cl) of the seawater that produces distinct variability in spectral reflectance of Landsat ETM+ image (Burrage and Wesson 2008). Salinity characteristics of seawater can be successfully estimated using spectral data that incorporated laboratory-derived coefficient values (Maged and Mazlan 2009). Seawater salinity interacts significantly with optical properties (spectral reflectance) of seawater and other suspended matters due to reflection or absorption at different wavelengths (McCarthy et al. 2015). Seawater salinity has produced various spectral reflectances based on salt concentration. The spectral reflectance of saline water is high in band 3 of the Landsat ETM + image; using the algorithm proposed by, seawater salinity is estimated with relatively accurate and the algorithm is expressed as follows:

$$\text{Salinity} = -102 + 9 \cdot 8 \times \text{ETM}_3 \tag{5.3}$$

5.4 Results and Discussion

Spectral variability of Landsat ETM+ bands plays a vital role for the estimation of phytoplankton (chl-a) concentration in the GoM region. The phytoplankton assemblage occurs in the coastal water and may result in unique spectral properties due to absorption (NIR) and reflection (Red) of EMR at different wavelengths. Based on this concept, the empirical algorithm is executed on all bands of ETM+ image for the estimation of phytoplankton concentration in pixel scale.

5.4.1 Landsat ETM Image Reflectance Response to Phytoplankton Concentration

Phytoplankton concentration in the coastal water consists of organic matters derived from the physical and climatological factors. The reflectance value of image pixel depends upon the concentration of phytoplankton (chl-a) at various stretches of wavelength that can alter the optical properties of the water column. The spectral reflectance of all ETM+ bands (VNIR and SWIR) clearly differentiate phytoplankton and other suspended matter (inorganic contents) from coastal water (see Fig. 5.3).

Landsat ETM+ image shows high spectral reflectance from coastal water which consists of higher concentration of phytoplankton (chl-a) in visible bands (blue, green and red bands) and lower reflectance in wavelength of NIR bands. The phytoplankton (chl-a) concentration has produced relatively nearest reflection

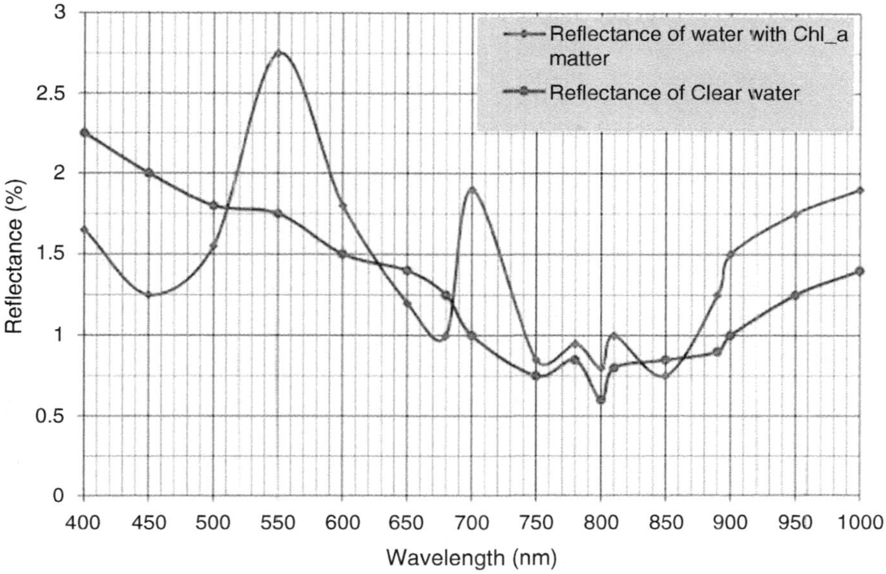

Fig. 5.3 Variability of spectral reflectance of phytoplankton (chl-a) assemblage and seawater from the Landsat ETM+ image bands of visible, NIR and SWIR

which ranges from 0.05% to 0.25% within the wavelength of 0.4–10.0 μm. The separation of spectral curves within this wavelength is more difficult to distinguish phytoplankton (chl-a) concentration from water content. This is due to the low amount of backscattering by phytoplankton (chl-a) concentration present in surface water. Highest phytoplankton (chl-a) concentration at 33.33 μg/L and above has been observed at the reflectance range from 0.17% to 0.25%. This analysis reveals that the response of reflectance in the image tends to increase with the increase of phytoplankton (chl-a) concentration in wavelength of all bands with a few mirror exception in the NIR spectrum. This is due to the wavelength of green (525–605 μm) and red (630–690 μm) bands. Comparative analysis of reflection and wavelength at different phytoplankton (chl-a) concentrations produces non-linear spectral profile, and that indicates the distribution of phytoplankton concentration in surface water; the spectral response of three different seasons are accurately demarcated (see Fig. 5.4). Phytoplankton (chl-a) concentration has often reflected more radiation in the visible portion than the infrared portion which includes NIR, mid-IR, SWIR and TIR bands. The variation in reflectance from individual band of ETM+ image is attributed to the estimation of phytoplankton (chl-a) concentration and dispersal. Assessment of seasonal variability of phytoplankton biomass is important for understanding the physical, chemical and environmental changes of marine ecosystem due to its influence of primary productivity in the coastal waters (Graf JR et al. 2015). The multi-temporal satellite images are primarily used for retrieving and mapping

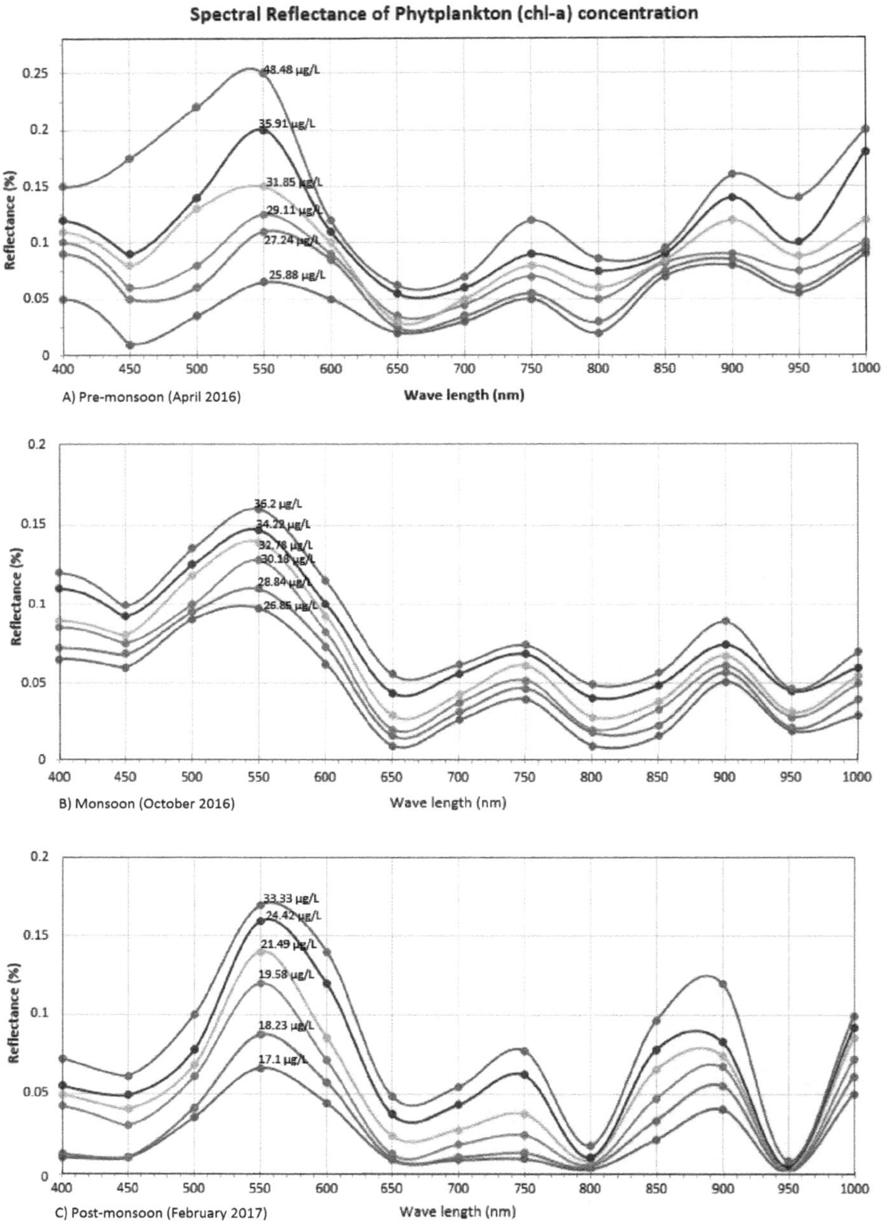

Fig. 5.4 Multispectral image reflectance response to phytoplankton concentration

the assemblage of phytoplankton known as phenology with its dispersal in seasonal scale.

5.4.2 Estimation of Phytoplankton Concentration

Landsat ETM+ image-derived phytoplankton (chl-a) concentration has been estimated using multivariate regression algorithm in GIS software environment. The phytoplankton (chl-a) concentration and its spatial dispersal have been estimated using multi-temporal Landsat ETM+ image acquired on 2016 (see Fig. 5.5). The result reveals that the phytoplankton (chl-a) concentration is estimated with ranges from 25.88 to 48.88 μg/L during pre-monsoon, 26.85–36.2 μg/L during monsoon and 17.1–33.33 μg/L during post-monsoon. Among them, coastal water surface that extends (8709 km^2) with the distance of 1 km from shoreline less than 10 m depth to seabed has been estimated with high phytoplankton (chl-a) concentration as 48.48 mg/L.

5.4.3 Spatio-Temporal Variation of Phytoplankton (chl-a) Biomass

Spatial dispersal of phytoplankton (chl-a) biomass and its seasonal variability have been demarcated for GoM region. The growth and productivity of phytoplankton (chl-a) biomass are highly sensitive to physical and chemical properties (SST and salinity) and hydrodynamic forces like waves and currents. Seasonal variability of plankton assemblage at various parts depends on waves and currents (Esaias et al. 1998; Burroughs 2007) and monsoonal changes in hydrodynamic processes which control their spatial dispersal. Phytoplankton (chl-a) concentration is estimated at higher rate (35.92–48.48 μg/L) in coastal water within the distance of less than 1 km from shore and depth of 10–100 m during pre-monsoon, and this is slightly decreased to 24.43–33.33 μg/L in monsoon and 34.23–36.20 μg/L in post-monsoon. The variation of salinity and SST influences growth and productivity of phytoplankton in coastal water column, and their assemblage varies at different places resulting from hydrodynamic processes. Quantitative measurement of phytoplankton (chl-a) and major influencing factors of SST and salinity in various spots are shown in Table 5.1.

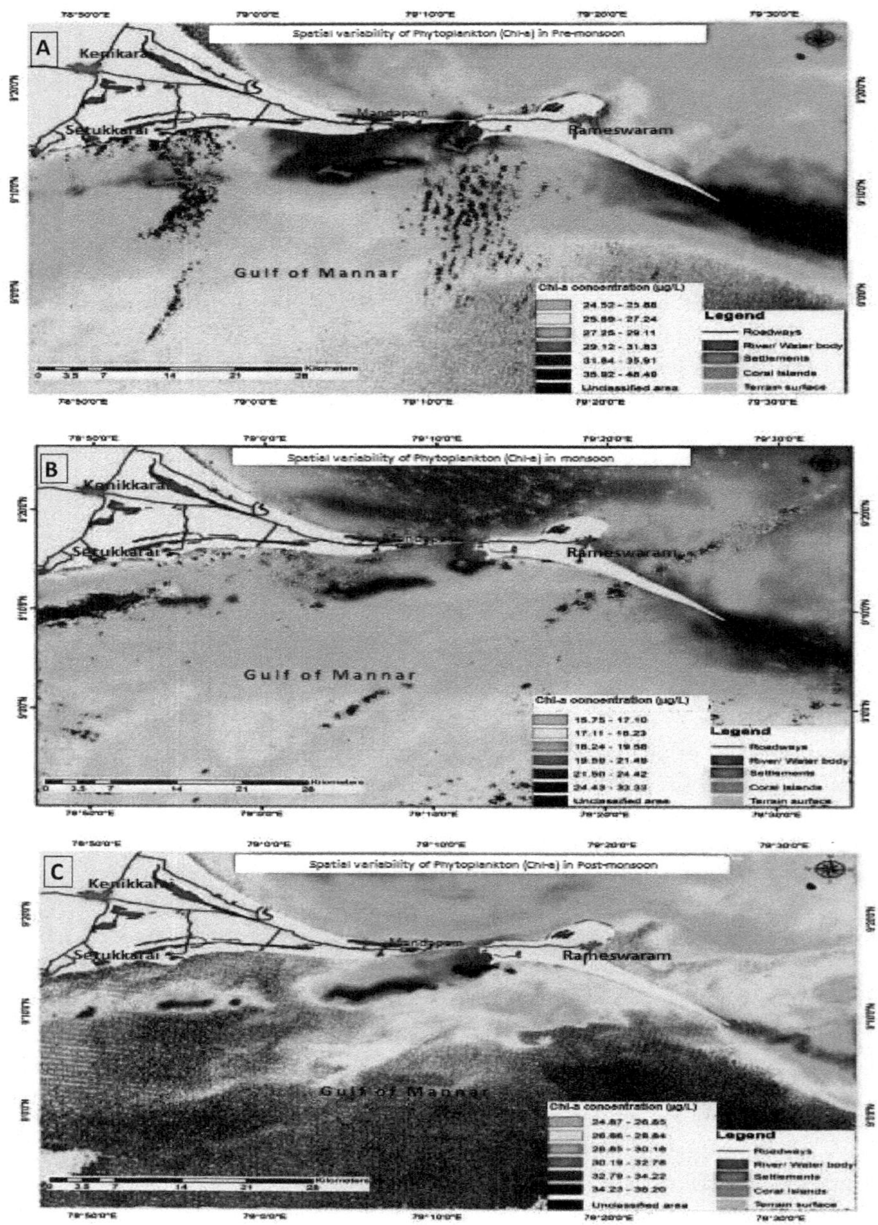

Fig. 5.5 Spatial assemblage of phytoplankton (chl-a) along the Gulf of Mannar at seasonal scale, pre-monsoon (**a**), monsoon (**b**), post-monsoon (**c**)

Table 5.1 Estimated rate of phytoplankton (chl-a) concentration using Landsat ETM+ images and the quantitative values of factors influencing during pre-monsoon (April 2016), monsoon (October 2016) and post-monsoon (February 2017)

Phytoplankton (chl-a) concentration (μg/L) Pre-monsoon	salinity (oo/o)	Temperature (°c)	Bathymetry (m)	Area (sq km)
→ 25.88	↓ 17.04	→ 19.53	→ 30	↑ 1288.333
→ 27.24	↓ 18.54	→ 19.67	→ 20	↑ 1859.58
→ 29.11	↓ 19.02	→ 19.91	→ 20	↑ 1325.7522
→ 31.85	→ 20.42	→ 21.12	↓ 10	→ 627.696
↑ 35.91	→ 22.19	→ 21.7	↓ 10	↓ 228.987
↑ 48.48	→ 23.67	→ 22.44	↓ 5	↓ 134.8659

Phytoplankton (chl-a) concentration (μg/L) Monsoon	salinity (oo/o)	Temperature (°c)	Bathymetry (m)	Area (sq km)
↓ 26.85	↓ 18.22	→ 20.65	↑ 30	↓ 651.465
↓ 28.84	↓ 17.56	→ 19.87	→ 20	↑ 2158.722
→ 30.18	→ 19.78	→ 24.27	↓ 15	↑ 1534.732
→ 32.78	→ 21.04	→ 22.37	↓ 10	→ 870.2496
↑ 34.22	↑ 22.17	↑ 26.17	↓ 10	↓ 425.3697
↑ 36.2	↑ 23.97	↑ 28.807	↓ 5	↓ 64.9287

Phytoplankton (chl-a) concentration (μg/L) Post-monsoon	salinity (oo/o)	Temperature (°c)	Bathymetry (m)	Area (sq km)
→ 17.1	↓ 11.36	↓ 19.27	↑ 30	↓ 109.9062
→ 18.23	↓ 12.74	↓ 19.31	→ 20	↓ 64.2249
→ 19.58	↓ 14.32	↓ 19.39	→ 20	↓ 421.7598
→ 21.49	→ 16.72	↓ 19.43	↓ 10	→ 454.2291
↑ 24.42	→ 19.52	↓ 19.77	↓ 7	↑ 1185.579
↑ 33.33	→ 23.17	↑ 21.7	↓ 5	→ 712.921

5.4.4 *Factor Influencing Phytoplankton (Chl-s) Dispersal*

5.4.4.1 Seawater Salinity

Seawater salinity is a dynamic entity and depends on chemical components of Na and Cl. Phytoplankton concentration reaches higher rate with lower salinity in coastal water. It describes that higher salinity may be found with lethal limit for growth and productivity of phytoplankton communities. Salinity of coastal water is being a medium for phytoplankton productivity by maintaining osmotic balance (Emery and Thomson 2001). Change of salinity causes osmotic stress on uptake or loss of nutrients by phytoplankton species. Seawater salinity is strong enough to control the growth of phytoplankton community and establish their spatial assemblage in specific sites. Seawater salinity can be estimated using Aqua/ MODIS, Landsat ETM and OCM images (Wong et al. 2007). The Landsat ETM+ image has been used to measure seawater salinity using standard regression algorithm proposed based on appropriate coefficient values. In the GoM region, the mean seawater salinity has been estimated with a range of 17–23 psu. Seasonal variations of seawater salinity were noted with a range of 17.04–28.67 psu in pre-monsoon; and during the periods, the phytoplankton concentration lever has reached at 25.88–48.88 μg/L. The seawater salinity has changed to 18.22–23.97 psu during monsoon, and phytoplankton concentration level is assessed at the rate of

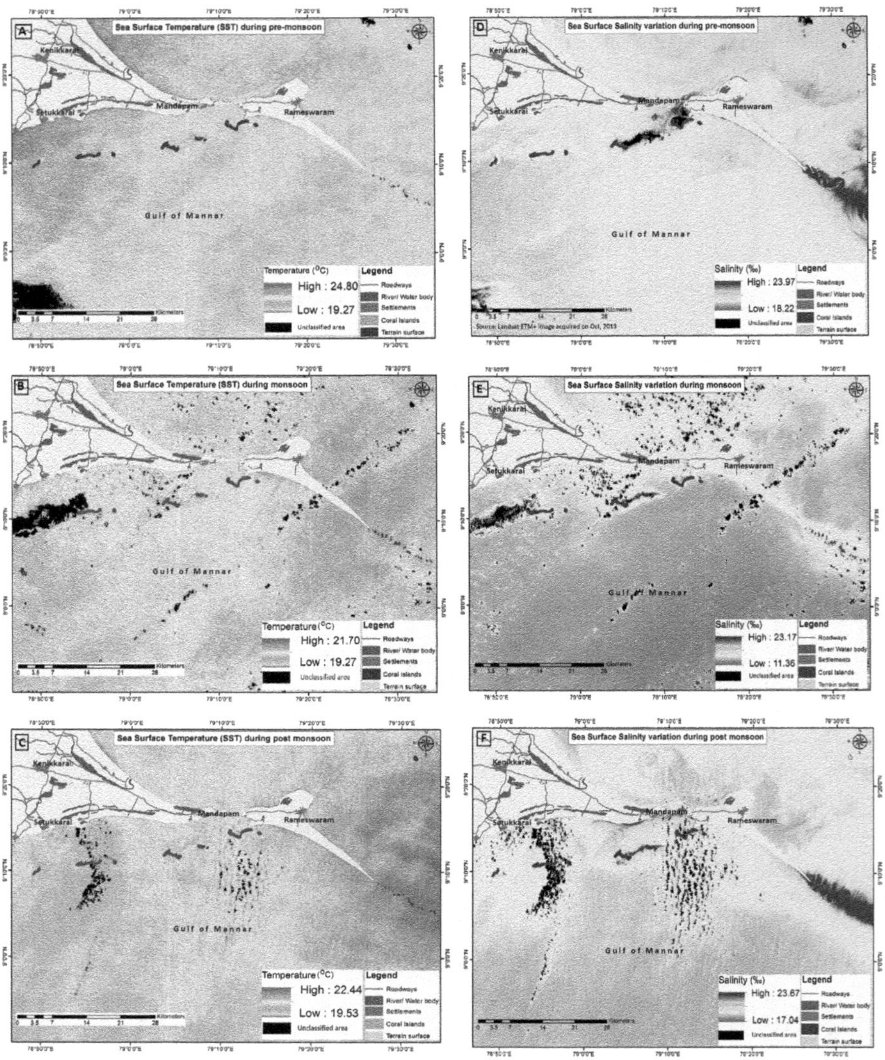

Fig. 5.6 Sea surface temperature (**a**, **b** and **c**) and salinity (**d**, **e** and **f**) in the Gulf of Mannar at seasonal scale where SST and salinity are represented as (**a** and **d**) is pre-monsoon, (**b** and **e**) is monsoon, (**c** and **f**) is post-monsoon, respectively

24.43–33.33 μg/L. However, seawater salinity prevails at the level of 11.36–23.17 psu during the post-monsoon and this has noticed as optimum salinity level for higher growth and productivity of phytoplankton and it estimated with the range of 34.23–38.20 μg/L while compared to rest of seasons in the GoM region (Fig. 5.6).

5.4.4.2 Sea Surface Temperature

Sea surface temperature (SST) varies from place to place and at vertical range of water column, which highly influences phytoplankton growth and productivity. Seasonal variation of SST determined phytoplankton assemblage and spatial patterns that shows the annual cycles of biomass variability and their recurrence strength (Walton et al. 1998; Wallhead et al. 2014). The SST variation in the GoM region is among the major determinants to influence phytoplankton growth rates and spatial and temporal distribution. In addition to that, the seasonal variability of SST regulates phytoplankton productivity by controlling photosynthesis that differently responds to different temperature regimes. The SST at pixel scale has been measured using TIR (band 6) of Landsat ETM+ image by applying the algorithm and the estimated SST at the range of 19.53–22.44 °C during pre-monsoon. The SST level has raised up to 20.65–28.81 °C in the monsoon period, and further the SST come down with range of 19.27–21.72 °C during post-monsoon. Meanwhile, the SST prevails during post-monsoon and is noticed as optimum temperature for constant productivity of phytoplankton communities, and the rate of concentration is estimated at 34.23–38.20 μg/L in coastal water of the GoM region.

5.4.4.3 Wave Height and Littoral Current

The wave height annually ranges from 2.7 to 5.155 m in the Gulf of Mannar. The dispersal of phytoplankton (phytoplankton (chl-a) concentration increases (33.33–48.48 μg/L) with the decrease in wave height at 2.7 m, and the dispersal of phytoplankton (phytoplankton (chl-a) concentration conversely decreases (17.1–25.8 μg/L) with increase in wave height at 5.155 m. The wave height is influenced corresponding to the wave direction, so this influences the wave height to increase along the deep sea bed and decrease along the shoreline. Spatial dispersal of phytoplankton communities depends on waves, wind and currents; however, their growth and productivity are regulated by SST and salinity. Wind-driven external factors transport phytoplankton assemblage towards the direction of waves and currents; wind-induced lateral processes have transported horizontally plankton assemblage towards shoreline and accumulated at lower depth of offshore. However, the movement of plankton assemblage at coastal water column depends on upwelling or downwelling processes resulting from induction of internal waves, and vertical distribution of phytoplankton is dynamically changed by seasonal hydrodynamic processes.

In the GoM region, the flow of littoral current influences phytoplankton concentration and its spatial dispersal in coastal water by controlling coastal system within the surf zone (Esaias et al. 1998; Dey and Singh 2003). In the Gulf of Mannar, the littoral current prevails with velocity range of 3.7–62.69 m/s, and this process has accumulated phytoplankton assemblage at higher level (33.33–48.48 μg/L) in coastal water of lower depth area within the distance of less than 1 km from the shoreline, whereas offshore area with a distance of more than 1 km has noticed littoral current velocity at 10.72 m/s, and the dispersal of phytoplankton (phytoplankton (chl-a) concentration decreases up to 17.1–25.8 μg/L. It is observed that the change of littoral current velocity subject to shoreline distance and depth of coastal water determines spatial dispersal of phytoplankton concentration along the GoM region.

5.5 Conclusion

Phytoplankton concentration and its seasonal variability in coastal water have been extracted using Landsat ETM+ image in GIS software environment. The spectral reflectance of VNIR and SWIR bands indicates the occurrence of phytoplankton (chl-a) in the coastal water. Band combination analysis of the total range of wavelength (0.45–1.0 μm) shows high phytoplankton (chl-a) concentration measured during pre-monsoon; however, constant productivity occurs in post-monsoon. Moreover, high phytoplankton (chl-a) is estimated along the shallow depth in the natural area, especially in the eastern part due to the large quality of river discharge materials shoaling by waves and current. Whereas the water surface away from the shore is found with low phytoplankton (chl-a) concentration, it tends to increase the depth and distance, and the low assemblage of them is only available to spread along the vast water surface area. The growth and productivity of phytoplankton depend on seasonal changes of SST and salinity, whereas their spatial dispersal is controlled by seasonal dynamics of waves and currents. The phytoplankton (chl-a) concentration in coastal water has higher rate in various parts during post-monsoon and lower level during pre-monsoon. It is observed that the assemblage of phytoplankton (chl-a) spread at high concentration in coastal water with lower depth and within the distance of 1 km to shoreline. It is concluded that phytoplankton (chl-a) concentration is inversely proportional to the depth and distance to the shoreline and has direct relationship with wave direction, littoral currents, SST and salinity prevailing over the GoM region. The phytoplankton (chl-a) concentration shows higher correlation with the logical relation of physical parameters like SST and salinity. Remote sensing-derived output of this study provides useful information for identification of potential fishing zone and marine ecosystem studies for researchers and management authorities.

Acknowledgement The corresponding author is thankful to the Director, ESSO – National Centre for Earth Science Studies, Ministry of Earth Sciences, Government of India, for continuous encouragement and support. The authors acknowledge the USGS EarthExplorer, NIOT, INCOIS and SOI for providing necessary data sources.

References

Ahn YH, Shanmugam P, Moon JE, Ryu JH (2008) Satellite remote sensing of a low salinity waters plume in the East China Sea. Ann Geophys 26:2019–2035

Balzano S, Ellis AV, Lan CL, Leterme SC (2015) Seasonal changes in phytoplankton on the north-eastern shelf of Kangaroo Island (South Australia) in 2012 and 2013. Oceanologia 57:251–262

Barale V (2010) Oceanography from space: revisited. Springer, Ispra, pp 237–238

Beaver JR, Crisman TL (1991) Temporal variability in algal biomass and primary productivity in Florida lakes relative to longitudinal gradients, organic colour and trophic states. Hydrobiologia 224:89–97

Behrenfeld MJ, Boss E (2006) Beam attenuation and phytoplankton (chl-a) concentration as alternative optical indices of phytoplankton biomass. J Mar Res 64:431–451

Bosart LF, Sprigg WA (1998) The meteorological buoy and coastal marine automated network for the United States. National Academies Press, Washington, DC

Burrage D, Wesson J (2008) Deriving sea surface salinity and density variations from satellite and aircraft microwave radiometer measurements: application to coastal plumes using STARRS. IEEE Trans Geosci Remote Sens 46(3):765–785

Burroughs JW (2007) Climate change: a multidisciplinary approach, 2nd edn. Cambridge University Press, Cambridge

Chauhan P, Mohan M, Sarngi RK, Kumari B, Nayak S, Matondkar SGP (2002) Surface phytoplankton (chl-a) a estimation in the Arabian Sea using IRS-P4 Ocean Colour Monitor (OCM) satellite data. Int J Remote Sens 23(8):1663–1676

Chen H, Quan W (2013) An improved algorithm for retrieving phytoplankton (chl-a)-a from the Yellow River Estuary using MODIS imagery. Environ Monit Assess 185:2243–2255

Desortiva B (1981) Relationship between phytoplankton (chl-a) concentration and phytoplankton biomass in several reservoirs in Czechoslovakia. Int Revue Ges Hydrobiol 66:153–169

Dey S, Singh RP (2003) Comparison of phytoplankton (chl-a) distributions in the northeastern Arabian Sea and southern Bay of Bengal using IRS P4 ocean color monitor data. Remote Sens Environ 85:424–428

El Hourany R et al (2017) Spatio-temporal variability of the phytoplankton biomass in the Levantine basin between 2002 and 2015 using MODIS products. Oceanologia 59:153 http://dx.doi.org/10.e1016/j.oceano.2016.12.002

Emery WJ, Thomson RE (2001) Data analysis methods in physical oceanography. Gulf Professional Publishing, Waltham

Esaias WE, Abbott MR, Barton I, Brown OB, Campbell JW, Carder KL, Clark DK, Evans RH, Hoge FE, Gordon HR, Balch WM, Letelier R, Minnett PJ (1998) An overview of MODIS capabilities for ocean science observations. IEEE Trans Geosci Remote Sens 36:1250–1265

Geider R, MacIntyre H, Kana T (1997) Dynamic model of phytoplankton growth and acclimation: responses of the balanced growth rate and the phytoplankton (chl-a) a: carbon ratio to light, nutrient-limitation and temperature. Mar Ecol Prog Ser 148:187–200

George DG, Heaney SI (1978) Factors influencing the spatial distribution of phytoplankton in a small productive lake. J Ecol 66:133–155

Graf JR et al (2015) Analytical phytoplankton carbon measurements spanning diverse ecosystems. Deep-Sea Res I(102):16–25

Kilpatrick KA, Podesta GP, Evans RH (2001) Overview of the NOAA/NASA pathfinder algorithm for sea surface temperature and associated matchup database. J Geophys Res 106:9179–9198

Laws EA (2013) Evaluation of in situ phytoplankton growth rates: a synthesis of data from varied approaches. Annu Rev Mar Sci 5:247–268

Maged M, Mazlan H (2009) Linear algorithm for salinity distribution modeling from MODIS data. In: CD-ROM proceeding of IGARSS 2009. Cape Town University, Cape Town, pp 1–4

McCarthy GD, Haigh ID, Hirschi JJM, Grist JP, Smeed DA (2015) Ocean impact on decadal Atlantic climate variability revealed by sea-level observations. Nature 521:508–510

Nagamani PV, Chauhan P, Dwivedi RM (2008) Development of chlorophyll-a algorithm for Ocean Colour Monitor onboard OCEANSAT-2 satellite. IEEE Geosci Remote Sens Lett 5(3):527–531
Nagamani PV, Shikhakolli R, Chauhan P (2011) Phytoplankton variability in the Bay of Bengal during winter monsoon using Oceansat-1Ocean Colour Monitor Data. J Indian Soc Remote Sens 39(1):117–126
Palmer SCJ, Odermatt D, Hunter PD, Brockmann C, Presing M, Balzter H, Toth VR (2015) Satellite remote sensing of phytoplankton phenology in Lake Balaton using 10 years of MERIS observations. Remote Sens Environ 158:441–452
Racault MF et al (2015) Phytoplankton phenology indices in coral reef ecosystems: application to ocean-color observations in the Red Sea. Remote Sens Environ 160:222–234
Reynolds RW, Smith TM (1994) Improved global sea surface temperature analysis using optimum interpolation. J Clim 7:929–948
Sarangi RK (2011) Impact of cyclones on the Bay of Bengal phytoplankton (chl-a) variability using remote sensing satellites. Indian J Geo-Mar Sci 40(6):794–801
Sarangi RK, Chauhan P, Nayak SR (2005) Inter-annual variability of phytoplankton blooms in the northern Arabian Sea during winter monsoon period (February-March) using IRS-P4 OCM data. Indian J Mar Sci 34(2):163–173
Sarangi RK, Nayak S, Panigrahy RC (2008) Monthly variability of phytoplankton (chl-a) and associated physical parameters in the southwest Bay of Bengal waters using remote sensing data. Indian J Mar Sci 37(3):256–266
Sathe PV, Jadhav N (2001) Retrieval of phytoplankton (chl-a) from the sea leaving radiance in the Arabian Sea. J Indian Soc Remote Sens 29(1–2):97–106
Selvavinayagam K, Surendran A, Ramachandran S (2003) Quantitative study on phytoplankton (chl-a) using IRS-P4 OCM data of Tuticorin coastal waters. J Indian Soc Remote Sens 31 (3):227–235
Siegel D et al (2013) Regional to global assessments of phytoplankton dynamics from the SeaWiFS mission. Remote Sens Environ 135:77–91
Singh RP, Chaturvedi P (2010) Comparison of phytoplankton (chl-a) concentration in the Bay of Bengal and the Arabian Sea using IRS-P4 OCM and MODIS Aqua. Indian J Mar Sci 39 (3):334–340
Solanki HU, Dwivedi RM, Nayak SR, Jadeja JV, Thakar DB, Dave HB, Patel MI (2001) Application of Ocean Colour Monitor phytoplankton (chl-a) and AVHRR SST for fishery forecast: preliminary validation results off Gujarat coast, northwest coast of India. Indian J Mar Sci 30:132–138
Vaillancourt RD, Brown CW, Guillard RRL, Balch WM (2004) Light backscattering properties of marine phytoplankton: relationships to cell size, chemical composition and taxonomy. J Plankton Res 26:191–212
Wallhead PJ, Garçon VC, Casey JR, Lomas MW (2014) Long-term variability of phytoplankton carbon biomass in the Sargasso Sea. Glob Biogeochem Cycles 28:825–841
Walton CC, Pichel WG, Sapper JF, May DA (1998) The development and operational application of nonlinear algorithms for the measurement of sea surface temperatures with the NOAA polar-orbiting environmental satellites. J Geophys Res 103:27999–28012
Wong M, Kwan SHL, Young JK, Nichol J, Zhangging L, Emerson N (2007) Modelling of suspended solids and sea surface salinity in Hong Kong using Aqua/ MODIS satellite images. Kor J Remote Sens 23(3):161–169

Part III
Global Modelling of Natural Hazard Risk Assessment

Chapter 6
Flood Inundation and Hazard Mapping of 2017 Floods in the Rapti River Basin Using Sentinel-1A Synthetic Aperture Radar Images

Rajesh Kumar

Abstract Globally, the flood magnitude and flood-induced damage are increasing. Hence, the geospatial technology has been used to minimise the adverse effects of floods and to plan the floodplain for the betterment of floodplain dwellers. One of the major causes of floods in the Rapti River basin is heavy rainfall induced by the break-in-monsoon condition. These days, geoscientists and planners use Sentinel-1A IW GRD synthetic-aperture radar (SAR) image for flood extent mapping. Gauge level and flood duration data recorded at Bhinga, Balrampur, Bansi, Regauli, Birdghat, Kakarahi, Uska Bazar and Trimohinighat sites provide the basis for the selection of SAR images. Extensive floods occurred in the Rapti River basin during August 13–September 01, 2017. The flood duration in the Rapti River basin varied from 3 (Bhinga) to 18 days (Birdghat) in 2017. The flood duration, normally, increases from the upstream to downstream along the Rapti River due to decreasing slope and discharges contributed by the tributaries. In this study, Sentinel-1A GRD SAR images of August 21 and 25, 2017, have been selected for flood mapping in the Indian part of the Rapti River basin. The water level of rivers was above the danger level (DL) at Bansi, Regauli, Birdghat, Kakarahi, Uska Bazar and Trimohinighat gauge and discharge (G/D) sites on August 21 and 25, 2017. The propagation of flood peaks and affected areas has been analysed using water level data and SAR images for the mentioned periods. The actual flooded areas covered 2046.7 km^2 area of the Indian part of the Rapti River basin during August 21–25, 2017. The validation of flooded areas has been done using GPS way points collected during field survey (November 2017) and Landsat 7 ETM+ images (August 24, 2017). Breach sites in flood-prone areas have been mapped using Sentinel-2A and B MSI images. The z-score method has been used for the standardisation of development block-wise flooded areas (km^2) and number of flood-affected villages. After standardisation, these two parameters have been added to formulate development

R. Kumar (✉)
Centre for the Study of Regional Development, Jawaharlal Nehru University, New Delhi, India
http://orcid.org/0000-0001-6092-2099

P. Kumar et al. (eds.), *Applications and Challenges of Geospatial Technology*,
https://doi.org/10.1007/978-3-319-99882-4_6

block-wise flood hazard index (FHI). High to very high FHI values have been observed in Siddharthnagar and Gorakhpur districts.

Keywords Sentinel-1A IW GRD SAR · Rapti River basin · Backscatter values · Danger level · Unprecedented flood · Flood hazard index

6.1 Introduction

Earth scientists and planners extensively use satellite images from active and passive remote sensing satellites for quick, accurate and real-time mapping of the flooded areas and hazard assessment in a river basin across the world. Active sensors utilise near-infrared and microwave region of the electromagnetic spectrum for terrestrial mapping (Jensen 2018). These sensors have the capacity to sense the Earth's objects during cloudy and light rainfall weather conditions. These sensors also work round the clock. However, the passive sensors do not detect the Earth's objects at night and during cloudy and light rainfall weather conditions. Hence, satellite images acquired by the active sensor are extensively used for flood mapping and its propagation downstream. But many researchers have used satellite images of the active and passive sensors to map the flood extent, depth, duration and turbidity (Kumar and Acharya 2016; Bhatt et al. 2016; Kumar 2016). Flooded or water pixel extraction from the top-of-atmosphere or surface reflectance images of the passive sensors, namely, Landsat operational land imager and thermal infrared sensor (OLI-TIRS), thematic mapper (TM), enhanced thematic mapper plus (ETM+) and multispectral scanner (MSS), are based on the normalised difference water index (NDWI), modified normalised difference water index (MNDWI), density slicing of near-infrared and shortwave infrared bands, visual interpretation and tasseled cap transformation (Crist and Cicone, 1984; McFeeters 1996; Jain et al. 2006; Xu 2006; Romshoo et al. 2018). Furthermore, the coarse spatial resolution images of Moderate Resolution Imaging Spectroradiometer (MODIS) and National Oceanic and Atmospheric Administration (NOAA) Advanced Very High Resolution Radiometer (AVHRR) sensors are used for large-scale mapping of the flooded areas due to their high temporal resolution (Ahamed and Bolten 2017; Islam and Sado 2000).

Presence of speckles in SAR image is a major disadvantage because it makes the texture analysis of image complicated (Lee 1981). Speckles in SAR images also influence the spatial characteristics of the backscattering coefficient of different Earth's objects that largely depends on surface roughness, dielectric constant, polarisation and incident angle of SAR waves (Senthilnath et al. 2013). Wind-induced ripples, the velocity of floodwaters, heavy rainfall and submerged vegetation and crops make the surface of the flood inundation rough that further creates a problem in image analysis and classification (Huang 2008; Eisuke 2012). Speckle filtering minimises the speckle noise, but it further degrades the resolution and object information of the SAR image (Sheng and Xia 1996; Manavalan 2017). Furthermore, the adaptive speckle filtering methods conserve the edges and texture information by computing the kernel mean and normalised standard deviation (Senthilnath et al. 2013). Different types of adaptive filters have been developed for the speckle suppression of SAR images. These are mean, median, refined Lee, sigma Lee and Lee (Lee 1981), Frost (Frost et al. 1982) and Gamma MAP (Lopès et al. 1990) filters. According to Qui et al. (2004), "In general, no filter consistently outperforms others. Each filter has its unique strengths and limitations". Hence, the selection of despeckle filter for SAR images depends on the nature of the study (Lee et al. 1994). Manavalan and Rao (2014) pointed out that the despeckle filtering is not necessary when the flood inundation surface is smooth. Flooded area mapping using SAR images is based on different methods like visual interpretation (Oberstadler et al. 1997), Ostu's method and thresholding of backscattering values (Hirose et al. 2001; Yamada 2001; Tan et al. 2004; Bhatt et al. 2013; Manjusree et al. 2015; Ban et al. 2017), band ratio and change detection (Giustarini et al. 2013; Schlaffer et al. 2015), SAR image-based supervised classification (DeRoo et al. 1999; Borghys et al. 2006), fuzzy rules (Pulvirenti et al. 2013), region growing (Mason et al. 2012a; Giustarini et al. 2013) active contour model (ACM) (Kass et al. 1988; Williams and Shah 1992; Horritt 1999), grey-level co-occurrence matrix (GLCM) (Song et al. 2007), segmentation and object-oriented image analysis techniques (Li et al. 2007). Many methods of extracting flooded areas from different SAR images are directly or indirectly based on finding an appropriate threshold range (Manavalan 2017).

Due to improvement in the spatial and temporal resolution, the Sentinel-1A SAR image has been used for detecting floods in rural areas (Clement et al. 2017). However, in urban areas, high density of buildings causes radar shadow and double bouncing effects that further creates a problem in detecting flooded areas (Clement et al. 2017). Therefore, the detection of flooding in urban areas requires fine resolution SAR images of COSMO-SkyMed, TerraSAR-X and RADARSAT-2 (Mason et al. 2012b; Pulvirenti et al. 2016).

In developing countries like India, the flood-prone areas are extensively used for agriculture and business activities. Despite huge investment on flood control measures in India, the flood-induced damage to houses, public utilities, the standing crops and the affected population has been increasing since 1980, inferring encroachment of high-value land use on chronically flood-affected areas (Sivasami 2002). Hence, flood hazard assessment using geospatial technology is a nonstructural measure for minimising the adverse effects of floods in flood-prone areas (Manjusree et al. 2015). Flood hazard assessment in Bihar has been analysed using flood layers (1998–2010), hazard area and number of floods in a year parameters, derived from RADARSAT-1 and RADARSAT-2 (C-band) HH polarisation images (Manjusree et al. 2015).

Uttar Pradesh has the highest flood-prone areas in India. The total flood-prone areas in Uttar Pradesh account for 73,400 km^2 (Planning Commission 2011). However, the total flood-prone areas in the Rapti River basin account for 4322.6 km^2 which is ~5.9% of the total flood-prone areas of Uttar Pradesh. The Rapti River, its tributaries and sub-tributaries experience severe floods in the monsoon season due to heavy downpour. Against the backdrop of above-mentioned methods and studies, the present study aims to map the flooding extent of 2017 floods in the Indian part of the Rapti River basin and to prepare a flood hazard map at development block level using the parameters, i.e. flooded areas and flood-affected hamlets.

6.2 Study Area

The entire Rapti River basin extends from 26° 18′ 00″ N to 28°33′06″ N and 81°33′00″ E to 83°45′06″ E and accounts for 25,793 km^2 of area, out of which 44% (11,401 km^2) lies in Nepal and 56% (14,392 km^2) in Uttar Pradesh (Fig. 6.1) (Rana et al. 2009; Kumar 2010). Administratively, the Rapti River basin covers Bahraich, Shrawasti, Balrampur, Siddharthnagar, Maharajganj, Gorakhpur, Kushinagar, Deoria, Sant Kabir Nagar and Basti districts of Uttar Pradesh.

The Rapti River originates from the Nepalese part of the lesser Himalaya where this river is known as the West Rapti. It is the largest tributary of Ghaghara River, which, in turn, is a major constituent of the Ganga (Rana et al. 2009; Kumar 2010; Kumar et al. 2013). The Jimruk, *Burhi* Rapti and its tributaries, Rohini and Gaura River are the major left bank tributaries of the Rapti. The right bank tributaries are Ami and Taraina River. In the Indian part of the basin, six gauge and discharge (G/D) sites, namely, Kakardhari, Bhinga, Balrampur, Bansi, Regauli and Birdghat (Gorakhpur), are located along the Rapti River. Kakarahi, Uska Bazar and

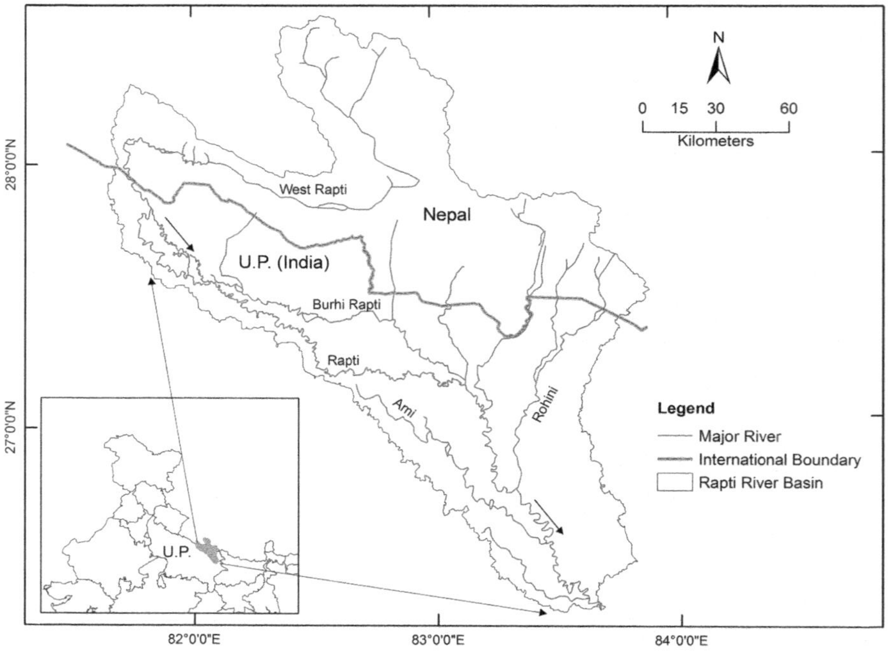

Fig. 6.1 Location map of the study area

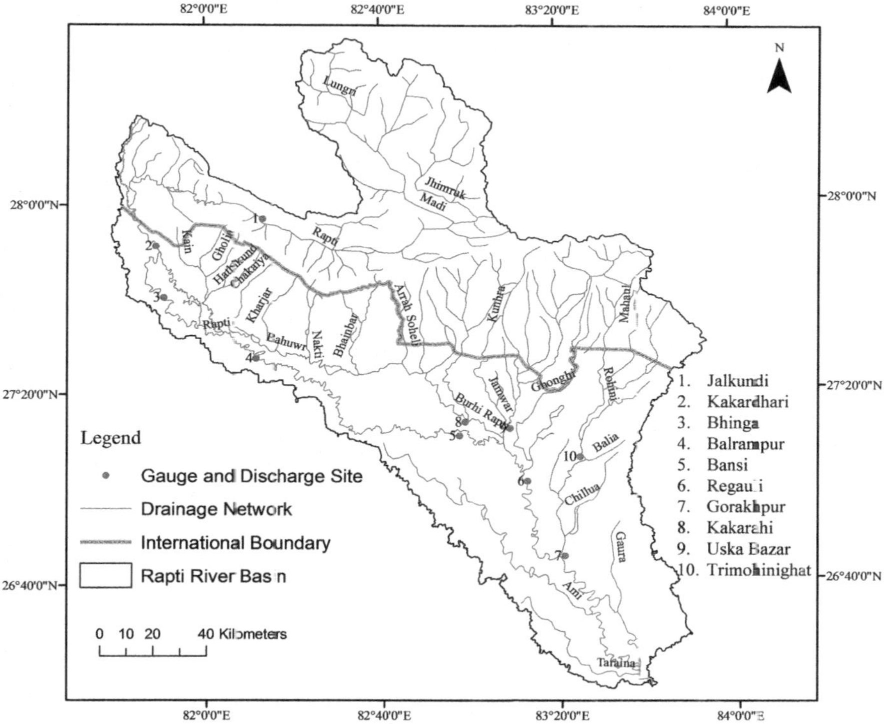

Fig. 6.2 Drainage map and gauge and discharge (G/D) sites of the Rapti River basin

Trimohinighat G/D sites are located along the *Burhi* Rapti, Kunhra and Rohini River, respectively (Fig. 6.2). The study region is a part of the vast alluvial plain of the Ganga where the depth of alluvium varies from ~4 to 6 km (Singh 1996). The major geomorphic features of the study area are piedmont alluvial plain, older alluvial plain, older floodplain and active floodplain (GSI and NRSC 2012). Furthermore, the active and older floodplains constitute the flood-prone areas of the Rapti River basin. In this study, areas bordered by the embankments are mapped as an active floodplain. The older floodplain is protected by the embankments. However, breaches in embankments cause flooding therein (Fig. 6.3). Agriculture is a general land use pattern in the study area (Yadav 1999).

6.3 Material and Methodology

Sentinel-1 consists of two-satellite constellation, namely, Sentinel-1A and Sentinel-1B. Sentinel-1A C-band SAR operated at a frequency of 5.405 GHz to detect the Earth's objects. Sentinel-1A and B missions were launched on April 3, 2014, and

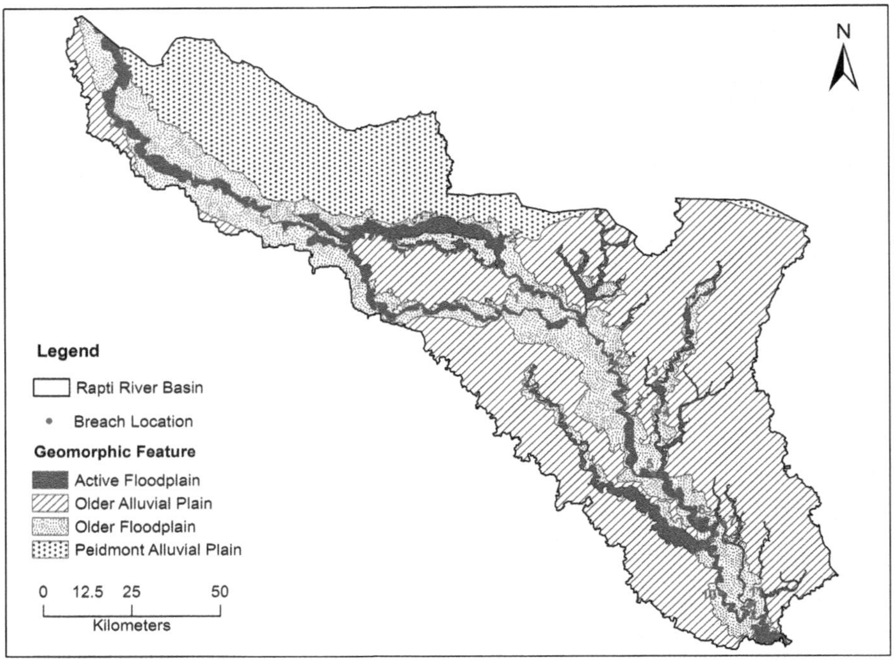

Fig. 6.3 Geomorphic features of the study area along with breach locations due to 2017 floods (Modified after GSI and NRSC 2012)

April 25, 2016, by the European Space Agency (ESA) under Europe's Copernicus programme, respectively, (ESA 2018a). The orbit height of Sentinel-1A and B is 693 km with a combined temporal resolution of 6 days. However, the temporal resolution of a single Sentinel-1 satellite is 12 days at the equator (ESA 2018a). The level-1 ground range detected (GRD) data products of Sentinel-1A have been used in this study. Interferometric wide swath (IW) GRD data product has dual polarisation modes, i.e. VV and VH. With the help of an Earth ellipsoid model, the GRD level-1 data have been detected, multi-looked and projected to the ground range (ESA 2018b). Sentinel-2 is a multispectral instrument (MSI). This is a two-satellite constellation, namely, Sentinel-2A and Sentinel-2B that were launched on June 23, 2015, and March 07, 2017, by the ESA, respectively (ESA 2018c). The temporal resolution of the combined Sentinel-2 satellites is 5 days at the equator, while it is 10 days for a single satellite (ESA 2018c). The height of the orbit of Sentinel-2A and B satellite is at 786 km (ESA 2018c). Sentinel-2 operates in visible, near-infrared and shortwave infrared part of the electromagnetic spectrum (Table 6.1) (ESA 2018d).

The Sentinel-1A SAR images have been used for mapping the flooded areas and pre-flood water bodies, while the Sentinel-2 images have been used for mapping of

Table 6.1 Spectral bands of the Sentinel-2A and B sensor

Band number	Band name	Sentinel-2A Central wavelength (nm)	Sentinel-2A Bandwidth (nm)	Sentinel-2B Central wavelength (nm)	Sentinel-2B Bandwidth (nm)	Spatial resolution (m)
1	Coastal aerosol	443.9	27	442.3	45	60
2	Blue	496.6	98	492.1	98	10
3	Green	560	45	559	46	10
4	Red	664.5	38	665	39	10
5	Vegetation red edge	703.9	19	703.8	20	20
6	Vegetation red edge	740.2	18	739.1	18	20
7	Vegetation red edge	782.5	28	779.7	28	20
8	NIR	835.1	145	833	133	10
8a	Vegetation red edge	864.8	33	864	32	20
9	Water vapour	945	26	943.2	27	60
10	Cirrus	1373.5	75	1376.9	76	60
11	SWIR	1613.7	143	1610.4	141	20
12	SWIR	2202.4	242	2185.7	238	20

Source: Sentinel online (2018)

the geomorphic features and breach sites. Landsat 7 ETM+ images of 30 m spatial resolution (EarthExplorer 2017) have been used for accuracy assessment of flooded and non-flooded areas extracted from the Sentinel-1 SAR images (Table 6.2). The Shuttle Radar Topography Mission (SRTM) digital elevation model (DEM) at 90 m spatial resolution (CGIAR-CSI 2008) has been used to extract the Rapti River basin boundary using hydrology module of the spatial analyst tool of ArcGIS 10. The development block boundaries have been digitised from the administrative atlas of Uttar Pradesh (Census of India 2001). The hamlet point feature data have been obtained from Mizushima Laboratory (2013). The geomorphic map has been obtained from Bhuvan, an Indian Geo-platform of ISRO (GSI and NRSC 2012). The rainfall and water level data have been obtained from the irrigation department of Uttar Pradesh (Irrigation & Water Resource Department 2017). Synoptic weather system information has been collected from all India weekly weather report (August 10–15, 2017), India Meteorological Department (IMD), Govt. of India (IMD 2017).

In this study, the Sentinel-1A IW GRD products of VV and VH polarisations have been processed using Sentinel Application Platform (SNAP) tool. The pre-processing steps using SNAP tool are given in Fig. 6.4. The standard procedures such as orbit correction, thermal noise removal, calibration (sigma0), speckle

Table 6.2 Description of satellite images

Scene ID	Date of acquisition	Track	Orbit
Sentinel-1A SAR scene IDs and location			
S1A_IW_GRDH_1SDV_20170602T122944_20170602T123009_016855_01C070_F427 (Naugarh-Regauli-Gorakhpur-Barhalganj)	June 02, 2017	158	16,855
S1A_IW_GRDH_1SDV_20170610T002751_20170610T002816_016964_01C3DB_561F ((Bansi-Regauli-Gorakhpur)	June 10, 2017	92	16,964
S1A_IW_GRDH_1SDV_20170610T002726_20170610T002751_016964_01C3DB_B609 (Kakardhari-Balrampur-Bansi)	June 10, 2017	92	16,964
S1A_IW_GRDH_1SDV_20170821T002730_20170821T002755_018014_01E3D1_6DE6 (Kakardhari-Balrampur-Bansi)	August 21, 2017	92	18,014
S1A_IW_GRDH_1SDV_20170821T002755_20170821T002820_018014_01E3D1_FABE (Bansi-Regauli-Gorakhpur)	August 21, 2017	92	18,014
S1A_IW_GRDH_1SDV_20170825T122949_20170825T123014_018080_01E5C1_50D7 (Naugarh-Regauli-Gorakhpur-Barhalganj)	August 25, 2017	158	18,080
Sentinel-2A and B (MSI)			
S2B_MSIL2A_20171016T050749_N0205_R019_T44RPR_20171016T051820.SAFE	October 12, 2017	N.A.	N.A.
S2B_MSIL2A_20171023T045829_N0206_R119_T44RQQ_20171023T084707.SAFE	October 23, 2017	N.A.	N.A.
S2A_MSIL2A_20171028T045911_N0206_R119_T44RQR_20171028T084742.SAFE	October 28, 2017	N.A.	N.A.
Landsat 7 ETM+		Path	Row
LE07_L1GT_142041_20170824_20170824_01_RT	August 24, 2017	142	41
LE07_L1TP_142042_20170824_20170824_01_RT	August 24, 2017	142	42

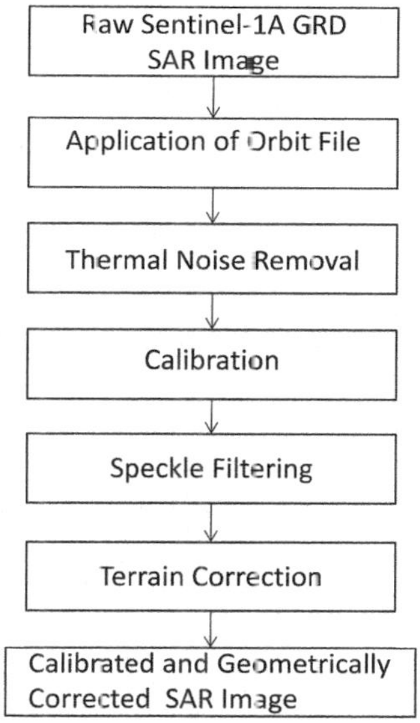

Fig. 6.4 Pre-processing steps of Sentinel-1A GRD SAR image

filtering and terrain correction have been applied on the raw SAR images to obtain the geometrically correct image along with backscattering values (sigma0) (Twele et al. 2016; Clement et al. 2017). A 7 × 7 Gamma MAP filtering method has been used for speckle suppression of the Sentinel-1A GRD SAR images. The sigma0 values have been converted into logarithmic scale, i.e. decibel (dB).

In this study, the selection of SAR images is based on the flood occurrence, duration and movement of the flood crest downstream. The statistics of training sites of flooded areas computed from VV and VH polarisation images of August 21, 2017 show unimodal distribution. Hence, the VV polarisation images of August 21, 2017 have been used for flood pixel extraction. Many researchers concluded that the VV polarisation of Sentinel-1A IW GRD product is appropriate for the flood detection (Twele et al. 2016). But VV polarisation is more affected by wind-induced ripples on the surface of the floodwaters than other polarisations (Manjusree et al. 2012). The VH polarisation image of August 25, 2017 provides better results for mapping the flooded areas than the VV polarisation using thresholding of the backscattered values. The histogram of training site statistics of the VV polarisation image of August 25, 2017, shows a polymodal distribution, while it is a unimodal distribution for VH polarisation image (Fig. 6.5). Hence, the computation of threshold value for the extraction of flooded areas from VH polarisation image is easy and appropriate.

After pre-processing, a thresholding procedure has been used to extract the flood pixels from the Sentinel-1A IW GRD SAR images. On the basis of statistics of the training sites of the flooded areas (minimum, mean, median and standard deviation),

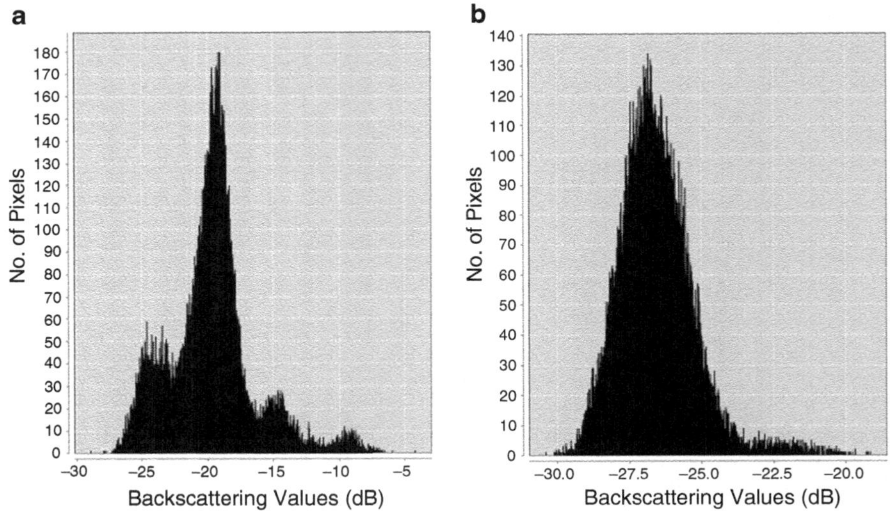

Fig. 6.5 Histogram of backscattering values (dB) of training sites of flooded areas (**a**) VV polarization and (**b**) VH polarisation of Sentinel-1A IW GRD SAR image (August 25, 2017)

lower and upper limit of the threshold range has been decided. Backscattering values (dB) of the flood areas are normally distributed for which the mean ± 3 standard deviation covers 99.7% of the distribution (Fig. 6.5) (Motulsky 2014). In this study, the lowest value of all training sites has been taken as the lower limit of the threshold range. The upper limit (UL) has been obtained using Eq. (6.1):

$$UL = [(\mu + 3\sigma)] \tag{6.1}$$

where σ stands for the standard deviation of training site backscattered values in decibel (dB) and μ is the mean backscattering values of training sites (flooded areas).

The actual flooded areas have been obtained by subtracting the pre-flood water bodies (June 02–10, 2017) from the total flooded areas of August 21–25, 2017. The accuracy assessment of actual flooded and non-flooded areas of August 25, 2017 has been done with the help of the Garmin eTrex global positioning system (GPS) waypoints and reference points selected from the Landsat 7 ETM+ images (August 24, 2017). The overall accuracy of flooded and non-flooded areas of August 25, 2017 is 91.6% with a high kappa coefficient value of 0.83. The user's accuracy for the flooded areas is 94.87% (Table 6.3).

The breach sites in the embankments have been mapped through visual interpretation of false colour composite (FCC) of Sentinel-2 MSI post-flood images (10 m spatial resolution). The development block-wise flood hazard assessment of 2017

Table 6.3 Accuracy assessment of flooded and non-flooded areas (August 25, 2017)

Class	Producer accuracy (%)	User accuracy (%)	Producer accuracy (pixels)	User accuracy (pixels)
Flooded Areas	88.1	94.87	37/42	37/39
Non-flooded areas	95.12	88.64	39/41	39/44

Overall accuracy, (76/83), 91.5% Kappa coefficient, 0.83

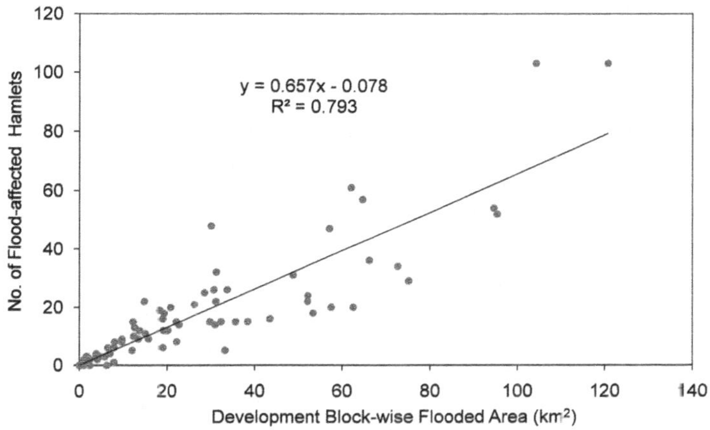

Fig. 6.6 Association between development block-wise flooded areas and flood-affected hamlets

floods has been based on two parameters, namely, development block-wise flooded areas and number of flood-affected hamlets. The linear association between these two parameters is strongly positive as the coefficient of determination (R^2) value is 0.79 (Fig. 6.6). Hence, the z-score (z) method has been applied for the standardisation of these two parameters (Eq.6.2) (Burt et al. 2009):

$$z = (\chi - \mu)/\sigma \tag{6.2}$$

where χ stands for a parameter. μ and σ are mean and standard deviation of a parameter, respectively.

The z-score values of these two parameters have been added linearly to make a composite index, i.e. flood hazard index (FHI). Furthermore, the development block-wise FHI values have been classified into four categories using natural break classification method.

6.4 Results and Discussion

6.4.1 Cause and Occurrence of 2017 Floods

The 2017 floods in the Rapti River basin were caused by the heavy rainfall during August 12–14 due to the break-in-monsoon system (IMD 2017). The highest rainfall of 303.6 mm was recorded at Kakarahi G/D site on August 14. Overall, the high rainfall was recorded at all the G/D sites of the basin on August 14 (Fig. 6.7). Hence, flooding was observed at all the G/D sites except Kakardhari. The flood peak which occurred at Bhinga on August 14 takes 2 days to reach Balrampur G/D site. The synchronisation of peak floods occurred at Bansi, Regauli and Birdghat. The unprecedented flood occurred at Bansi on August 21, 2017, when the water level crossed the previous highest flood level (1998) by 0.06 m. The flood peaks at Trimohinighat, Kakarahi and Uska Bazar occurred on August 16, 18 and 20, 2017, respectively. Deviation of the maximum water level (MWL) from the danger level (DL) at Bansi, Regauli, Birdhat, Kakarahi, Trimohinighat and Uska Bazar was large and ranged between 1.53 and 3.14 m. Such a large deviation of MWL indicates that the major floods occurred at these G/D sites in 2017 (Table 6.4). When the water level in a river remains above the DL by 1 m or more, it is defined as major floods (Dhar and Nandargi 2003).

The flood duration in the basin varied from 3 (Bhinga) to 18 days (Birdghat) during 2017 floods. The flood duration along the Rapti River in the basin showed an increasing trend from upstream to downstream due to decrease in slope and discharges contributed by the tributaries. Hence, the extreme downstream Birdghat G/D site recorded the highest flood duration of 18 days from August 15 to September 01, 2017 (Table 6.5).

Due to the high water level in Rapti, Burhi Rapti, Gaura and Rohini River, breaches in embankments occurred during 2017 floods. A breach of 130 m length occurred in the right embankment of the Burhi Rapti River (breach ID 1) (Table 6.6).

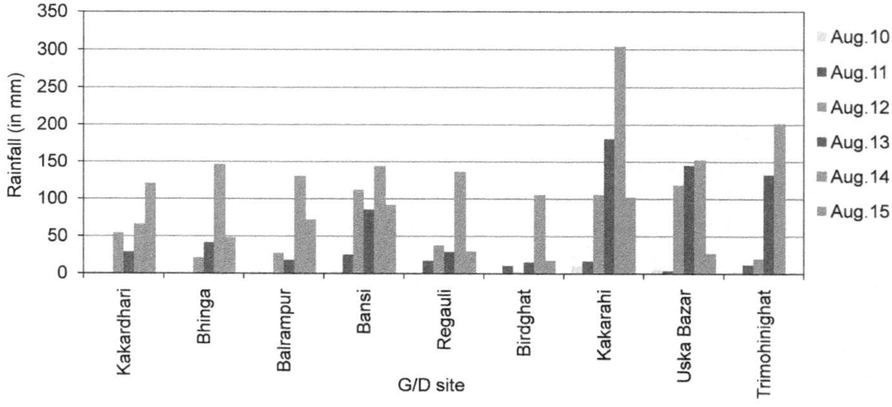

Fig. 6.7 Rainfall recorded at different G/D sites during August 10–15, 2017

Table 6.4 Maximum water level (MWL), danger level (DL) and highest flood level (HFL) in Rapti River basin

G/D Site	DL (m)	MWL, 2017 (m)	Date	Deviation of MWL from DL (m)	HFL (year)
Kakardhari	131	130.77	August 13, 2017	−0.23	132.37 (2014)
Bhinga	119.5	120.11	August 14, 2017	0.61	120.3 (2014)
Balrampur	104.62	105.52	August 16, 2017	0.90	105.51 (2014)
Bansi	84.9	85.88	August 21, 2017	0.98	85.82 (1998)
Regauli	80.3	81.84	August 21, 2017	1.54	82.12 (2000)
Birdghat	74.98	77.22	August 21, 2017	2.24	77.54 (1998)
Kakarahi	85.65	88.79	August 18, 2017	3.14	88.97 (1998)
Uska Bazar	83.52	85.05	August 20, 2017	1.53	85.62 (1998)
Trimohinighat	82.44	85.23	August 16, 2017	2.79	85.43 (2001)

Source: Irrigation and Water Resource Department (2017)

Table 6.5 Flood duration in Rapti River basin during 2017

G/D site	River	Days above DL	Date
Kakardhari	Rapti	0	N.A.
Bhinga	Rapti	3	August 13–15
Balrampur	Rapti	8	August 13–20
Bansi	Rapti	11	August 16–26
Regauli	Rapti	15	August 15–29
Birdghat	Rapti	18	August 15–September 01
Kakarahi	Burhi Rapti	14	August 14–27
Uska Bazar	Kunhra	12	August 15–26
Trimohinighat	Rohini	11	August 13–23

Source: Irrigation and Water Resource Department (2017)

Two breach sites (ID 3 & 4) were located along the Rohini River (Fig. 6.3). Two breaches (ID 7 & 11) were also mapped along the right bank embankment of the Gaura River. The rest of the breach sites were located along the Rapti River. The largest breach (ID 9) of 132 m occurred in the left bank embankment of the Rapti River. Such large breaches in embankments indicate that the structural measure is not a permanent solution for flood control in the basin. Breaches in embankment often produce more coarse to medium sand in the floodwaters that settle in the nearby embankment-protected arable land. Such deposition of coarse and medium sand in the arable land makes it infertile for many years (Kumar 2010).

Table 6.6 Breach in embankments during 2017 floods

Breach site ID	River name	Length (m)
1	Burhi Rapti	130
2	Rapti	40
3	Rohini	50
4	Rohini	80
5	Rapti	110
6	Rapti	36
7	Gaura	99.3
8	Rapti	71.9
9	Rapti	132
10	Rapti	23.9
11	Gaura	113
12	Rapti	94.4

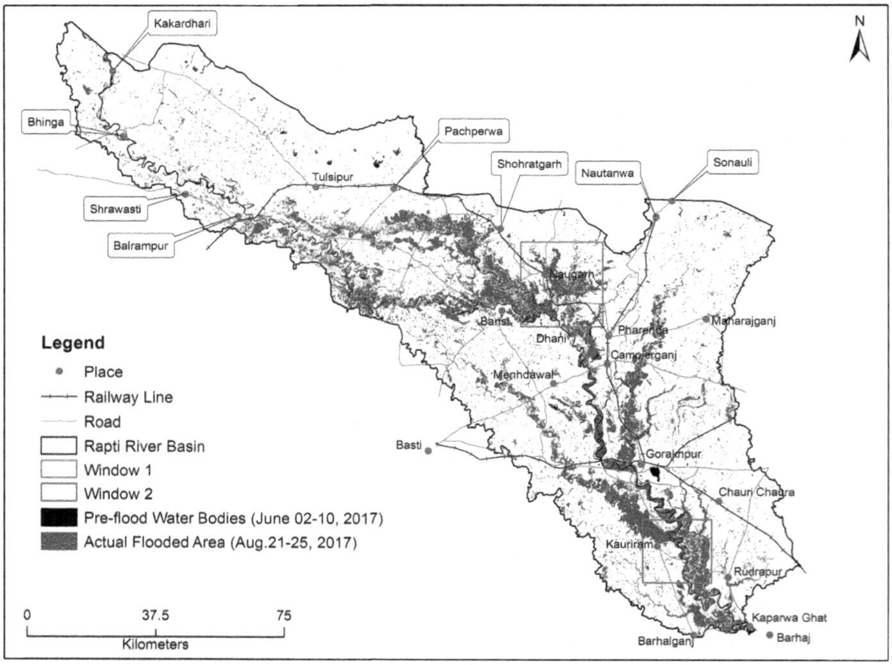

Fig. 6.8 Actual flooded areas and pre-flood water bodies of the Rapti River basin (Indian part)

6.4.2 *Flooded Area of the Rapti River Basin in 2017*

The actual flooded area in the basin is 2046.7 km^2, while the pre-flood water bodies account for 68.6 km^2 only (Fig. 6.8). The flood-prone area covers 30% geographical area of the Rapti River basin (UP). During 2017 floods, the actual flooded areas account for 38.8% geographical area of the total flood-prone area. The actual flooded areas of the flood-prone region cover 80% of the total actual flooded areas of the basin.

6.4.3 Downstream Movement of 2017 Floods

Downstream movement of the flood has been analysed with the help of flood extent in window 1 and window 2. The actual flooded area in window 1 on August 21, 2017, was 205 km^2, while it was 131 km^2 on August 25, 2017 (Fig. 6.9). The water level recorded at Kakarahi and Uska Bazar G/D site was above the DL with a decreasing trend during August 21–25, 2017 (Table 6.7). The window 2 largely covers the older floodplain between the Rapti and Gaura River downstream of Birdghat G/D site. The actual flooded area in window 2 was 91 km^2 on August 21, 2017, while it was increased to 143 km^2 on August 25, 2017 (Fig. 6.10). The main causal factor for an increase in flooded area was breaches in embankments along the Rapti and Gaura River. Such breaches in the embankments in window 2 were caused by the flood flow from the upstream reaches.

6.4.4 Development Block-Wise Flood Hazard Assessment

Flood hazard assessment in a populated river basin is basically based on certain key parameters such as elevation, flooded areas, turbidity, flood frequency, depth,

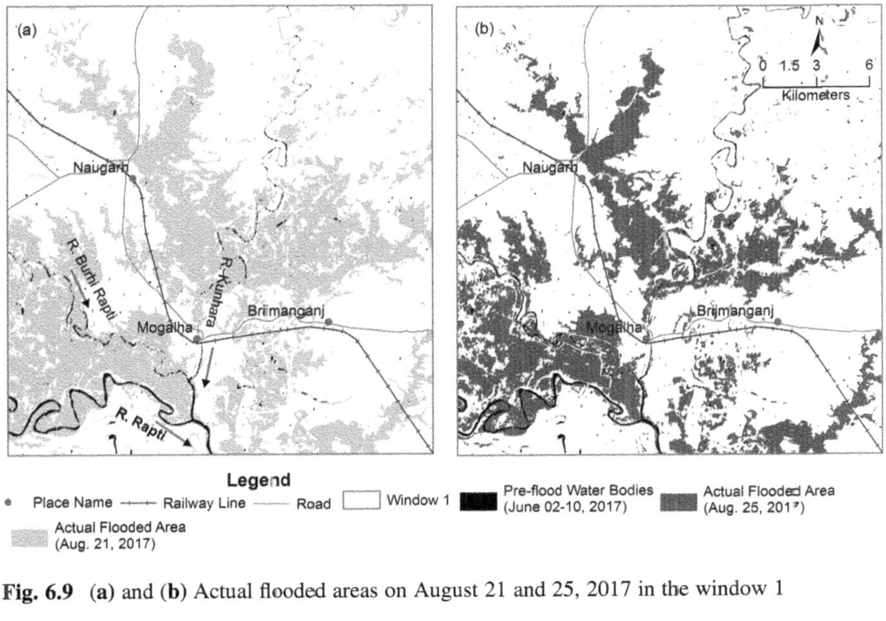

Fig. 6.9 (a) and (b) Actual flooded areas on August 21 and 25, 2017 in the window 1

Table 6.7 Danger level (DL), water level (WL) and deviation of WL from DL in Rapti River basin during August 21 and 25, 2017

G/D Site	Danger level (m)	WL (August 21, 2017) (m)	WL (August 25, 2017) (m)	Deviation of WL from DL (m) (August 21, 2017)	Deviation of WL from DL (m) (August 25, 2017)
Kakardhari	131	129.01	128.83	−1.99	−2.17
Bhinga	119.5	118.06	118.11	−1.44	−1.39
Balrampur	104.62	104.51	103.98	−0.11	−0.64
Bansi	84.9	85.88	85.33	0.98	0.43
Regauli	80.3	81.84	81.4	1.54	1.1
Birdghat	74.98	77.22	76.6	2.24	1.62
Kakarahi	85.65	87.71	86.93	2.06	1.28
Uska Bazar	83.52	84.99	84.09	1.47	0.57
Trimohinighat	82.44	83.92	81.63	1.48	−0.81

Source: Irrigation and Water Resource Department (2017)

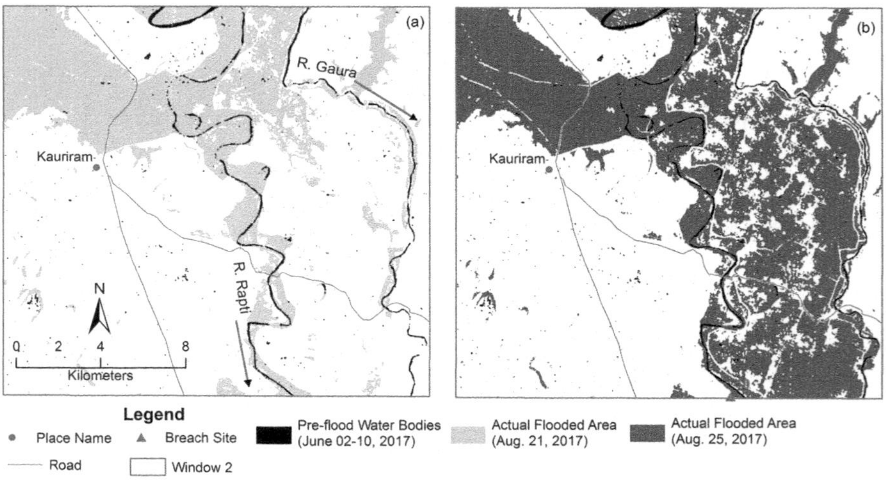

Fig. 6.10 (**a**) and (**b**) Actual flooded areas on August 21 and 25, 2017 in the window 2 along with breach locations

duration, flood-affected population and hamlets (Sanyal and Lu 2006; Manjusree et al. 2015; Kumar and Acharya 2016; Kumar 2016; Kumar et al. 2016). In this study, the development block-wise flooded areas show the extent of flooding, while the number of flood-affected hamlets indicates the flood-induced damage. Therefore, a composite index (FHI) has been computed using these parameters.

In this study, development block-wise low flood hazard is observed in all districts except Siddharthnagar (Fig. 6.11). Generally, the low flood hazard zone

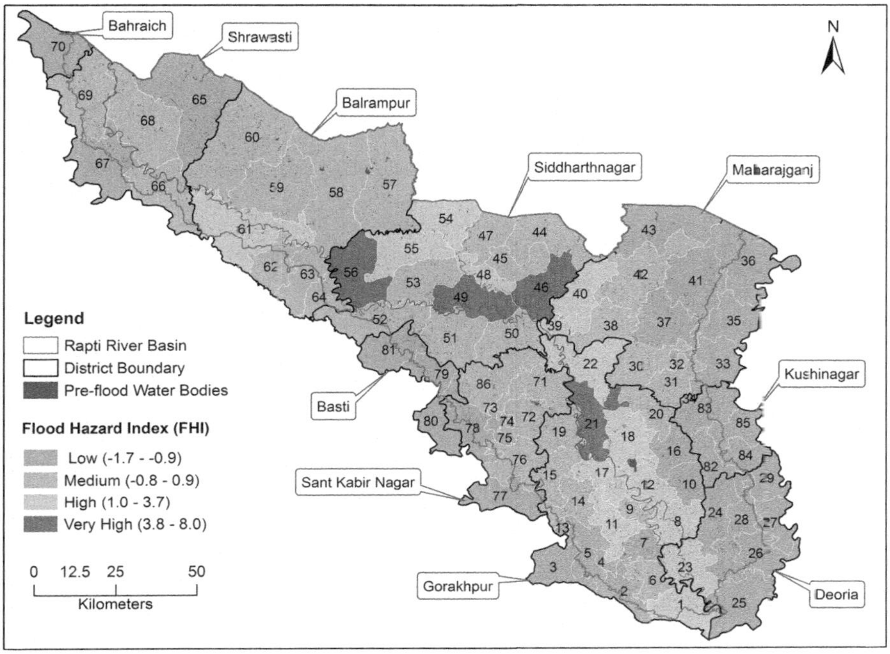

Fig. 6.11 Development block-wise flood hazard categories with label IDs. Name of the districts is given in callouts

Table 6.8a District-wise flood-affected development blocks in low flood hazard zone

District	Affected development block (label ID)
Bahraich	Nawabganj (70)
Shrawasti	Sirasia (65) and Gilaula (67)
Maharajganj	Part of Siswa Bazar (32), Ghughuli (33), part of Partawal (34), Siswa Bazar (35), Nichlaul (36), Maharajganj (37), Mithaura (41) and Nautanwa (43),
Kushinagar	Sukrauli (82), Captanganj (83), Hata (84) and Motichak (85)
Deoria	Gauri Bazar (24), Barhaj (25), Deoria (26), Rampur Karkhana (27), Baitalpur (28) and Desai Deoria (29)
Gorakhpur	Gola (2), Belghat (3), part of Gagaha (4), Urua (5), part of Kauriram (7) and part of Belghat (13)
Sant Kabir Nagar	Part of Semariyawan (73, 74 & 78) and Baghauli (75)
Basti	Rudhauli (79), Saughat (80) and Ramnagar (81)

lies in the older alluvial plain that comes under flood-free zone or *Bangar* land (Table 6.8a).

Medium flood hazard zone is found in the development blocks of Shrawasti, Balrampur, Siddharthnagr, Maharajganj, Gorakhpur and Sant Kabir Nagar districts

Table 6.8b District-wise flood-affected development blocks in medium flood hazard zone

District	Affected development block (Label ID)
Shrawasti	Ikauna (66), Hariharpur (68) and Jamunaha (69)
Balrampur	Pachperwa (57), Gainsari (58), Tulasipur (59), Harraiya Satgharwa (60), Shridattganj (62), Utraula (63) and Gaindas Buzurg (64)
Siddharthnagar	Birdpur (44), Naugarh (45), Shohratgarh (47), Kheserha (50), Mithwal (51) and Domariyaganj (52)
Maharajganj	Paniara (30), Partawal (31), Pharenda (38) and Lakshmipur (42)
Gorakhpur	Gagaha (6), Kauriram (9), Khajani (14), Shahjanwa (15) and Pali (19)
Sant Kabir Nagar	Mehdawal (71), Semariyawan (72), Khalilabad (76) and Santha (86)

Table 6.8c District-wise flood-affected development blocks in high flood hazard zone

District	Affected development block (Label ID)
Balrampur	Balrampur (61)
Siddharthnagar	Jogiya Khas (48), Barhni Bazar (54) and Itwa (55)
Maharajganj	Dhani (39) and Brimanganj (40)
Deoria	Rudrapur (23)
Gorakhpur	Barhalganj (1), Brahmpur (8), Bansgaon (11), Khorabar (12), Piprauli (17), Chargawan (18) and Campierganj (22)

Table 6.8d District-wise flood-affected development blocks in very high flood hazard zone

District	Affected development block (Label ID)
Siddharthnagar	Uska Bazar (46), Bansi (49) and Bhanwapur (56)
Gorakhpur	Jungal Kaudia (21)

(Table 6.8b). In general, this flood zone falls in the topographically low-lying areas of older alluvial plain and older floodplain. Such low-lying areas comprise oxbow lakes, abandoned and minor channels. In the local language, these lakes and channels are called as *chaurs* that often get flooded during the monsoon season. The high flood hazard zone falls in the development blocks of Balrampur, Siddharthnagar, Maharajganj, Deoria and Gorakhpur districts (Table 6.8c). This zone covers the topographically low-lying areas of older alluvial plain, older and active floodplain.

In Siddharthnagar district, a very high flood zone is observed in Uska Bazar, Bansi and Bhanwapur development blocks that fall in the active floodplain of Kunhra, Rapti and Burhi Rapti River (Fig. 6.11). Jungle Kaudia development block of Gorakhpur district also falls in the very high flood zone (Table 6.8d). Jungle Kaudia block comes under the older and active floodplain of Rapti and Rohini River. The topographically low-lying areas of the older floodplain of this block are protected by embankments. Breaches in embankments along the Rohini and Rapti Rivers often cause floods in Jungle Kaudia block.

6.5 Conclusions

The 2017 floods in the Rapti River basin have been caused by the heavy rainfall due to the break-in-monsoon condition. Breaches in embankments are another reason for flooding in embankment-protected areas, i.e. older floodplain. Major floods occurred at Regauli, Birdghat, Kakarahi, Uska Bazar and Trimohinighat G/D sites in 2017.

Siddharthnagar and Gorakhpur were the most flood-affected districts in 2017. The confluence of the *Burhi* Rapti and Rohini River with the Rapti comes under the very high flood hazard zone due to backwater effect. The development block-wise flood hazard zone of a major flood in the Rapti River basin has immense importance to the district administration for the execution of flood management strategies at the block level. The geospatial approach used in this study for flood hazard assessment can be improved with the help of amount of compensation money given to the floodplain dwellers whose houses were damaged by the 2017 floods.

Methods discussed in this study are easy to detect flood-affected areas during the monsoon season using freely available Sentinel-1A IW GRD SAR images. Sentinel-1A SAR detects the Earth's objects round the clock and during cloudy and light rainfall weather conditions. Hence, Sentinel-1 GRD SAR images have immense potential for the implementation of flood fighting measures during floods.

Acknowledgements The author thankfully acknowledges the IMD and Irrigation and Water Resource Department of Uttar Pradesh for providing synoptic weather condition, daily rainfall and water level data of 2017 free of cost. He also expresses gratitude to ESA Copernicus Open Access Hub and USGS EarthExplorer web portals for providing free access to the Sentinel-1A GRD and Landsat 7 ETM+ images, respectively. He also acknowledges ESA science toolbox exploitation platform for providing SNAP tool (ver.6.0.0).

References

Ahamed A, Bolten JD (2017) A MODIS-based automated flood monitoring system for southeast asia. Int J Appl Earth Obs Geoinf 61:104–117. https://doi.org/10.1016/j.jag.2017.05.006

Ban HJ, Kwon YJ, Shin H, Ryu HS, Hong S (2017) Flood monitoring using satellite-based RGB composite imagery and refractive index retrieval in visible and near-infrared bands. Remote Sens. https://doi.org/10.3390/rs9040313

Bhatt CM, Rao GS, Begum A, Manjusree P, Sharma SVSP, Prasanna L, Bhanumurthy V (2013) Satellite images for extraction of flood disaster footprints and assessing the disaster impact: Brahmaputra floods of June–July 2012, Assam, India. Curr Sci 104(12):1692–1700

Bhatt CM, Rao GS, Farooq M, Manjusree P, Shukla A, Sharma SVSP, Kulkarni SS, Begum A, Bhanumurthy V, Diwakar PG, Dadhwal VK (2016) Satellite based assessment of the catastrophic Jhelum floods of September 2014, Jammu & Kashmir, India. Geomat Nat Haz Risk. https://doi.org/10.1080/19475705.2016.1218943

Borghys D, Yvinec Y, Perneel C, Pizurica A, Philips W (2006) Supervised feature-based classification of multi-channel SAR images. Pattern Recogn Lett 27(4):252–258

Burt JE, Barber GM, Rigby DL (2009) Elementary statistics for geographers, 3rd edn. The Guilford Press, New York

Census of India (2001) Administrative atlas of Uttar Pradesh. Office of the Registrar General & Census Commissioner, New Delhi

CGIAR-CSI (2008) SRTM 90m digital elevation data. http://srtm.csi.cgiar.org/SELECTION/inputCoord.asp. Accessed on 01 Jan 2008

Clement MA, Kilsby CG, Moore P (2017) Multi-temporal synthetic aperture radar flood mapping using change detection. J Flood Risk Manage. https://doi.org/10.1111/jfr3.12303

Crist EP, Cicone RC (1984) A physically-based transformation of thematic mapper data—the TM tasseled cap. IEEE Trans Geosci Remote Sens 22:256–263. https://doi.org/10.1109/TGRS.1984.350619

De Roo A, Van der Knijff J, Horritt MS, Schmuck G, De Jong S (1999) Assessing flood damages of the 1997 Oder flood and the 1995 Meuse flood. Proceedings of the 2nd international symposium on operationalisation of remote sensing, Enschede, The Netherlands, pp 3459–3465

Dhar ON, Nandargi S (2003) Hydrometeorological aspects of floods in India. Nat Hazards 28:1–33

EarthExplorer (2017) Landsat collection 1 level-1, Landsat7 ETM+ C1 level-1. https://earthexplorer.usgs.gov/. Accessed on 30 Aug 2018

Eisuke K (2012) Basic principles of Synthetic Aperture Radar (SAR). 8th Sentinel Asia System (SAS) Operation Training, Bangkok

ESA (2018a) Sentinel-1 mission details. https://earth.esa.int/web/guest/missions/esa-operational-eo-missions/sentinel-1. Accessed on 23 Jan 2018

ESA (2018b) Level-1 GRD Products. https://sentinel.esa.int/web/sentinel/technical-guides/sentinel-1-sar/products-algorithms/level-1-algorithms/ground-range-detected. Accessed on 23 Jan 2018

ESA (2018c) Sentinel-2 mission details. https://earth.esa.int/web/guest/missions/esa-operational-eo-missions/sentinel-2. Accessed on 23 Jan 2018

ESA (2018d) Multispectral instrument (MSI) overview. https://earth.esa.int/web/sentinel/technical-guides/sentinel-2-msi/msi-instrument. Accessed on 23 Jan 2018

Frost VS, Stiles JA, Shanmugan KS, Holtzman JC (1982) A model for radar images and its application to adaptive digital filtering of multiplicative noise. IEEE Trans Pattern Anal Mach Intell 4(2):157–166

Giustarini L, Hostache R, Matgen P, Schumann GJP (2013) A change detection approach to flood mapping in urban areas using TerraSAR-X. IEEE Trans Geosci Remote Sens 51(4):2417–2430. https://doi.org/10.1109/TGRS.2012.2210901

GSI and NRSC (2012) Geomorphology and Lineament NGLM, Landform, 50K, Uttar Pradesh. http://bhuvan.nrsc.gov.in/gis/thematic/index.php. Accessed on 01 Jan 2012

Hirose K, Maruyama Y, Quy D-V, Tsukada M, Shiokawa Y (2001) Visualization of flood monitoring in the lower reaches of the Mekong River. Proceedings of 22nd Asian Conference on Remote Sensing, vol 1, Nov. 5–9 2001, Singapore, pp 314–319

Horritt S (1999) A statistical active contour model for SAR image segmentation. Image Vis Comput 17(3–4):213–224. https://doi.org/10.1016/S0262-8856(98)00101-2

Huang SQ (2008) Change mechanism analysis and integration change detection method on SAR images. The international archives of the photogrammetry, remote sensing and spatial information Sciences. 37(B7). Beijing

IMD (2017) Weekly weather report. http://www.imd.gov.in/pages/weeklyweatherreport.php. Accessed on 10 Sept 2017

Irrigation & Water Resource Department (2017) Flood bulletin for the year 2017. Irrigation & Water Resource Department, U.P. http://idup.gov.in/pages/en/topmenu/dept.-activities/civil/floods/en-flood-bulletin. Accessed on 10 Sept 2017

Islam MM, Sado K (2000) Development of flood hazard maps of Bangladesh using NOAA-AVHRR images with GIS. Hydrol Sci J 45:337–355. https://doi.org/10.1080/02626660009492334

Jain SK, Saraf AK, Goswami A, Ahmad T (2006) Flood inundation mapping using NOAA AVHRR data. Water Res Manage 20:949–959. https://doi.org/10.1007/s11269-006-9016-4

Jensen JR (2018) Introductory digital image processing: a remote sensing perspective. 4th edn. Pearson India Education Services Pvt. Ltd, Noida
Kass M, Witkin A, Terzopoulos D (1988) Snakes: active contour models. Int J Comput Vis 1 (4):321–331
Kumar R (2010) Fluvial processes in lower Rapti River basin: a case study of impacts on arable land. Unpublished Ph.D thesis. Centre for the Study of Regional Development, Jawaharlal Nehru University, New Delhi
Kumar R (2016) Flood hazard assessment of 2014 floods in Sonawari sub-district of Bandipore district (Jammu and Kashmir): an application of geoinformatics. Remote Sens Appl Soc Environ 4:188–203
Kumar R, Acharya P (2016) Flood hazard and risk assessment of 2014 floods in Kashmir Valley: a space-based multisensor approach. Nat Hazards 84(1):437–464
Kumar R, Kamal V, Singh RK (2013) Geomorphic effects of 2011 floods on channel belt parameters of Rapti River: a remote sensing and GIS approach. Corona J Sci Technol 2(2):4–12
Kumar R, Kumar S, Pandey PC (2016) Delineation and zonation of flood prone area using geo-hydrological parameters: a case study of lower Ghaghara River Valley. In: Srivastava PK, Pandey PC, Kumar P, Raghubanshi AS, Han D (eds) Geospatial technology for water resource applications. CRC Press, Taylor & Francis Group, Boca Raton, pp 78–100
Lee J-S (1981) Speckle analysis and smoothing of synthetic aperature radar images. Comput Graphics Image Process 17(1):24–32
Lee J-S, Jurkevich I, Dewaele P, Wambacq P, Oosterlinck A (1994) Speckle filtering of synthetic aperture radar images: a review. Remote Sens Rev 8:313–340. https://doi.org/10.1080/02757259409532206
Li J, Chen W, Touzi R (2007) Optimum RADARSAT-1 configurations for wetlands discrimination: a case study of the Mer Bleue peat bog. Can J Remote Sens 33(Suppl 1):S46–S55. https://doi.org/10.5589/m07-046
Lopès A, Nezry E, Touzi R and Laur H (1990) Maximum aposteriori speckle filtering and first order texture models in SAR Images. Proceedings International Geoscience and Remote Sensing Symposium (IGARSS'90). Ref IEEE90CH2825–8. Washington DC, USA, pp 2409–2412
Manavalan R (2017) SAR image analysis techniques for flood area mapping-literature survey. Earth Sci Inf 10(1):1–14. https://doi.org/10.1007/s12145-016-0274-2
Manavalan, Rao (2014) DEM and SAR image based flood feature extraction techniques to map the deep and shallow flood inundated regions of known as well as remote disaster regions. Geocarto Int 29(7):745–757. https://doi.org/10.1080/10106049.2013.838310
Manjusree R, Kumar LP, Bhatt CM, Rao GS, Bhanumurthy V (2012) Optimization of threshold ranges for rapid flood inundation mapping by evaluating backscatter profiles of high incidence angle SAR images. Int J Disaster Risk Sci 3(2):113–122. https://doi.org/10.1007/s13753-012-0011-5
Manjusree P, Bhatt CM, Begum A, Rao GS, Bhanumurthy V (2015) A decadal historical satellite data nalysis for flood hazard evaluation: a case study of Bihar (North India). Singap J Trop Geogr 36:308–323
Mason DC, Schumann GJP, Neal JC, Garcia-Pintado J, Bates PD (2012a) Automatic near real-time selection of flood water levels from high resolution Synthetic Aperture Radar images for assimilation into hydraulic models: a case study. Remote Sens Environ 124:705–716
Mason DC, Davenport IJ, Neal JC, Schumann GJP, Bates PD (2012b) Near real-time flood detection in Urban and rural areas using high-resolution synthetic aperture radar images. IEEE Trans Geosci Remote Sens 50(8):3041–3052
McFeeters SK (1996) The use of the normalized difference water index (NDWI) in the delineation of open water features. Int J Remote Sens 17:1425–1432
Mizushima Laboratory (2013) India Place Finder: http://india.csis.u-tokyo.ac.jp/, Department of Oriental History, Graduate School of Humanities and Sociology, The University of Tokyo
Motulsky H (2014) Intuitive biostatistics: a nonmathematical guide to statistical thinking, 3rd edn. Oxford University Press, New York

Oberstadler R, Hoè Nsch H, Huth D (1997) Assessment of the mapping capabilities of ERS-1 SAR data for flood mapping: a case study in Germany. Hydrol Process 11:1415–1425. https://doi.org/10.1002/(SICI)1099-1085(199708)11:10<1415::AID-HYP532>3.0.CO;2-2
Planning Commission (2011) Report of working group on food management and region specific issues for XII plan. Govt of India, New Delhi
Pulvirenti L, Pierdicca N, Guerriero L (2013) Monitoring flood evolution in vegetated areas using COSMO-skyMed data: the Tuscany 2009 case study. IEEE J Sel Top Appl Earth Observ Remote Sens 6(4):1939–1404
Pulvirenti L, Chini M, Pierdicca N, Boni G (2016) Use of SAR for detecting floodwater in urban and agricultural areas: the role of the interferometric coherence. IEEE Trans Geosci Remote Sens 54(3):1532–1544. https://doi.org/10.1109/TGRS.2015.2482001
Qui F, Berglund J, Jensen JR, Thakkar P, Ren D (2004) Speckle noise reduction in SAR imagery using a local adaptive filter. GISci Remote Sens 41(3):244–266
Rana NK, Kumar R, Kumar D (2009) Nature of channel shift of a foothill-fed river in the alluvial setting: a case study of River Rapti, India. Indian J Geomorphol 13(1&2):83–98
Romshoo SA, Sadaff A, Rashid I, Dar RA (2018) Climatic, geomorphic and anthropogenic drivers of the 2014 extreme flooding in the Jhelum basin of Kashmir, India. Geomat Nat Haz Risk 9 (1):224–248. https://doi.org/10.1080/19475705.2017.1417332
Sanyal J, LU X (2006) GIS-based flood hazard mapping at different administrative scales: a case study in Gangetic West Bengal, India. Singap J Trop Geogr 27:207–220. https://doi.org/10.1111/j.1467-9493.2006.00254.x
Schlaffer S, Matgen P, Hollaus M, Wagner W (2015) Flood detection from multi-temporal SAR data using harmonic analysis and change detection. Int J Appl Earth Obs Geoinf 38:15–24. https://doi.org/10.1016/j.jag.2014.12.001
Senthilnath J, Shenoy HV, Rajendra R, Omkar SN, Mani V, Diwakar PG (2013) Integration of speckle de-noising and image segmentation using Synthetic Aperature Radar image for flood extent extraction. J Earth Syst Sci 122(3):559–572. https://doi.org/10.1007/s12040-013-0305-z
Sheng Y, Xia Z (1996) A comprehensive evaluation of filters for radar speckle suppression. Proceed IGARSS 2:1559–1561
Singh IB (1996) Geological evolution of Ganga plain: an overview. J Palaeontol Soc India 41:99–137
Sivasami KS (2002) Environmental effect due to floods and reservoirs. In: Subramanian V (ed) Environmental hazard in South Asia. Capital, New Delhi, pp 65–82
Song Y-S, Sohn H-G, Park C-H (2007) Efficient water area classification using RADARSAT-1 SAR Imagery in a high relief mountainous environment. Photogramm Eng Remote Sens 73 (3):285–296
Tan Q, Bi S, Hu J, Liu Z (2004) Measuring lake water level using multisource remote sensing images combined with hydrological statistical data. IEEE Geoscience and Remote Sensing Symposium, IGARSS-2004, 7, 4885–4888. Anchorage, AK, Sept. 20–24, 2004. https://doi.org/10.1109/IGARSS.2004.1370258
Twele A, Cao WX, Plank S, Martinis S (2016) Sentinel-1-based flood mapping: a fully automated processing chain. Int J Remote Sens 37(13):2990–3004. https://doi.org/10.1080/01431161.2016.1192304
Williams DJ, Shah M (1992) A fast algorithm for active contours and curvature estimation. CVGIP: Image Underst 55(1):14–26
Xu H (2006) Modification of normalised difference water index (NDWI) to enhance open water features in remotely sensed imagery. Int J Remote Sens 27:3025–3033. https://doi.org/10.1080/01431160600589179
Yadav RP (1999) Floods in eastern U.P. Radha Publication, New Delhi-2
Yamada Y (2001) Detection of Flood-inundated area and relation between the area and Micro-Geomorphology using SAR and GIS. IEEE Geoscience and Remote Sensing Symposium, IGARSS. 7: 3282–3284. Sydney, NSW, July 9–13, 200. https://doi.org/10.1109/IGARSS.2001.978329

Chapter 7
Application of ASTER Remote Sensing for Lithological Mapping in the Udaipur District of Rajasthan, India

S. S. Salaj, S. K. Srivastava, Rahul Dugal, Richa Upadhyay, D. S. Suresh Babu, and S. Kaliraj

Abstract Remote sensing applications for earth studies such as lithological discrimination, geological mapping and potential mineral exploration have shown great success worldwide. Advanced Spaceborne Thermal Emission and Reflection Radiometer (ASTER) Level-1B image includes visible and near-infrared (VNIR) and shortwave infrared (SWIR) bands that have been analysed to discriminate lithology features in meta-sedimentary terrains of Aravalli Supergroup in Udaipur area of Rajasthan, India. The area comprises various types of geological settings and rock types composed of economic valuable deposits of lead, zinc, copper micas and marbles; they show spectral reflectance distinctly in bands of VNIR and SWIR. The unique spectral signature reflected by lithological unit shows effectiveness in lithological mapping. The reflectance spectra of various rock types, namely, phyllitic dolomite, siliceous dolomite, metagreywacke, quartzite and gneiss, were collected in situ using spectroradiometer and used as reference of ASTER image for the preparation of spectral signature of different lithological units. The image is applied to analysis atmospheric correction using Fast Line-of-sight Atmospheric Analysis of Hypercubes (FLAASH) and empirical line calibration techniques to convert pixel radiance values into reflectance. A minimum noise fraction (MNF) transform is applied to identify the inherent variance of spectral reflectance and effectively discriminates various lithological units. The different types of lithological units are clearly discriminated using MNF method. Spectral Angle Mapper (SAM) classification is an effective tool for differentiating rock types and its distinct mineralogical

S. S. Salaj · D. S. Suresh Babu · S. Kaliraj (✉)
Central Geomatics Laboratory (CGL), ESSO-National Centre for Earth Science Studies (NCESS), Ministry of Earth Sciences, Government of India, Thiruvananthapuram, India
e-mail: s.kaliraj@ncess.gov.in

S. K. Srivastava · R. Upadhyay
Indian Institute of Remote Sensing, Dehradun, India

R. Dugal
University of Pune, Pune, India

P. Kumar et al. (eds.), *Applications and Challenges of Geospatial Technology*,
https://doi.org/10.1007/978-3-319-99882-4_7

composition from associated terrains. Spectral Angle Mapper (SAM) classification uses field-derived spectral signature to demarcate various lithological features with its spatial extent. The result shows different lithological units under Aravalli Supergroup, Banded Gneissic Complex and intrusive formations that are composed of meta-arkose, conglomerate, phyllite, mica schist, dolomite, metagreywacke and migmatites in various locations. The extracted geological features using ASTER image show strong resampling with the district resource map and validated using ground truth verification. The overall accuracy of SAM-classified map of lithological units is 73.39% and Kappa coefficient of 0.59. Mapping the lithological features using ASTER image, data coupled with MNF and SAM techniques provides relatively accurate result, and this study may be used for discrimination of lithological units with its spatial characteristics.

Keywords ASTER · Lithological mapping · FLAASH · Minimum noise fraction · Spectral Angle Mapper · Remote sensing and GIS

7.1 Introduction

Application of remote sensing in geological studies shows a great success in lithological discrimination, mineral exploration and identification of hydrothermally altered rock types (Chandan Kumar et al. 2015). Remotely sensed multispectral images involve earth exploration studies especially delineation of geological outcrops, different rock types and soils and minerals (Kruse and Dietz 1991; Rowan and Mars 2003; Tangestani et al. 2011). Multispectral images including AVIRIS, MODIS, MERIS, ASTER and ETM+ have great utility in geological mapping at different scales; several studies dealt with the utilization of ASTER image which includes spectral bands of VNIR, SWIR and TIR for lithological discrimination and detection of rocks alteration products (Boardman and Kruse 1994; Di Tommaso and Rubinstein 2007; Gabr et al. 2010; Hosseinjani and Tangestani 2011; Pour and Hashim 2012). Multispectral images of ASTER and ETM+ with higher spectral and spatial resolution have been well employed for mapping of lithological features and lineaments (Gomez et al. 2005; Gad and Kusky 2007; Rajendran et al. 2012). Spectral reflectance of various rock types is used as endmember for discriminating location and spatial extent of them along with their mineral compositions. The spectral combination analysis of ASTER image provides distinct reflectance in the spectral reflectance of visible, NIR bands that is used for discriminating the main hydrothermal alteration mineral zones (Kruse and Dietz 1991; Rowan et al. 2003). ASTER image shows spatial extends of lithological features and rock types and used for various geological applications. Imaging spectroscopy allows geologists to explore vast territories for surface features that may indicate the presence of minerals in a certain area, using specific absorption features, caused by chemical bonds in materials (Borengasser et al. 2008). ASTER image with continuous spectrum of

VNIR and SWIR bands is utilized for effective lithological mapping and provides new prospective of investigating earth surface mineral compositions (Van der Meer et al. 2012). Spectral reflectance of ASTER sensor acquired in the VNIR, SWIR and TIR regions offers unique spectral signature for surface-exposed lithological features (Rajendran et al. 2012) minerals (Chandan Kumar et al. 2015), hydrothermally altered rocks include propylitic, argillic, phyllic and potassic, etc. with relatively high accuracy (Zhang et al. 2007). ASTER image comprises wide range of spectral coverage to enhance specific rock for lithological mapping from other features (Hewson et al. 2005; Gad and Kusky 2007). Advanced development of image processing methods is applied to ASTER data as a tool for mapping lithological features and mineral zones associated with their rock types. The development of image processing techniques, i.e. SAM, MNF, ANN, etc., can provide complementary mapping for lithological, mineral and altered rocks using ASTER image (Gomez et al. 2005; Hosseinjani and Tangestani 2011). Advanced image processing techniques like principal component analysis (PCA), minimum noise fraction (MNF), band ratios (BRs), band combinations (BCs) and spectral indices (SIs) are well employed on ASTER image for discrimination of various lithological units. The target spectra (endmember)-based classification algorithms such as Spectral Angle Mapper (SAM), Spectral Feature Fitting (SFF), Matched Filter (MF), Constrained Energy Minimization (CEM), Linear Spectral Unmixing (LSU) and Mixture-Tuned Matched Filtering (MTMF) are provided subpixel level classification of lithological features along with their abundant mineral compositions. Minimum noise fraction (MNF) is a statistical method widely used to demarcate the inherent properties of spectral dimensionality from lithological features based on segregation of minimum noise and identify various types of lithological features along with their associated features with spatial extent on the ground surface.

Spectral Angle Mapper (SAM) is a standard classification method that is successfully used for mapping lithological features compared to other classification systems. SAM classification is used to determine spectral similarity between image spectra and reference reflectance spectra (field-observed spectra also known as endmember). This is highly suitable for classification of lithological features from multispectral image with medium and high spatial resolution (Boardman and Kruse 1994; Borengasser et al. 2008; Pour and Hashim 2011). SAM classification of ASTER image has been used worldwide for mapping geological features and alteration minerals and for identifying potential mineral exploration sites with the use of field reflectance spectra (i.e. USGS/Japan Spectral Library) as reference spectrum (Tangestani et al. 2011). This classification provides a pixel-wise feature extraction for improved lithological mapping and is quiet an advanced one compared to other methods because this executes spectral properties at each pixel of images (Rowan and Mars 2003; Zhang et al. 2009; Pour and Hashim 2012). The ASTER image is classified using SAM algorithm to quantify spectral similarity of lithological features based on endmember (reference spectra) of known rock types observed in the field (Qiu et al. 2006). SAM algorithm is used as spectral matching method for identification of lithological features using multispectral or hyperspectral images (Sabins 1999; Tangestani et al. 2011). ASTER image provides clear discrimination of mineral composition for mapping alteration minerals with high accuracy. Many classification techniques are widely used for feature extraction from the image based on spectral reflectance, and SAM is one of the most suitable classification methods for lithological discrimination using multispectral images such as Landsat TM, ETM +, OLI and ASTER and hyperspectral images (Hyperion and AVIRIS) that provide mineralogical compositions. The endmember applied on SAM algorithm can execute image classification at subpixel level to demarcate different rock types based on field-observed reference spectra collected from various rocks (Youssef et al. 2009). SAM model quantifies spectral similarity between spectrum of image pixel and field-observed spectra from known components (rock types), and the model's algorithm executes spectral matching method and hence widely used for identification of lithological type using multispectral images (Pour and Hashim 2012). The present study investigates ASTER image for lithological mapping in the Udaipur district, Rajasthan, India, using field-observed spectral signatures; the various types of lithological features have been effectively discriminated and mapped using advanced image processing techniques such as minimum noise fraction (MNF) and Spectral Angle Mapper (SAM) methods in remote sensing and GIS environment.

7.2 Study Area and Geological Setting

ASTER image-based lithological features have been discriminated in the south-eastern part of the Udaipur district of Rajasthan, India. The surface-exposed rock types comprise different geological settings associated with various rock types due to formation of Archean to recent alluvium formations. Figure 7.1 shows geographical location of the study area. The geographical extent of the study area lies between

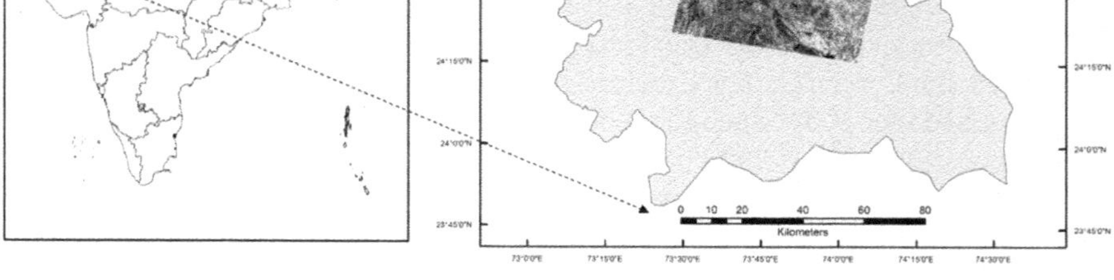

Fig. 7.1 Location map of the study area

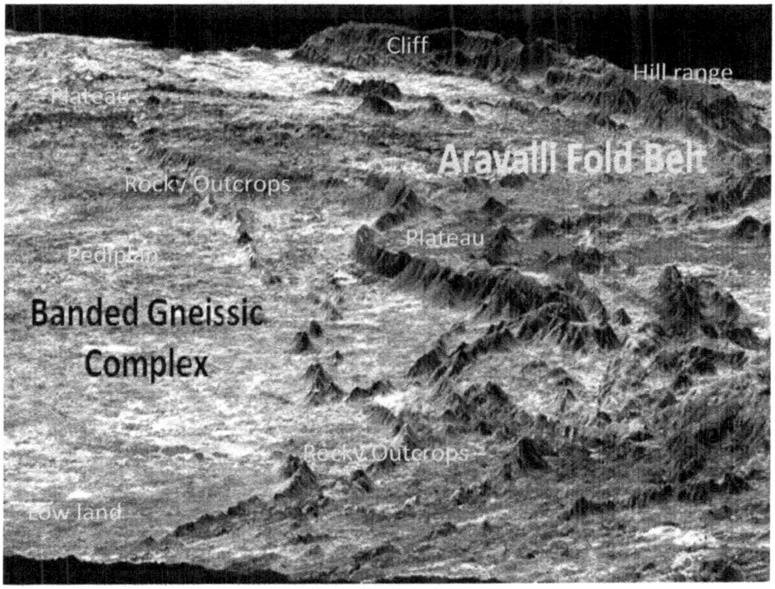

Fig. 7.2 Physiography of the study area

the longitude of 74° 14′ 13″E – 73° 24′ 38″E and latitude of 24° 54′ 44″N – 24° 14′ 40″N with the coverage of total area which is approximately 3600 km^2. The terrain achieved the present-day topography largely due to long period of erosion. Aravalli Mountains are highly folded and show undulating topography with ridges and peaks intervened by low-lying valleys. Figure 7.2 shows the 3D view of physiography of study area using ASTER image. The study area consists of rocky-exposed terrains which prevail during semiarid climatic condition that are generally composed of economically valuable mineral resources including iron, lead, zinc, copper, rock phosphate, soapstone, limestone, barites, marble, etc. The Udaipur district of Rajasthan state holds enriched mineral resources, whereas the Aravalli Supergroup of rocks hosts some of the best mineral belts being worked in the district (Roy and Paliwal 1981; Sharma 2009). The geological setting of the study area is demarcated in Fig. 7.3. Lithological formations occur during orogenic cycles, and the terrain is classified as (i) Aravalli Supergroup (includes sub-groups of Jharol, Bari Lake (basic volcanics, metaconglomerate and phyllite), Udaipur group (overlain by Zawar formation in some areas)), (ii) Banded Gneissic Complex (Bhilwara Supergroup), (iii) Debari group and (iv) synorogenic granite and gneiss (intrusive rocks). Aravalli Supergroup (ASG) shows two contrasting lithofacies composed of surface facies of shale-sand-carbonate assemblage and deepwater facies of carbonate-free shale-arenite association; within this, the ultramafic rocks (serpentinites) are emplaced as formation of Aravalli fold belt (Sharma 2009; Chandan Kumar et al. 2015). Debari group and Udaipur group represent platform facies of the surface. Debari group consisting of coarse clastics, carbonates and pelites overlies the basement Banded

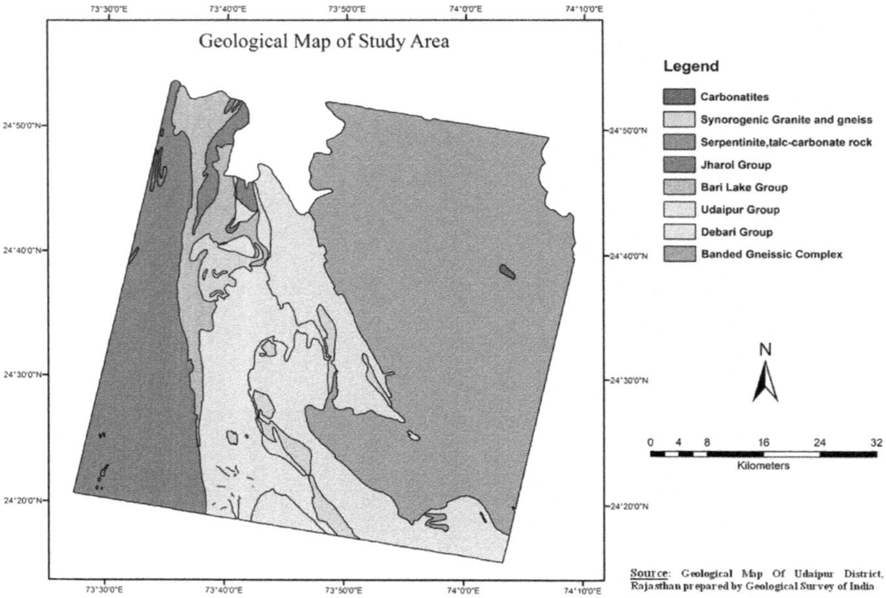

Fig. 7.3 Geological map of the study area

Gneissic Complex. Debari group overlay by Udaipur group comprising greywacke and phyllite. A carbonate sequence named as Zawar formation hosting lead-zinc mineralization overlies Udaipur group. Jharol group represents deep-sea facies and ultramafics mainly consisting of pelite-dominant sequences, serpentinites, chlorite schist and quartzite with phosphoritic dolomite (Roy et al. 1988). The metamorphic formation occurs in Aravalli sequence which extends from east to west. The major area covers rock types of gneiss, migmatites, phyllite, quartzite, metagreywacke and dolomite. Dolomite in Zawar formation constitutes the principal host rock for sulphide mineralization. The rock types generally formed between Archaean to mid-Proterozoic period (Sinha-Roy et al. 1998). Meta-sedimentary rocks of Aravalli Supergroup occur in the plateau of Aravalli hills. The Proterozoic rock formation has deposited as supracrustal forms, and that is called as Archaean basement gneiss complex under Banded Gneissic Complex (BGC). This formation has predominantly occurred as poly-metamorphosed, multi-deformed rock-suite of tonalite-trondhjemite gneiss, amphibolite, migmatite and granitoids (Sharma 2009). The rocks of Aravalli Supergroup are formed as phyllite, greywacke, quartzite, dolomite and intra-formational conglomerate, and these formations were deposited in Pre-cambrian Aravalli sea during 2.3 Ga (Roy and Paliwal 1981; Sinha-Roy et al. 1998). The sedimentary deposits are formed as folded outcrops with mildly metamorphosed rocks. Synorogenic intrusive formation comprises with granite and gneiss that are formed as isolated patches in various parts of the study area.

7.3 Materials and Methodology

Geological remote sensing has been achieved for mapping of lithological features in arid to semiarid environments. Terra onboard ASTER sensor has been launched in December 1999 by NASA's Earth Observing System (EOS) AM-1. ASTER consists of three separate instrument subsystems to observe the three different spectral bands including visible and near-infrared (VNIR), shortwave infrared (SWIR) and thermal infrared (TIR). ASTER image comprises long-narrow spectral reflectance of an object in SWIR band that mostly used for discriminating lithological features and mineral mapping. Hydrothermally altered mineral assemblages can be identified using their diagnostic spectral absorption from VNIR through SWIR (0.4–2.5 μm) and TIR (8.0–14.0 μm) wavelength regions. Significant spectral properties of ASTER have been commonly used for geological applications including (i) discrimination and identification of rocks, minerals and lithological features and (ii) identification of iron oxide minerals in surface and mapping of carbonates and silicates. The methodology flowchart that shows systematic geoprocessing operations used for this study is shown in Fig. 7.4.

ASTER Level-1B image of VNIR and SWIR bands which cover the Udaipur region has been analysed for this study. ASTER image consist of three bands including visible/near-infrared (VNIR) and six spectral bands in shortwave infrared (SWIR) regions. The image with enough spatial resolution provides a new

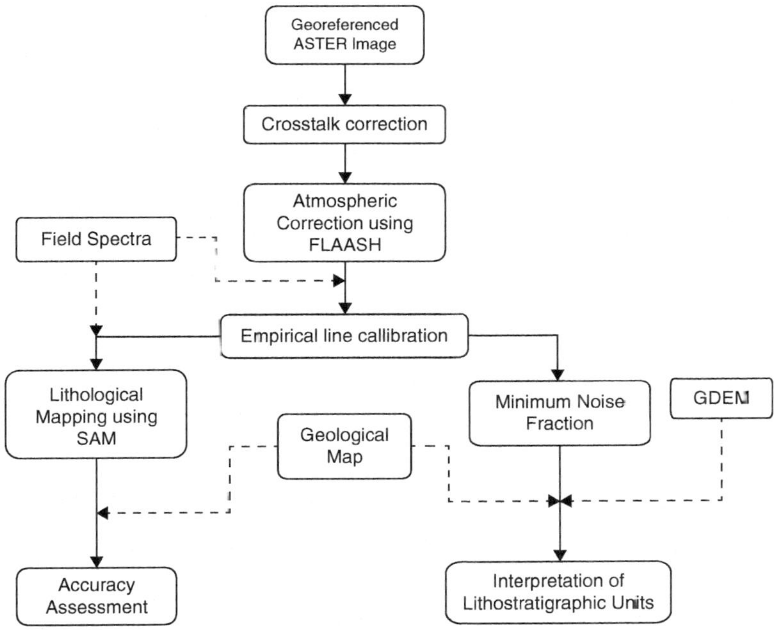

Fig. 7.4 Methodological flowchart shows geoprocessing functionalities involved in this study

prospective for exploration studies on the earth surface. The VNIR region with three bands (0.52–0.86 μm) provides distinct spectral reflectance from the ores of transition metals such as iron, manganese, chromium, copper, etc. The six bands of SWIR region (1.60–2.43 μm) have been used for identification of carbonate, hydrate and hydroxide minerals. The TIR band covers wavelength of 8.125–11.65 μm that have perfectly provided spectral reflectance of silicates and its composed matter (Chandan Kumar et al. 2015). Multispectral bands of ASTER image have been used for identification of various lithological features based on different band combination analyses. The spectral reflectance of VNIR and SWIR bands in ASTER image has scientifically been proved for geological applications such as lithological discrimination, mineral mapping and identification of altered rock types.

7.3.1 *Field Spectra Reflectance Data Collection*

The objects that occur on the earth surface, near the sub-surface, reflect unique spectral reflectance that indicates their spectral signature based on shape, size, pattern and mineral composition in nature. The earth surface features reflect or emit unique portion of spectral reflectance in a VNIR and SWIR wavelength depending on their mineral composition and textural properties such as grain size, packing, mixing and physical state that are involved for lithological mapping. Various rock types exposed on the surface in the semiarid terrain of Udaipur in Rajasthan and the spectral reflectance collected as endmember from various rocks can be used to demarcate lithology and hydrothermal-altered minerals. Field-observed spectral reflectance describes material's characteristics in continuous electromagnetic spectrum interact with them. The spectroradiometer instrument (Analytical Spectral Device) is used for collection of reference spectra (endmembers) with narrow-continuous bands that are having high-intensity contact with the rock types to generate spectral signature. The spectral sampling locations were marked using handheld GPS system. The ASD spectroradiometer is used to collect continuous spectra with the wavelength range of 325–1075 nm, in which VNIR and SWIR bands cover 0.35–2.5 nm (VNIR), 10 nm (SWIR) spectral resolutions and 1 nm spectral sampling. The ASD field spectra were collected mainly from dominant rock types, viz. gneiss, quartzite, phyllitic dolomite, siliceous dolomite and metagreywacke, across the Udaipur district in Rajasthan state. Figure 7.5 shows the field-observed ASD spectra (endmember spectra collection) from different rock types of the study area. The spectral reflectance collected from various rock types at different angles is used for empirical line correction and reference for reflectance spectrum of ASTER image for mapping the lithological features.

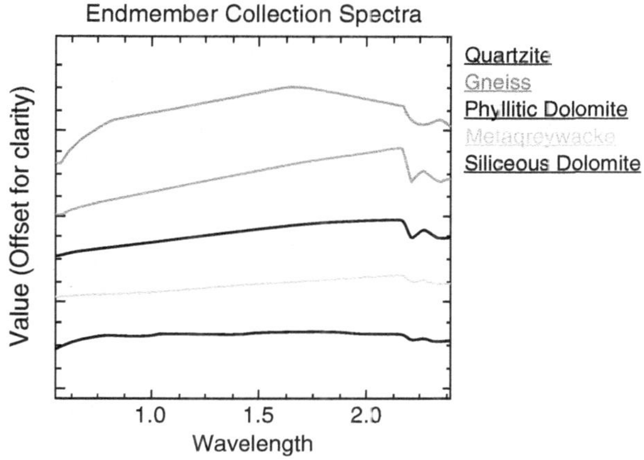

Fig. 7.5 Field-observed reflectance spectra (endmember collection spectra) from different rock types of the study area

7.3.2 ASTER Pre-processing and Image Calibration

A systematic calibration requires an ASTER image for removing atmospheric and radiometric errors and other sensor defects, before performing reliable image analysis. The pixel values (digital number) of an image may not represent surface reflectance and that required transferring into spectral reflectance. ASTER image can maximize by calibrating it for atmospheric and radiometric correction to represent actual spectral reflectance of objects as it is in the earth surface. Image atmospheric correction plays an important role for converting DN value to spectral reflectance. FLAASH (Fast Line-of-sight Atmospheric Analysis of Spectral Hypercubes) model is applied to ensure atmospherically corrected ASTER image to produce true surface reflectance. This model incorporates MODTRAN4 radiation transfer code, which is applied on individual bands of ASTER image to correct atmospheric scattering due to water vapour and aerosol on a pixel-by-pixel basis. FLAASH uses the algorithm and is expressed as

$$L = \left(\frac{A\rho}{1-\rho_e S}\right) + \left(\frac{B\rho_e}{1-\rho_e S}\right) + L_a \qquad (7.1)$$

where ρ is the pixel surface reflectance, ρ_e is an average surface reflectance for the pixel and a surrounding region, S is the spherical albedo of the atmosphere, L_a is the radiance back scattered by the atmosphere and A and B are coefficients that depend on atmospheric and geometric conditions but not on the surface. Image atmospheric correction is used as mathematical equation (Eq. 7.1) for removing the atmospheric disturbances (effects) that involve conversion of the image radiance to ground truth reflectance (GTR). FLASSH tool has been calibrated using ASTER image comprising VNIR and SWIR bands to improve spectral reflectance of the objects. Moreover, the process of conversion of absolute radiance (DN value) to ground truth Reflectance involves the classification of earth features from the image using field spectral

Table 7.1 The rock types and their locations of ASD field spectra collection for ASTER image empirical line calibration

Field sample	Latitude	Longitude
Phyllitic dolomite	24°21′08.5″N	73°43′18.2″E
Siliceous dolomite 1	24°21′38″N	73°43′44.3″E
Siliceous dolomite 2	24°21′43.2″N	73°42′00.61″E
Metagreywacke	24°21′41.7″N	73°43′45.1″E
Gneiss	24°24′32.6″N	73°39′30.3″E
Quartzite	24°.1″N	73°44′15″E

reflectance, mainly used for discrimination of earth features (Chandan Kumar et al. 2015).

7.3.3 Empirical Line Calibration (ELC)

Table 7.1 shows the rock types and its locations of ASD field spectra collection for ASTER image empirical line calibration. This model is executed for image-measured spectra or field spectral reflectance (endmember) for enhance spectral properties of each objects that provides as an alternative to radioactive transfer modelling for lithological mapping. This method is executed multivariate regression equation, and this is represented as

$$DN_k = \rho_\lambda A_k + B_k \tag{7.2}$$

where DN_k is the DN value for a particular pixel in band (k), $\rho_\lambda$equals to scaled surface reflectance of object within the remote sensor IFOV at a specific wavelength (λ), $A_k$is gain and $B_k$is offset.

In an ASTER image after FLASSH model-based atmospheric corrections, some residual errors like path radiance remain in the image data, but it can be corrected using empirical line calibration (ELC) based on regression statistical method. Image pixel spectra is calibrated to match ground spectra (ASD field spectral reflectance data) using certain ground control points from contrasting albedo areas. This model calculates linear relationship between image-derived spectra and ground-measured reflectance spectra of objects with a range of contrasting albedo to calculate total reflectance gains and total reflectance offsets and to convert image pixel values into reluctance factor. Field spectral reflectance of the selected targets is used for ELC for ASTER bands of VNIR and SWIR to produce independent reflectivity of image measurement.

7.3.4 Minimum Noise Fraction (MNF) Transformation

The minimum noise fraction (MNF) is a standard transformation model that computes normalized linear combinations of spectral reflectance of each band, which maximizes signal (reflectance) to noise ratio for filtering noise and inverting reflectance value in a pixel of an image (Rajendran et al. 2013). Output of MNF transformation produces ternary coloured MNF image that shows distinct variability of different rock types in their spectral signature for lithological compositional mapping. MNF transformation was applied on ASTER image with VNIR and SWIR bands using ENVI 5.4 software to implement cascaded analysis such as PCA and MNF transformation. The selection of endmembers from the MNF image using matched filtering produces enough spectral variability as a function of different mineralogies in the visible and near-infrared and shortwave-infrared ranges that compare successfully with the existing geological map. MNF algorithm analyses the ASTER VNIR and SWIR bands to determine inherent dimensionality, to segregate noise in the data and to reduce the computational requirements for subsequent image processing. This analysis executes MNF transformation as a first step, which is used to segregates noise in the image in which the noise has unit variance and no band-to-band correlations. Secondly, principal components analysis (PCA) is executed on VNIR and SWIR bands to transform the pixels with noise-whitened data into new uncorrelated bands that have maximum variance in spectral reflectance values (DN value) to highlight uncorrelated features for exploring the rock types exposed on the surface.

7.3.5 Spectral Angle Mapper Classification

Spectral Angle Mapper (SAM) is a standard classification model, which is applied to classify ASTER image to experimentally map the different rock types using field reflectance spectra (endmember) derived from ASD spectroradiometer instrument (350–2500 nm). SAM classification executes pixel-based analysis of ASTER image using endmember or reference spectra (spectral signatures of the objects measured directly in the field or laboratory) to classify various features and ponderation of their values assigned as 0 (low spectral similarity) and 1 (high spectral similarity). This model takes inherent property of rocks for further magnified as they acquired in remote sensing image based on spectral variability. Two methods were executed to derive endmember classes (reflectance spectra) used for SAM classification for the preparation of improved lithological mapping using ASTER image; (i) calculate mean spectral derivatives using sum of spectral values from field-observed spectra and image-measured original spectra (spectral value of pixel). It is described as the combination of image-derived mean spectrum and field-observed mean spectral derivative used in the SAM classification to improve class separability (spectral variability) for a corresponding rock type; (ii) use of multiple endmembers (reflectance spectra of many rock samples) into SAM classification to accommodate spectral variability among various types of rocks. In the present study, the SAM method is used for the classification of rock types using ASTER image by calculating spectral similarity of image spectra to reference reflectance spectra (Kruse et al. 1993). Spectral reflectance of different rock samples was collected using field measurements or extracted directly from the image as an endmember or reference spectra used for SAM classification analysis. The algorithm of SAM classifier is used to measure spectral similarity of an object compared to endmember by calculating spectral angle between image spectra and field-observed spectra and treating them as vectors in n-dimensional space (Kruse et al. 1993; Rowan and Mars 2003). The statistical correlation of spectra showing narrow angle between above spectra indicates occurrence of higher similarity for demarcating the rock types, whereas the wider spectral angle indicates low similarity and causes poor classification results. The SAM model is independent from spectral, radiometric errors due to solar illumination factors, because this model mainly calculates spectral angle of two different spectra of the same features and it is independent from vector's length (Kruse et al. 1993). The SAM algorithm calculates spectral similarity of image spectra and endmember spectra by computing arccosine value at dot product for test spectrum 't' and reference spectrum 'r' involved in the analysis (Kruse et al. 1993). The algorithm computes spectral similarity at each pixel of an image based on comparative analysis between spectral angles between reference spectrum (endmember) and image-derived spectrum, and the output shows a vector in n-dimensional space, in which the narrow spectral angles represent high similarity and indicates closer matches to reference spectrum derived from the field (Kruse et al. 1993). In this analysis, SAM classification has been executed on ASTER image with VNIR and SWIR bands using field-observed spectra for mapping lithological units. The field-observed spectra is used as a reference (endmember), which have been collected from different rocks at various locations. There are 436 sample points that are comparatively analysed with stratified random distribution for accuracy assessment using error matrix method (Congalton 1991). SAM-classified lithological map was comparatively verified using existing geological features on district resource map published by the Geological Survey of India (GSI 1997).

7.4 Results and Discussion

ASTER remote sensing provides for spectrally unmixed products of various rock types that may be useful for mapping lithological features with improved geological discrimination details using MNF and SAM classification techniques (Tangestani et al. 2011). Field-observed reflectance spectra from different rock types have been used as reference spectrum for classifying specific lithological units with their pattern, association and spatial extent as it is on the earth surface. ASTER image with VNIR and SWIR bands (spatial resolution 30 m) has been used for discriminating different lithological features from various parts of the study area. The output of lithological map derived from ASTER image shows Gneissic Complex formation of gneiss in major parts of the study that have exposed as rocky outcrops. Migmatites and granitic rocks combined with shale-sand-carbonate assemblage, deepwater facies of carbonate-free shale-arenite association and intrusive formation of synorogenic granite and gneiss. The lithological map with location and spatial extent of mineral resources may be helpful for planning socio-economic development activities.

7.4.1 ASTER Image Spectral Characteristics of Rock Types

The spectral reflectance of ASTER image shows unique signature of various lithological units. Figure 7.5 shows spectral variability reflected by different rock types at wavelength of VNIR and SWIR bands of ASTER image. The field spectra collected from various rock samples using spectroradiometer are highly resembled with ASTER image spectra for discriminating the rock types. Mapping of lithological features has been performed using band combination analysis of ASTER image, i.e. (i) band composition analysis of band 4 (SWIR 1.60–1.70 μm), band 5 (SWIR 2.145–2.185 μm) and band 7 (2.235–2.285 μm), (ii) MNF analysis of VNIR and SWIR bands and (iii) band ratio of image of (6/8:R; 4/8:G and 11/14:B). Massive sheared granodiorites and diorites have been discriminated as light and dark grey image signatures, respectively, on the 4/8 band ratio of ASTER image. The formation of hard rock types consists of granites and dykes that have spectrally demarcated by their dark grey signature on image using the 6/8 band ratio analysis. The altered rocky formations of hornblende, biotite and plagioclases (chlorite, sericite and clay minerals) may be responsible for absorption features at the wavelength of 1.65 μm and 2.35 μm regions. The TIR bands of ASTER image may be attributed unique emissivity to different lithological units; the band ratio of 11/14 shows different lithological units like diorite, granodiorites and post-tectonic granites; and these features have been discriminated based on bright, grey and dark image signatures, respectively. The different rock types reflect their unique spectral signature in ASTER image of VNIR and SWIR bands, and these signature are derived as endmember classes (reference spectra) for SAM classification for mapping the lithological features.

7.4.2 Discrimination of Lithological Feature Using MNF Analysis

Figure 7.6 shows the ASTER image-derived lithological features that have categorized into three major groups of lithological formations such as (i) Aravalli Supergroup (ASG) which occurred with the assemblage of shale-sand-carbonate, carbonate-free deepwater facies and shale-arenite association, in which ultramafic rocks (serpentinites) have emplaced as group or isolated bathes along the Aravalli fold belt; (ii) Banded Gneissic Complex (BGC) which comprises the formation of gneiss, migmatites and granitic rocks; and (iii) intrusive formation of synorogenic granite and gneiss. Figure 7.7 shows the minimum noise fraction analysis of

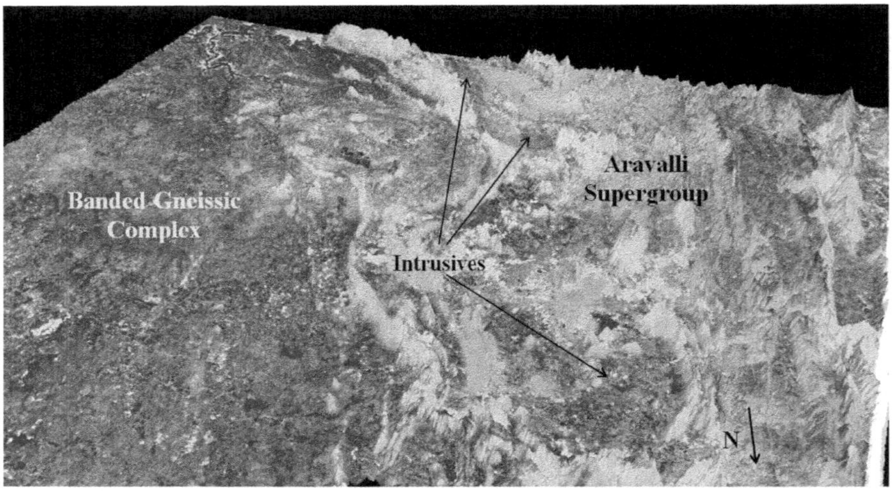

Fig. 7.6 MNF colour composite (RGB: 312) draped over DEM viewed from the north. The major groups of lithological formation in the study area: (i) Aravalli Supergroup of shale-sand-carbonate assemblage and deepwater facies of carbonate-free shale-arenite association, in which ultramafic rocks (serpentinites) emplaced along the Aravalli fold belt, (ii) Banded Gneissic Complex formation of gneiss, migmatites and granitic rocks; (iii) intrusive formation of synorogenic granite and gneiss

lithological discrimination; Locality 1 shows lithological features: Debari group of meta-arkose and conglomerate (A); phyllite and mica schist (B); dolomite, phyllite and mica schist (C); phyllite and metagreywacke (D); composition of metagreywacke, dolomite and phyllite (E); intrusive formation of granite and gneiss (F); Banded Gneissic Complex of gneiss, migmatites and granitic rocks. Locality 2 shows the different lithological features such as synorogenic granite and gneiss formation (1); metaconglomerate (2); pebbly arkose (3); Udaipur group of metagreywacke and phyllite (4); Debari group of phyllite and mica schist formation (5); and Jharol group of phyllite, chlorite and mica schist. Aravalli Supergroup formations are dominated with heterogeneous assemblage of amphibolite facies of metamorphites associated with migmatites, composite gneisses, feldspathic mica schist, sillimanite-kyanite, mica schist, hornblende schist, granite gneiss and amphibolite with mixture of minor carbonates. BGC layers overlay on ASG formations such as self facies of mafic, volcanics, coarse clastics and carbonates that are densely assembled in the eastern parts, and carbonate-free deep-sea facies formed dominated phyllites with quartzite in the western parts. Major parts of self facies have been further subdivided into (i) volcanic formation of Delwara group and (ii) volcanic-free Debari group, in which the Delwara group are found as the lowermost sequence layer under the ASG formations, and they are mainly composed of mafic volcanic, quartzite, quartz-pebble conglomerate and banded iron formation. Debari group of rocks are formed as middle layer of ASG that comprised coarse clastics, carbonates and phyllites. Udaipur group consists of greywacke and phyllite overlain

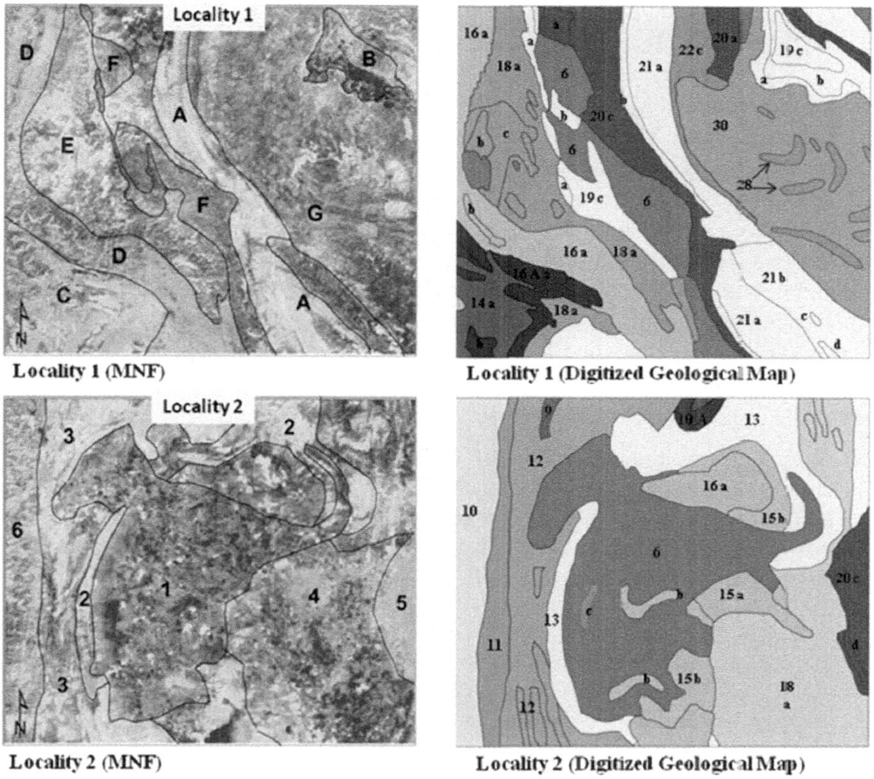

Fig. 7.7 The minimum noise fraction analysis of lithological discrimination; Locality 1 shows lithological features: Debari group of meta-arkose and conglomerate (A); phyllite and mica schist (B); dolomite, phyllite and mica schist (C); phyllite and metagreywacke (D); composition of metagreywacke, dolomite and phyllite (E); intrusive formation of granite and gneiss (F); Banded Gneissic Complex of gneiss, migmatites and granitic rocks. Locality 2 shows lithological features: synorogenic granite and gneiss formation (1); metaconglomerate (2); pebbly arkose (3); Udaipur group of metagreywacke and phyllite (4); Debari group of phyllite and mica schist formation (5); and Jharol group of phyllite, chlorite and mica schist

by carbonate sequence composed of Pb-Zn minerals, and it is known as Zawar formation. The Jharol group of ASG consist of carbonate-free and pelite dominant sequence arenite bands and are mainly composed of phyllites, chlorite schist and garnetiferous mica schist. Assemblage of lithological features shows that majority of them have undergone severe metamorphic processes, and they are formed as hydrothermally altered rocks with economically valuable deposits in various parts of the study area.

7.4.2.1 Locality 1: 73° 53′ E to 74° 40′ E and 24° 20′ N to 24° 30′ N (Area of Kotra, Jhamarkotra and Zawar)

Since the variance of MNF bands is very high, their combinations highlight different zones in the image. These zones are uncorrelated areas which arise due to compositional and topographic variations in the terrain. The major geological formations in the area can be demarcated using MNF colour composites. The highlighted zones are correlated with the existing geological map to identify the formations. As the extent of the study area is very large, two localities exposing maximum lithological variations in the MNF colour composite image (RGB: 312) have been selected as representatives for comparison with the digitized geological map (Figs. 7.7 and 7.8). The ASTER image-classified lithological map shows major parts of the area covered: Debari group, Udaipur group, Synorogenic intrusive and Banded Gneissic Complex.

(a) *Debari group*: It is represented by yellow-coloured zone (A) in the central part and a prominent blue-coloured patch ‘B’ in the upper right (Fig. 7.7a). Zone A is highly correlated to a meta-arkose and conglomerate formations. The lower part this zone, the layer noticed by bluish patch parallel to upland, mainly consists of phyllite and mica schist. Zone ‘B’ is another formation belonging to Debari

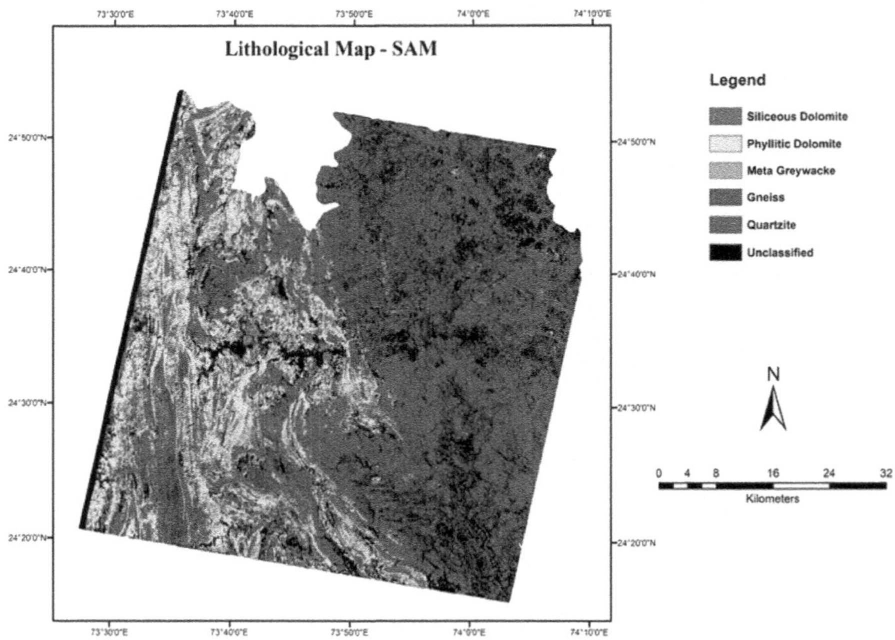

Fig. 7.8 The Spectral Angle Mapper (SAM) classification of lithological features (map shows various lithological features with its spatial extent that are discriminated based on spectral reflectance variability)

group that comprises phyllite and mica schist, and the narrow patches of these formations also occur in the western side to meta-arkose formation.

(b) *Udaipur group*: This formation comprises three types of rocks represented as C, D and E in Fig. 7.7. Zone C is found in southwestern parts that mainly contains dolomite, phyllite and mica schist. In between this zone, there are some narrow reddish patches, which can correlate with Zawar formation consisting of sulphide-bearing dolomite. Zone D occurs with the formation comprised of phyllite and metagreywacke, whereas phyllitic rocks appear in blue-coloured patches, and metagreywacke have shown in greenish colour. Zone E represented as shade of yellow and red colour that mainly contains metagreywacke, dolomite and phyllite. This is another formation in Udaipur group that is noted in middle-western parts.

(c) *Synorogenic intrusive*: Zone F denoted occurrence of this formation that comprises of granite and gneiss intrusive rocks; they are clearly demarcated by reddish-blue-coloured patches in the north-western parts.

(d) *Banded Gneissic Complex*: Zone G represented spatial distribution of this formation in middle-eastern parts, and this is noted as reddish-coloured patches. BGC forms basement for the Aravalli rocks and mainly contains gneiss, migmatites and granitic rocks that consist of Bhilwara Supergroup.

7.4.2.2 Locality 2: 73° 35′ E to 73° 44′ E and 24° 36′ N to 24° 42′ N (Area of Thur, Badgaon, Fatehpura and a Part of National Highway 27)

Locality 2 shows the major formations of synorogenic intrusive, Bari Lake group, Udaipur group, Debari group and Jharol group. These formations consist of the lithological units such as synorogenic granite and gneiss formation (1); metaconglomerate (2); pebbly arkose (3); Udaipur group of metagreywacke and phyllite (4); Debari group of phyllite and mica schist formation (5); and Jharol group of phyllite, chlorite and mica schist.

(a) *Synorogenic intrusive*: Zone 1 denoted the occurrence of this formation predominantly in middle parts that correlated with synorogenic granite and gneiss in geological map. The intrusion of granite and gneiss marked the end of Aravalli orogeny. The exposed part of intrusive is well distinguished in MNF image.

(b) *Bari Lake group*: Zones 2 and 3 belong to this group. These formations appear in bale greenish yellow patches of enriched metaconglomerate associated with some pebbly arkose in the northern parts and distributed NW and NE direction. Spatial pattern of this formation is clearly highlighted in MNF image. The Bari Lake group overlay with Udaipur group and Jharol group in the study area.

(c) *Udaipur group*: Zone 4 shows Udaipur group formation that comprises of metagreywacke and phyllite, and this formation are marked by a bluish region southeast to the intrusive. These formation covers in larger area that extend from middle-eastern part to southeast parts.

(d) *Debari group*: Zone 5 shows occurrence of phyllite and mica schist formation under Debari group that shows light green colour in middle-eastern parts parallel associated with Udaipur group formation.
(e) *Jharol Group*: Zone 6 shows occurrence of the Jharol group that runs parallel to the western parts and extends from north to south direction. This formation represented as shade of bluish green patches that consists of phyllitic and chlorite-mica schist rock types.

ASTER-derived lithological mapping explores various rock types along with their spatial distribution of different lithological units. The result shows relative positions of various geological formations based on their topographic zones, such as planar region of BGC basement, upland deposition of Aravalli Supergroup of rocks that are highly folded and synorogenic intrusive formation at lower position of the study area.

7.4.3 SAM Classification of ASTER for Lithological Mapping

The study area comprises two major lithological formations such as phyllitic and gneissic terrains that have predominantly distributed in the western and eastern parts of the area. Figure 7.8 shows the Spectral Angle Mapper (SAM) classification of lithological features such as siliceous dolomite, phyllitic dolomite, metagreywacke, gneiss and quartzite. The phyllitic dolomite predominantly distributed in western parts that mainly associated with siliceous dolomite and metagreywacke (Roy et al. 1988; Congalton 1991; Chandan Kumar et al. 2015). The eastern parts of the area are found predominantly with gneissic formation that constituted batches of quartzite in various locations. The error matrix of SAM-classified lithological features used for accuracy assessment is shown in Table 7.2. SAM-classified image of overall accuracy is estimated as 73.39% with Kappa coefficient 0.59. The result is comparatively high correlation with GSI-published geological map for Udaipur district. The gneissic rocks formed as Banded Gneissic Complex (BGC) and synorogenic intrusive, and siliceous dolomite formation occurs in the Zawar area. Phyllitic dolomite rocks occur in major parts of Aravalli Supergroup formation. Metagreywacke are found as thick layers in some locations of Udaipur group formation. Quartzite rock types have formed in Jharol group that mainly are associated with gneissic terrain in western parts.

7.5 Conclusion

Lithological mapping using ASTER image provides various lithological units with their spatial characteristics. Atmospherically corrected image shows actual spectral reflectance of rock types. Analysis of multispectral bands of the image using MNF

Table 7.2 Accuracy assessment of SAM classification for lithological mapping (error matrix of the SAM-classified lithological features)

Classified data	Siliceous dolomite	Metagreywacke	Phyllitic dolomite	Gneiss	Quartzite	Row total	User's accuracy
Siliceous dolomite	22	0	10	0	0	32	68.75%
Metagreywacke	0	10	2	0	4	18	55.56%
Phyllitic dolomite	0	0	62	0	0	86	72.09%
Gneiss	2	2	18	202	14	270	74.81%
Quartzite	0	0	0	2	24	30	80.00%
Column total	24	12	92	204	42	436	
Producer's accuracy	91.67%	83.33%	67.39%	99.02%	57.14%		

and SAM methods shows spatial extent of various lithological features. The SAM classification of ASTER image using field-observed spectra produces lithological mapping successfully. The experimental result of SAM method shows spectral variability of different lithological classes with mean reference spectrum. Accuracy assessment of ASTER-derived lithological mapping proved the rock types and other lithological features as similar as GSI-published map for the study area. Lithological mapping using ASTER image may provide basic information for exploring mineral resources and helpful for planning socio-economic developmental activities.

Acknowledgement The authors are grateful to the directors of the Indian Institute of Remote Sensing (IIRS), Dehradun, and National Centre for Earth Science Studies (NCESS), Thiruvananthapuram, for constant support and encouragement. Thanks are also due to Dr. Rabi N. Sahoo, Indian Agricultural Research Institute (IARI), New Delhi, for lab support and Dr. T. N. Prakash, NCESS, for extending XRD facilities. We are also grateful to Dr. R. R. Chowdhary, Dr. P. R. Golani, Director of GSI FTC, Zawar and Col. Kakkad for their support during fieldwork.

References

Boardman JW, Kruse FA (1994) Automated spectral analysis: a geologic example using AVIRIS data, north Grapevine Mountains, Nevada. In: Proceedings, Tenth Thematic Conference on Geologic Remote Sensing, Environmental Research Institute of Michigan, Ann Arbor, MI, I-407–I-418

Borengasser M, Hungate WS, Watkins R (2008) Hyperspectral remote sensing principles and applications. Taylor & Francis, Boca Raton, pp 119–130

Chandan Kumar, Amba Shetty, Simit Raval, Richa Sharma PK, Champati Ray (2015) Lithological discrimination and mapping using ASTER SWIR data in the Udaipur area of Rajasthan, India. Proc Earth Planet Sci 11:180–188

Congalton RG (1991) A review of assessing the accuracy of classifications of remotely sensed data. Remote Sens Environ 37:35–46

Di Tommaso I, Rubinstein N (2007) Hydrothermal alteration mapping using ASTER data in the Infiernillo porphyry deposit, Argentina. Ore Geol Rev 32(1):275–290

Gabr S, Ghulam A, Kusky T (2010) Detecting areas of high-potential gold mineralization using ASTER data. Ore Geol Rev 38(1):59–69

Gad S, Kusky T (2007) ASTER spectral ratioing for lithological mapping in the Arabian–Nubian shield, the Neoproterozoic Wadi Kid area, Sinai, Egypt. Gondwana Res 11(3):326–335

Gomez C, Delacourt C, Allemand P, Ledru P, Wackerle R (2005) Using ASTER remote sensing data set for geological mapping, in Namibia. Phys Chem Earth, Parts A/B/C 30(1):97–108

GSI (1997) District resource map of Udaipur District, Rajasthan on 1:250,000 scale. Geological Survey of India, Calcutta

Hewson RD, Cudahy TJ, Mizujiko S, Ueda K, Mauger AL (2005) Seamless geological map generation using ASTER in the Broken Hill-Curnmona province of Australia. Remote Sens Environ 99:159–172

Hosseinjani M, Tangestani MH (2011) Mapping alteration minerals using sub-pixel unmixing of ASTER data in the Sarduiyeh area, SE Kerman, Iran. Int J Digital Earth 4(6):487–504

Kruse FA, Lefkoff AB, Boardman JB, Heidebreicht KB, Shapiro AT, Barloon PJ, Goetz AFH (1993) The Spectral Image Processing System (SIPS) – interactive visualization and analysis of imaging spectrometer data. Remote Sens Environ 44:145–163

Kruse, FA, Dietz JB (1991) Integration of visible-through microwave-range multispectral image data sets for geologic mapping. In: Proceedings of the Cinquieme Colloque International,

Mesures Physiques et Signatures En Teledetection, 14–18 January, Couchevel, France, European Space Agency, Paris, France, ESA SP-319, vol 2, pp 481–486

Pour AB, Hashim M (2011) Identification of hydrothermal alteration minerals for exploring of porphyry copper deposit using ASTER data, SE Iran. J Asian Earth Sci 42(6):1309–1323

Pour AB, Hashim M (2012) Identifying areas of high economic-potential copper mineralization using ASTER data in the Urumieh – Dokhtar Volcanic Belt, Iran. Adv Space Res 49 (4):753–769

Qiu F, Abdelsalam M, Thakkar P (2006) Spectral analysis of ASTER data covering part of the Neoproterozoic Allaqi-Heiani suture, Southern Egypt. J Afr Earth Sci 44(2):169–180

Rajendran S, Al-Khirbash S, Pracejus B, Nasir S, Al-Abri AH, Kusky TM, Ghulam A (2012) ASTER detection of chromite bearing mineralized zones in Semail Ophiolite Massifs of the northern Oman Mountains: exploration strategy. Ore Geol Rev 44:121–135

Rajendran S, Nasir S, Kusky TM, Ghulam A, Gabr S, El-Ghali MA (2013) Detection of hydrothermal mineralized zones associated with listwaenites in Central Oman using ASTER data. Ore Geol Rev 53:470–488

Rowan L, Hook SJ, Abrams MJ, Mars JC (2003) Mapping hydrothermally altered rocks at Cuprite, Nevada, using the Advanced Spaceborne thermal emission and reflection radiometer (ASTER), a new satellite-imaging system. Econ Geol Bull Soc Econ Geol 98(5):1019–1027

Rowan LC, Mars JC (2003) Lithologic mapping in the Mountain Pass, California area using Advanced Spaceborne Thermal Emission and Reflection Radiometer (ASTER) data. Remote Sens Environ 84(3):350–366

Roy AB, Paliwal BS (1981) Evolution of lower Proterozoic epicontinental deposits: stromatolite-bearing Aravalli rocks of Udaipur, Rajasthan, India. Precamb Res 14:49–74

Roy SS, Malthottra G, Mohanty M (1988) Geology of Rajasthan. Geology Society of India, Bangalore

Sabins FF (1999) Remote sensing for mineral exploration. Ore Geol Rev 14(3):157–183

Sharma RS (2009) Cratons and Fold Belts of India, Lecture notes in earth sciences. Springer-Verlag, Berlin, pp 304–320

Sinha-Roy S, Malhotra G, Mohanty M (1998) Geology of Rajasthan, 1st edn. Geological Society of India, Bangalore, pp 278–290

Tangestani MH, Jaffari L, Vincent RK, Sridhar BM (2011) Spectral characterization and ASTER-based lithological mapping of an ophiolite complex: a case study from Neyriz ophiolite, SW Iran. Remote Sens Environ 115(9):2243–2254

Van der Meer FD, Van der Werff HM, van Ruitenbeek FJ, Hecker CA, Bakker WH, Noomen MF, Woldai T (2012) Multi-and hyperspectral geologic remote sensing: a review. Int J Appl Earth Observ Geoinfor 14(1):112–128

Youssef AM, Hassan AM, EL-Haddad AAA (2009) Mapping of Prerift – Synrift sedimentary units using Enhanced Thematic Mapper Plus (ETM+): Sidri-Ferian Area, Southwestern Sinai Peninsula, Egypt. J Indian Soc Remote Sens 37:377–393

Zhang X, Pazner M, Duke N (2007) Lithologic and mineral information extraction for gold exploration using ASTER data in the south Chocolate Mountains (California). ISPRS J Photogramm Remote Sens 62(4):271–282

Zhang X, Yang H, Shuai T (2009) Comparison of FLAASH versus Empirical Line Approach for Atmospheric Correction of OMIS-II Imagery. In: Proceedings of 30th Asian Conference on Remote Sensing (ACRS), Beijing, China

Chapter 8
Interactive Approach for Earthquake Scenario Development and Hazards Resource Estimation

B. S. Chaudhary, Ram Kumar Singh, Nupur Bhatia, Ravi Mishra, Md Ataullah Raza Khan, Juhi Yadav, and Shashikanta Patairiya

Abstract Indian subcontinent attained present physical form due to vast tectonic movements that resulted into large number of earthquakes. In studies it has been found that more than 50% area in the country is prone to damaging earthquakes. The northeastern part of India as well as the entire Himalayan belt is susceptible to earthquake of magnitude more than 8.0. The present study is principally aimed at understanding the intricate seismological processes in the study area, Sikkim which is on hilly terrain of Eastern Himalayas. Sikkim is situated in a region where major cause of earthquake is displacement of the Indian plate toward the Eurasian plate having complex geology. Remote sensing and GIS model builder and syntax were proven for hazard and vulnerable map creation used in earthquake scenarios development, planning, management, and resource estimation. In this study the preliminary factors including geology, topography, slope, relief, land use/cover, major roads, and historical epicenter were used with mechanical weightage, and overlay categorization was used for hazard index map and zone identification.

Keywords Earthquake scenarios management · Hazard index map · Overlay analysis

B. S. Chaudhary
Department of Geophysics, Kurukshetra University, Kurukshetra, Haryana, India
e-mail: bschaudhary@kuk.ac.in

R. K. Singh
Department of Natural Resources, TERI School of Advanced Studies, New Delhi, India

N. Bhatia · R. Mishra · M. A. R. Khan · J. Yadav (✉)
Remote Sensing and GIS, Kumaun University, Almora, Uttarakhand, India

S. Patairiya
Gurugram Metropolitan Development Authority, Gurugram, Haryana, India

P. Kumar et al. (eds.), *Applications and Challenges of Geospatial Technology*,
https://doi.org/10.1007/978-3-319-99882-4_8

8.1 Introduction

India has faced several serious world's enormous earthquakes, and about more than 50% of the country region is said to be vulnerable of earthquake (Sengar et al. 2013). The most vulnerable region of earthquake is the northeastern regions of India, including the entire Himalayan (Kaila and Narayan 1976). Earthquake is identified as natural sudden shaking of the ground earth due to movement on earth's rocks and surface (Armas 2012; Armas and Gavris 2013). Earthquake occurs when the plate tectonic is unexpectedly separated to form a fault or fold and releases energy in form of seismic waves and tremble ground surface (Pavel and Vacareanu 2016; Chemenda et al. 1995; Fekete 2009; Walker et al. 2014). Seismograph or Seismometer is used to detect and record the seismic waves. The major cause of earthquake is a sudden release of energy due to sudden breaks in the underlying rocks along the faults that leads to procreation of seismic waves. Plate tectonic refers to the explanation of the movements of the earth surface in an existing place or the past experienced places. The spot under the ground where the earthquake occurs is known as focus, and the place perpendicularly above the focus is termed as the epicenter (Chen et al. 2011; Pavel et al. 2016; Rufat 2009). The focus of the earthquake is an underground point where rocks first begin to move and the seismic waves travel outward from the earthquake's focus. The epicenter is the point on the earth's surface which is directly above the focus point. Lower magnitude of earthquake even causes mass destruction even more than an atom bomb, so hazard resource estimation and scenarios development are very important (Skarlatoudis et al. 2004).

The major cause of earthquake is the shifting of tectonic plates which forms the folds and faults feature. Another reason could be due to the volcanic inflation, explosions, deep penetrations bombs, geothermal energy, stress transfer, groundwater extraction which decreases the pore pressure, surface quarrying, and many reasons more (Allen and Kanamori 2003; Ansal et al. 2004; Strasser et al. 2008; Toma-Danila et al. 2015). Earthquake scenarios help in management of hazardous resources which refers to the series of actions for maintaining the influence areas or zones to reduce the disaster (Champatiray et al. 2005). It does not completely reduce the influence of disaster, but it brings out the outcome or plans to decrease the risk factor. There is no such planning or arrangements for the disaster management that would lead to the worst-case scenario and include the loss of life as well as damage to the properties (Roustaei et al. 2005; Trifunac and Todorovska 1997; Lillesand et al. 2008). India has faced many powerful impacts of the world's greatest earthquakes; therefore, greater amount of area of the country is evaluated to be prone to earthquakes (Zhou et al. 2009).

The spectacular geomorphic-tectonic features as exhibited by the Indian lithosphere plate include the continental mountain systems, plateaus, and uplands (Glennie et al. 1990; Wilson 1965; Ouma et al. 2011; Lang et al. 2012). The reason behind the occurrence of earthquake in this particular region is due to the displacement of the Indian plate which is shifting toward the Eurasian plate and also collides with it which leads to the creation of the convergent boundaries. The Indian plate is bounded by some major and minor plates from all direction: the African plate in the west, the

Arabian and Iranian microplates in the north-west, the Eurasian plate toward north, the Chinese plate toward east and southeast, and Antarctic plate in the south.

Geographic information system is a unified application of the scientific knowledge which refers to computer science, topography, technology, and most importantly geography (Gong Jianya 2004). GIS uses all input factors for analyzing and assigning different weightage priority and various mechanisms in handling and inspecting the spatial data (Bai Yungang et al. 2004). It has been proven successfully to an enormous field including geography (Chen et al. 2006). This conception is being used widely throughout the world for various purposes which include the identification of the risk regions. It is capable for identifying risk factors and vulnerable prone areas for earthquake (Morrow 1999; Zhou et al. 2009). The utilization of GIS for earthquake disaster is substantially focused on the hazards emerging from the earthquakes commence from the active fault locations frequently termed from the inter plate earthquakes.

The representation of these maps has been done by using two software, that is, ERDAS Imagine and ArcGIS. The GIS technology is being used widely for the seismic fields including analyzing approach for seismic forecasting, resistance to earthquake, prevention of disaster, and mitigation and emergency command (Weiwei et al. 2009). GIS is growing in a vast approach toward analyzing the vulnerability and risk zones (Sun et al., 2008). GIS is used for mapping, land use, planning, water conservancy, and civil engineering as every connection is linked toward the earthquake disaster and these processes are interlinked to the geographical location.

8.2 Study Area

Sikkim, one of the states of India, is situated in the Eastern Himalayas which comes under the world's third highest mountain Khangchendzonga which is of 8585 m, restrained by the people of Sikkim as their protective deity. Sikkim is parted from the Singalila range from Nepal in the west, Chola range from Tibet and Bhutan, and also the Rangeet and Rango rivers from the borders. Sikkim is located at the 27.33°N and 88.62°E; therefore the total area covered by is about 7096 Sq. Km^2 (Fig. 8.1). The elevation of Sikkim ranges from 300 m to about 8583 m above the mean sea level, Mt. Khangchendzonga being the highest. It aggregates 82.31% (Sikkim ISFR 2017) of the total land which is covered with forest, consisting of the 8 species of the tree ferns, 11 species of the oak trees, along with the 20 species of bamboos, 300 species of ferns, and for about 400 species of the flowering plants. Its population is around 6.11 Lakhs (Census of India 2011). It is one of the states in India which receives the regular snowfall and the heavy rainfall, and hence there is an increased chance for landslides (Onagh et al. 2012; Lin et al. 2006). The land area of Sikkim is not fit for agricultural activities, but still most of its hill slopes are converted in the form of terrace farming land. The hills of the state mainly comprise of the gneiss rock and the schist rocks which are likely to erode due to the process of weathering and erosion. The geology

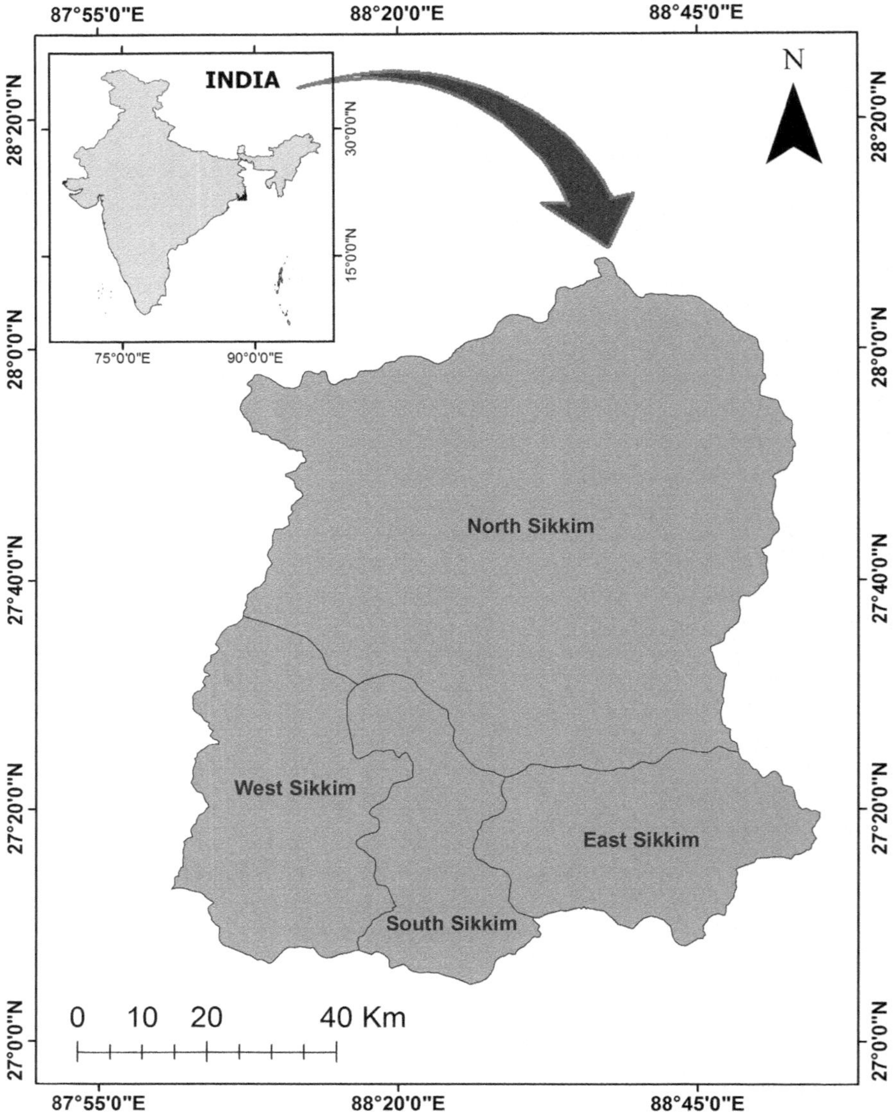

Fig. 8.1 Study area map

of Sikkim largely comprises of the infertile soil, inappropriate for the agricultural purposes. Sikkim comprises of the three important rocks named half-schistose, gneiss-ose, and the Precambrian. Precambrian rock is established and is significant part of Sikkim geology that is made up of schists and phyllites. The existence of the schist's and phyllites forms the gradients of the region that are prone to weathering and erosion. Earthquakes in India show some striking aspects which can be implemented for planning the earthquake disaster management. Sikkim falls under the category of zone IV having high damage risk zone. The study area Sikkim, India, is chosen due to its earth structure, located at highly terrain of Eastern Himalayas, displacement of the Indian plate toward the Eurasian plate, moderate population in open area, and statistical earthquake, falling into earthquake zone IV (Sarkar and Nandy 1974).

8.3 Material and Methodology

Consolidating geospatial technology is accompanied in conception of informations and experiences during earthquake for constructing a hazard map for risk-prone regions that will be helpful for future planning and strategy making (Chen et al. 2006). Therefore, the limitation envisages by using GIS, and maps are classified into three classes. After formative criteria assessment ranges, the ranking of the class values from 1 to 3 is assigned with respect to scientific codes and postulation.

The foremost factors used for the study and categorization (Table 8.1) for geology, soil, slope, land use, and road are also vital mechanisms of dwindling modeling development and are together mapped. In this study classification is done to organize the factors used for categorizing the map so as to help the decision-makers understand the situation and distribute the mechanism rapidly by visualizing the map. The whole procedure for GIS data and investigation is carried out by using the model builder tool. In this way a mechanical GIS system may also be a significant element for classification-based researches. The weight and individual features (Table 8.1) are made flexible to transform through a user interface.

Material used for study Landsat-8 Operational Land Imager (OLI) Image, Cartosat Stereo pairs, major roads source DIVA-GIS, and historical earthquake data are used to derive useful hazard index.

8.3.1 Land Use/Land Cover

The land cover figure documents how any area is enclosed by forests, wetlands, and impervious surfaces, agricultural together with the other land and water types, whereas the land use represents how people utilize the landscape whether for growth, preservation, or diverse uses. Land cover can be resoluted by analyzing satellite and the aerial imagery (Metternicht et al. 2005). Land cover maps endow with the statistics to assist the managers paramount and recognize the existing landscape (Fig. 8.2). The term land use and land cover are used interchangeably. Using Landsat 8 OLI, image was used for land use-land cover classification using maximum likelihood classifier to classify into nine classes specifically agricultural, snow and glacier, alpine scrub, alpine meadow, oak forest, conifer forest, roads, alpine thicket, and the alder forest.

Table 8.1 Representing weightage rank for five basic factors related to earthquake

S. No	Themes	Weightage	Individual factors	Rank/Value for individual entity	Weightage* rank assigned
1	Geology	0.15	Metamorphic	1	0.15
			Gabbros	2	0.3
			Chert/limestone facies	3	0.45
			Gabbros and Ultrabasics	4	0.6
			Fluviatile deposits	5	0.75
			Ultrabasics-peridotite, Precambrian rock	6	0.9
2	Soil	0.1	Torrifluvents, Torriorthents and rock outcrops, clay-rich soil	1	0.1
			Rock outcrops, torripsamments, and calciorthids	2	0.2
3	Slope (in percentage)	0.15	Low (0–30)	1	0.15
			Medium (30–60)	2	0.3
			High (>60)	3	0.45
4	Land use	0.1	Green areas	1	0.1
			Water bodies	2	0.2
			Scrub land	3	0.3
			Rock outcrop, dunes, and crushers	4	0.4
			Built-up	5	0.5
5	Road (m)	0.05	1500	1	0.05
			1000	2	0.1
			500	3	0.15

8.3.2 Cartosat DEM

Cartosat is an exclusive stereoscopic undertaking with the ability of attaining the along track stereo imagery with the unchanging B/H ratio. It holds two payloads which are the fore camera and the aft camera. These two panchromatic cameras with the 2.5 m spatial resolution are to obtain two descriptions concurrently. ERDAS IMAGINE Photogrammetry is used to create elevation model of 15 m.

8.3.3 Slope Approach

A slope map represents the variations in altitude which simultaneously shows the aspect and degree of slope for a terrain. Slope map is made up of Cartosat 15 m

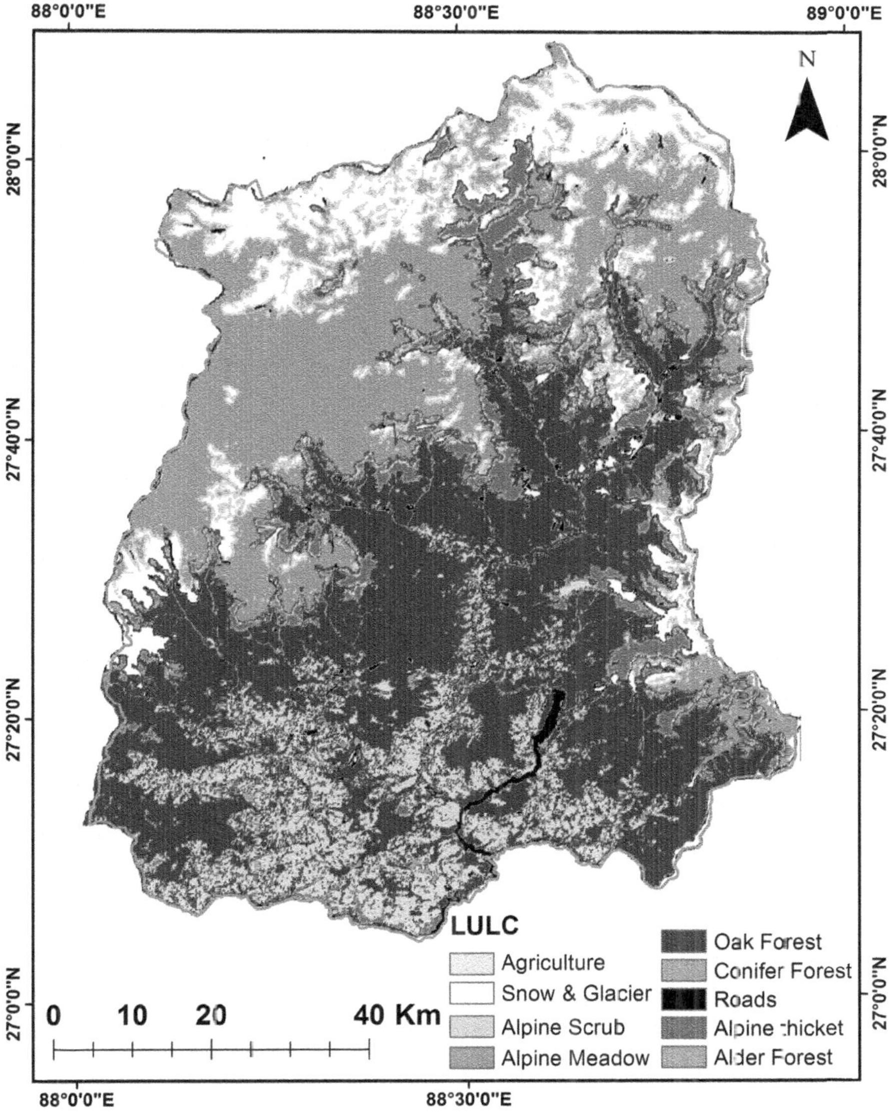

Fig. 8.2 Land use and land cover of the study area

DEM. It represents the direction and steepness of the slope for a terrain for continuous surfaces. They are characterized as a symbolized such as the combination of the hues such as the red, yellow, orange, and green.

8.3.4 Relief

Relief map basically displays the altitude and topography. They are more better-quality adaptation of the topographic maps as the topographic map utilizes equivalent elevation to generate two-dimensional models. The relief map attempts to depict the physical appearance of the ground based on its shape and height. The relief of topography has an important impact on the earthquake response of a region. The climate and weather of any region largely determines the variation in altitude which influences the rock and soil type found in the region.

8.3.5 Weightage Assignments

Weightage assignment is taken as important preliminary factors to set up parameters to study any seismic study. The most important authority of geologic environment on both strong trembling description and consequential damage pattern are integrated in this study. The relationship between gray scales, statistical sample gray value, and earthquake intensity has been established and analyzed for affected area from the change image and published report. Model the probability distribution of the gray value in different intensity zones by Eq. (8.1).

$$F_j(x_i) = p\text{norm}(x_i, \mu_i, \sigma_i) \tag{8.1}$$

$F_j(x_i)$ denotes the probability density function of the gray value in intensity j = 5, 6, 7, 8. . ., and x_i is the gray value of the change image. μ_i, σ_i are the mean and standard deviation of statistical sample gray value in intensity j, respectively. A map projection conversion is directed with a suitable location of the origin of coordinates or false northing and false easting to make all XY coordinates positive, and then the contours were generated.

Take $a_j = \max(x_{ij})$, $b_j = \min(x_{ij})$, c = max (y_{ij}), and d = min (y_{ij}), where I is for identification of the sample point, j is for intensity, e = min (j), f = max (j), and x_{ij},y_{ij} are the coordinates of point i with intensity j. The circle contour can be stated by the following (Eq. 8.2).

$$R_J^2 = (x_j - x_0)^2 + (y_j - y_0)^2 \tag{8.2}$$

$$x_0 = \frac{1}{2(f - e + 1)} \sum\nolimits_{j=e}^{f} (\mathrm{a_j} + \mathrm{b_j})$$

$$y_0 = \frac{1}{2(f - e + 1)} \sum\nolimits_{j=e}^{f} (\mathrm{c_j} + \mathrm{d_j})$$

$$R_j = \frac{1}{2}\sqrt{(a_j - b_j)^2 + (c_j - d_j)^2}$$

where (x_0,y_0) is the coordinate of a circle center and R_j is the equivalent radius of the region with intensity j. If the rupture direction is recognized (in 0–90 degree), an ellipse can be determined in Eq. 8.3.

$$\frac{((x_j - x_0)\alpha_j + (y_j - y_0)\beta_j)^2}{A_j^2} + \frac{((y_j - y_0)\alpha_j + (x_j - x_0)\beta_j)^2}{B_j^2} = 1 \tag{8.3}$$

$$A_j = R_j, B_j = \max\left(\sqrt{(a_j - b_j)^2 + (c_j - d_j)^2}\right)$$

$$\alpha_j = \cos\varnothing_j, \beta_j = \sin\varnothing_j, \qquad \varphi_j = \arctan\left(\frac{c_j - d_j}{a_j - b_j}\right)$$

where A_j is the radius along the major axis and B_j is the radius along the major axis.

8.3.6 Trend Surface Analysis

The trend surface analysis is the most widely applicable global surface-fitting procedure. It is a global procedure with an inexact interpolation used for obtaining smooth and approximate value points with known values using a polynomial equation (Chang, 2006). A cubic polynomial surface such as in Eq. (8.4) can be used to model the trend of the spatial distribution of the RS variables. After global polynomial interpolation by Eq. (8.4), the contours without a given shape can be generated by horizontal cuts. The mapped data are approximated by a polynomial expansion of the geographic coordinates of the control points, and the coefficients of the polynomial function are found by the method of least squares, insuring that the sum of the squared deviations from the trend surface is a minimum. Each original observation is considered to be the sum of a deterministic polynomial function of the geographic coordinates plus a random error.

$$Z(x,y) = b_0 + b_1x + b_2y + b_3x^2 + b_4xy + b_5y^2 + b_6x^3 + b_7x^2y + b_8xy^2 + b_9y^3 \tag{8.4}$$

where Z(x, y) is the surface equation and $b_0, b_1, b_2, b_3, b_4, \ldots\ldots\ldots b_9$ are coefficients to be fitted in the regression.

8.3.7 *Co-Kriging*

Co-Kriging is a geo-statistical analyst module that is considered as a point interpolation. It needs point data as input which is used to yield raster data with approximations of an optionally inaccurate data. It is a multivariate variant of the usual kriging operation that predicts for a poorly tested variable (predictor) with support of a well-sampled co-variable, where they should be highly correlated using Eq. 8.5. This paper considers intensity as the primary variable and statistical sample gray value as the secondary variable.

$$P^*(u_0) = \sum_{k=1}^{n} \lambda_k * p(u_k) + \sum_{k=1}^{n} \square_l * q(u_l) \tag{8.5}$$

Where $P^*(u_0)$ is for the co-kriging estimate at pointu_0, n is for the number of p(.) points, and m is for the number of q(.). The $p(u_k)$ is for the intensity at point u_k and q (u_l) for the statistical sample gray value at point u_l, respectively. The corresponding weights λ_k,$\square_l$are from the solution of co-kriging equations of intensity semivariogram, statistical sample gray value semivariogram and cross-covariance of intensity, and statistical sample gray value.

8.3.8 *Isoseismic Line*

Concentric loops are the most implied isoseismic lines but are very coarse. Ellipses are better with two structures but are still not compliant since the form gets fixed. The lines with different roman numerals, V, VI, VII, IX, and X, respectively, denote intensity zones. Both the co-kriging and trend surface methods can generate isoseismic lines quite well. Isoseismic lines acquired from the trend surface model are very smooth, and those from the co-kriging model must be further leveled. The larger the numbers of "n" and "m," the smoother the isoseismic lines, and it can be smoothed using other mathematical tools.

8.4 Results and Discussion

8.4.1 *Land Use/Land Cover*

The land use and land cover map is a chief principle for regulating the earthquake risk zones. In the present study, nine land use and land cover classes specifically agricultural, snow and glacier, alpine scrub, alpine meadow, oak forest, coniferous forest, roads, alpine, teaks, and the alder forest have been classified from the LANDSAT-8 image using Maximum likelihood classifier. The supervised classification was done along with the image analysis keys such as the tone, association, texture, etc. (Fig. 8.2).

8.4.2 Cartosat DEM

The Cartosat digital elevation model (DEM) of 15-meter resolution was used for generating the slope and relief map for the study area (Fig. 8.3). The map represents the digital elevation model of Sikkim, and the value of the map varies from 188 (low) to 8387 (higher).

8.4.3 Slope Map

A slope map represents the variations of altitude within a distance. Slope map is made up of Cartosat DEM (Fig. 8.3). It represents the direction and steepness of the slope for a terrain for continuous surfaces. Slope map is categorized on the basis of combination of the hues such as the red, yellow, orange, green, etc. (Fig. 8.4).A slope map specifies in the direction of the topography of a region alongside with a study of topographic description as they have inclined and may persist to authority the enlargement of land. The slope map is determined of steepness characteristic comparative to the horizontal plane. The surface topography is one of the most influential determinants for the site effects during an earthquake. The slope resolution is classified into five classes for slope degree in between 0 and 87.5 values (Fig. 8.4).

8.4.4 Relief

The relief of topography has an important impact on the earthquake response of a region. The variation in altitude is greatly affected by the climate and the weather which determines the rock and soil type found in the region. The relief map was extracted from the Cartosat DEM classified into five classes ranging from 0 to 8340 m (Fig. 8.5).

8.4.5 Earthquake Mapping

Earthquake transpire when the plate tectonic unpredictably disengage to form a fault or fold during the time of seismic activity and makes ground to tremble (Fig. 8.6). An earthquake map is prepared by contemplating the past fault and the earthquake

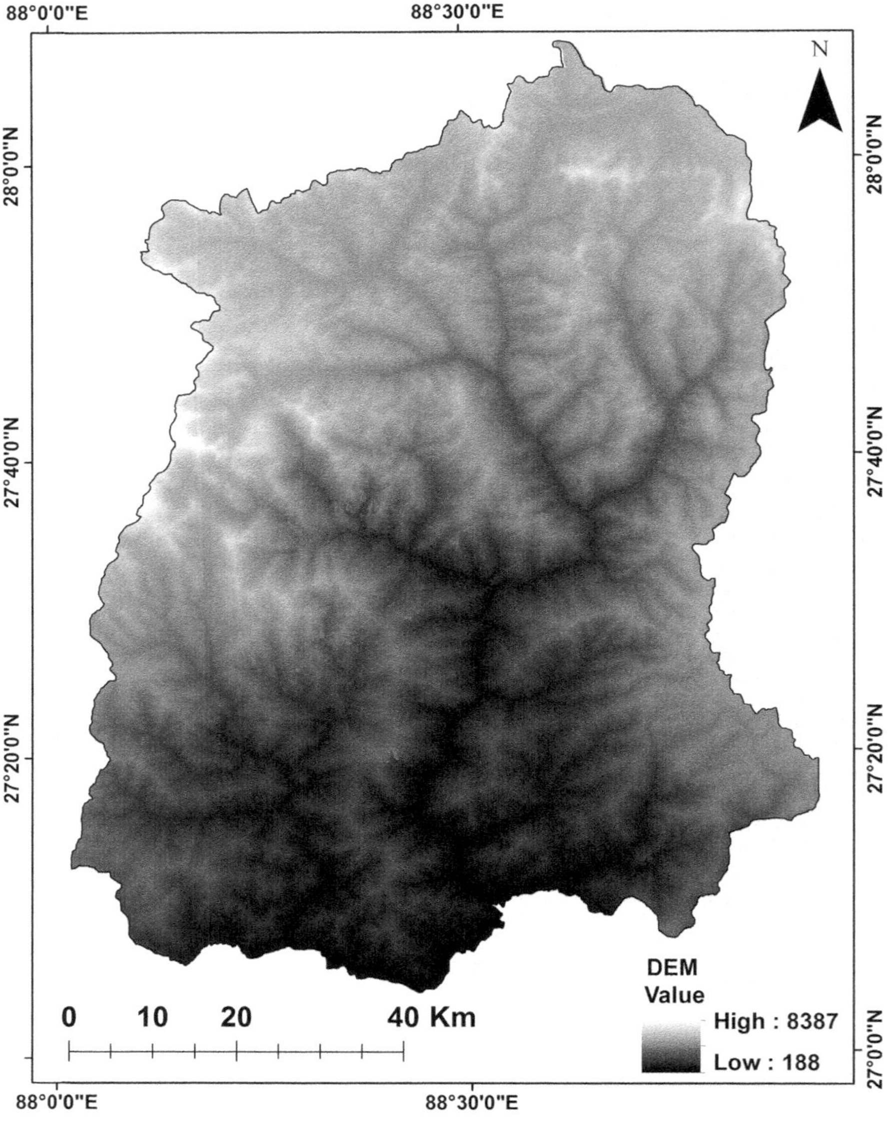

Fig. 8.3 Cartosat digital elevation model (DEM) of the study area

activities as well as the performance of the seismic waves traveling in different direction. An earthquake map can be utilized for the development of a land use, alleviation, and the crisis response. The earthquake map represents different probabilities selected to provide an overview of relative range of seismic activities. It shows areas that were affected by the earthquake along with the hazard intensity. The intensity of earthquake varies from lowest 6 to 8.5 which are of highest intensity (Fig. 8.6).

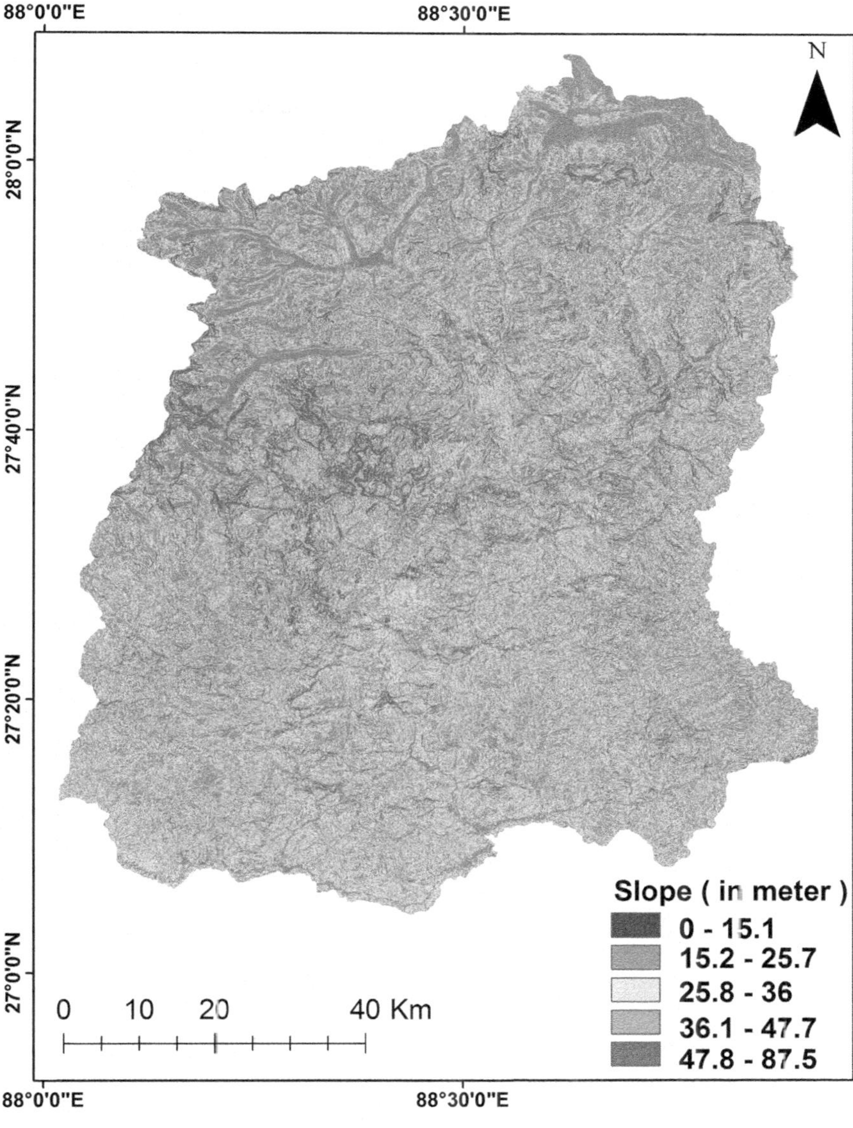

Fig. 8.4 Slope map of the study area

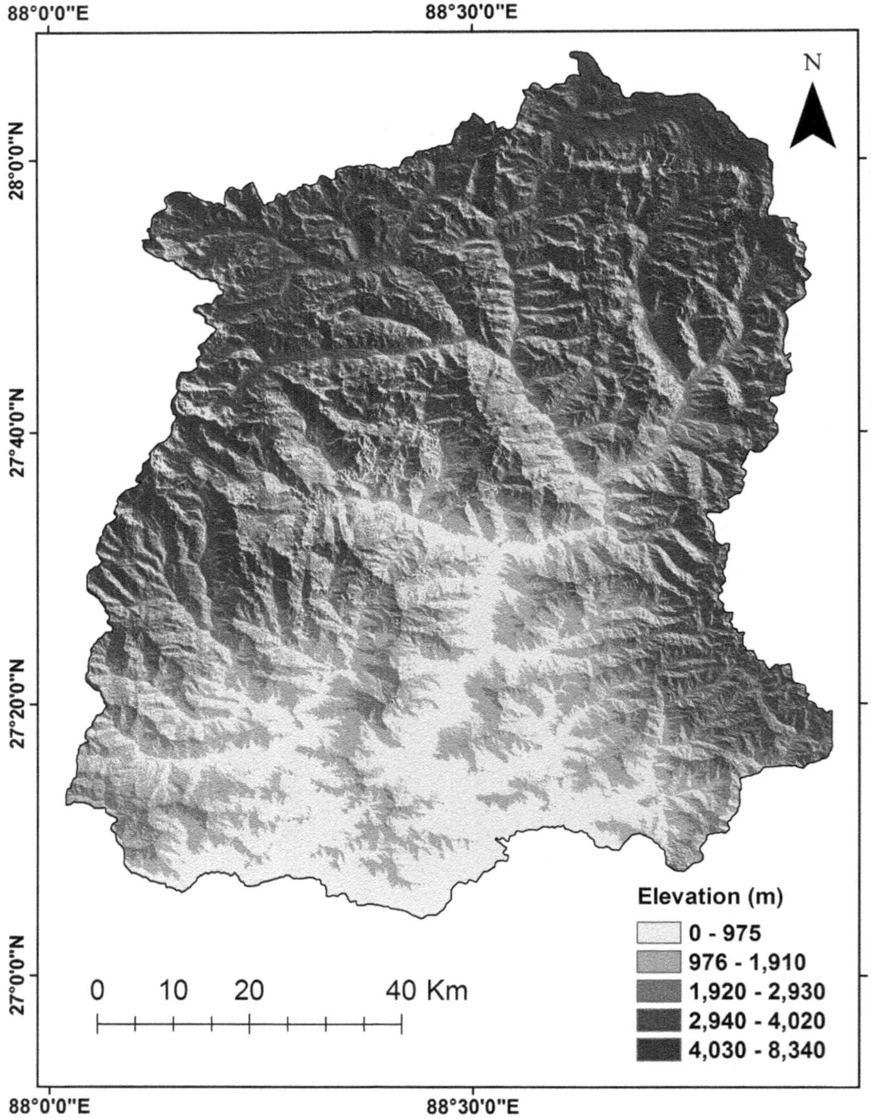

Fig. 8.5 Relief map of the study area

8.4.6 Seismic Hazard Map

Based on the parameters such as geology, slope, soil, land use and land cover pattern, and the earthquake mapping (USGS 2014), the potential risk zones of the earthquake vulnerable areas were studied (Fig. 8.7). Map describes that complication

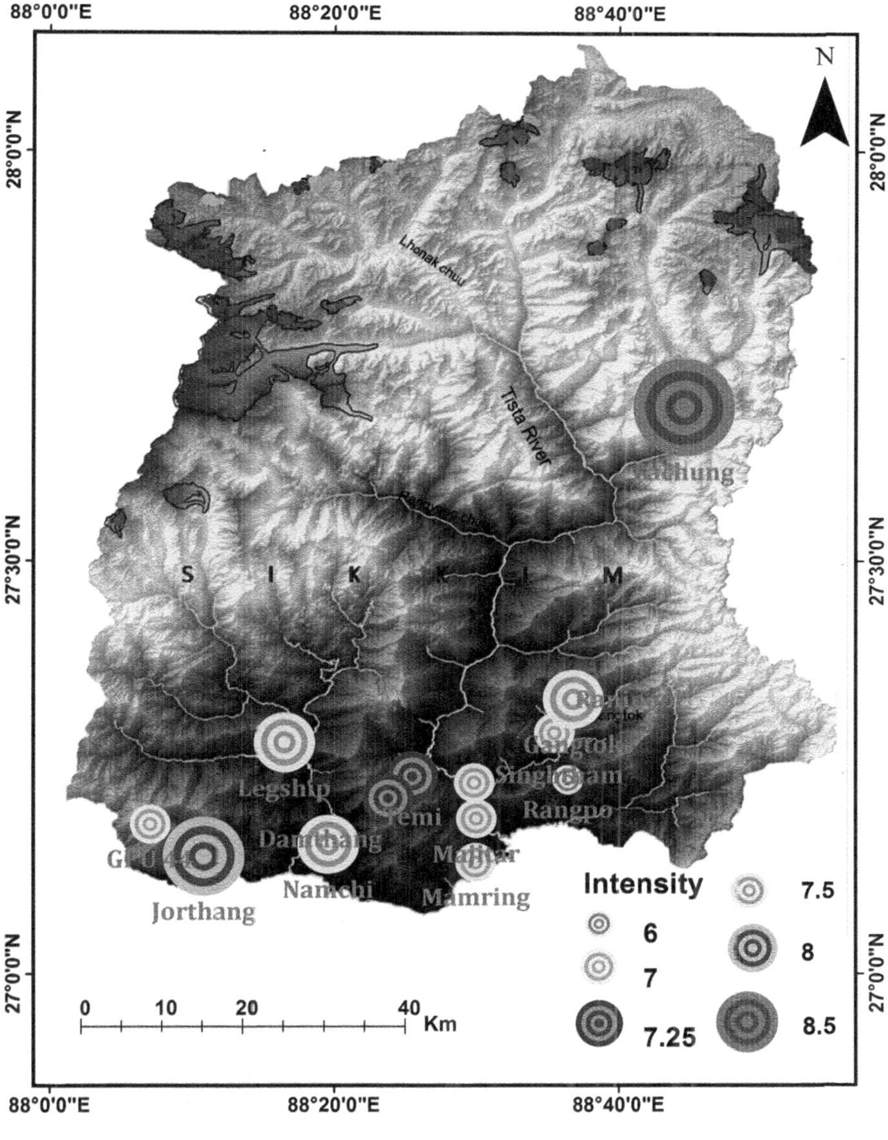

Fig. 8.6 Earthquake mapping of the study area

of the risk is not fully restricted to the state of the potentially exposure and the socially tempered acknowledgement. Therefore risk is considered to be the feature of the urban society that can be an admittance through an arrangement of the ecological aspect that are related with the physical environment of the geographic space where the urban society is situated or the social situation of the population in that region

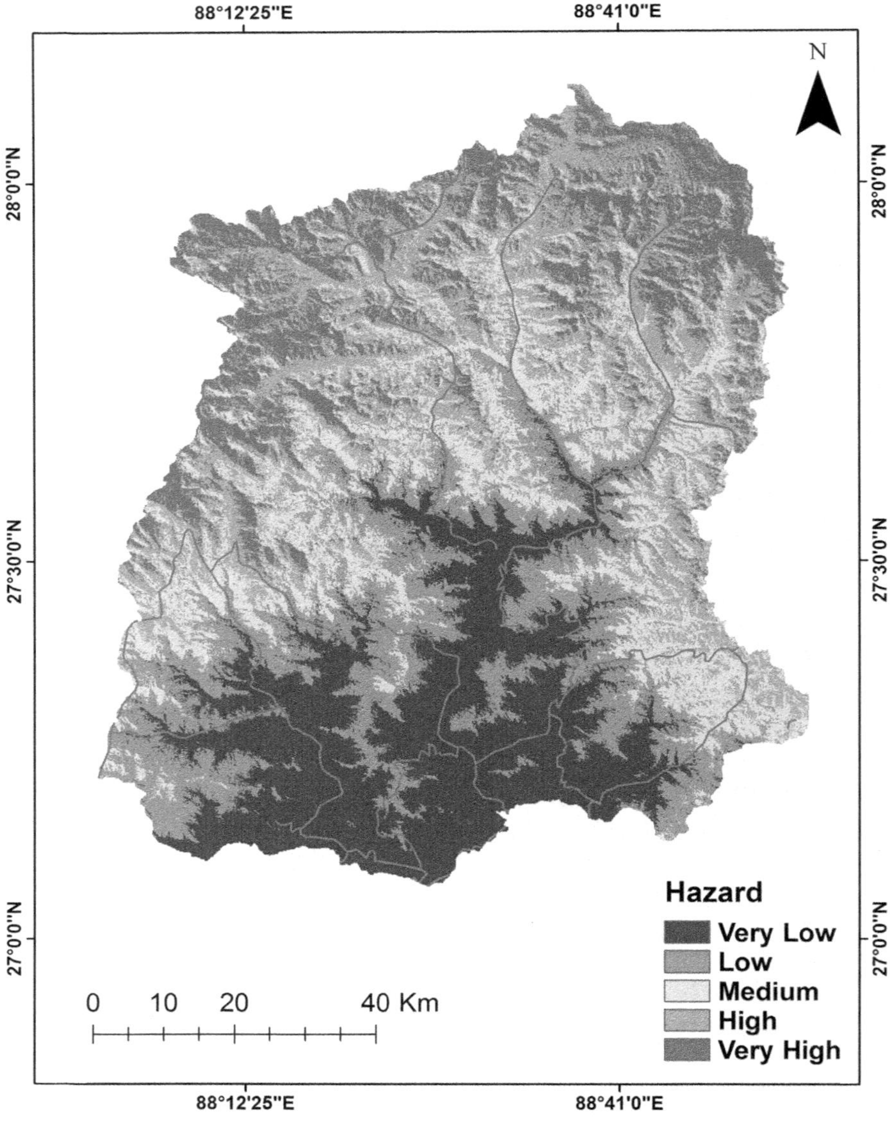

Fig. 8.7 Hazard map of the study area

(Fischhoff et al. 1981).Through the study of risk or the vulnerability zones, the highly prone areas are easily identified and differentiated from the other regions within the study area. The risk zone is considered to be the spatial problem which involves the observant confirmation of the risk zones based on the weighted analysis

(Keeney 1995). By the use of GIS-based spatial multi-measurement, the problem toward the risk analysis can be tackled.

Exploration and analysis are still continued to be done and grow on the earthquake mapping and the hazard mapping in a large-scale format. The shift in the hazard to disasters has continued, and the spotlight on human vulnerability has been maintained (White et al. 2001). The affiliation between a hazard and its prospect can be used to finish off the on the whole level of risk. The surroundings can be expensive in economic and social term; a direct hazard to life is the most severe threat. Risk is occasionally taken as identical with hazard, but risk has extra implication of the possibility of exact hazard actually happening. Hazard is paramount viewed as a physically taking place or human-induced process or occasion, with impending to generate loss. Risk is the actual exposure of something of human value to a hazard and is often regarded as the product of probability and loss (Okrent 1980).

8.5 Conclusion

The research characterizes the attempt to study and mapping of the earthquake for the state of Sikkim which lies in the zone-IV category classified as the great vertical extend zone. The research emphasizes on the areas that were affected by the earthquake or the areas that are vulnerable to earthquake. This has been calculated via various algorithms and with the use of ArcGIS and ERDAS Imagine. The concluded map is the overlay analysis map representing all the combination of factor maps that slope map, relief map, digital elevation model map, and land use and land cover map. The research was taken into account to analyze the five criteria that are geology, soil, slope, road, and the land use/land cover using the GIS to create the earthquake hazard map. This map is concluded to increase the awareness, management, and preparedness among the people invulnerable areas in Sikkim.

References

Allen RM, Kanamori H (2003) The potential for earthquake early warning in southern California. Science 300:786–789

Ansal A, Biro Y, Erken A, Gulerce U (2004) Seismic microzonation: a case study. In: Ansal A (ed) Recent advances in earth quake geotechnical engineering and microzonation. Kluwer Academic Publishers, The Netherlands, pp 253–266

Armas I (2012) Multi-criteria vulnerability analysis to earthquake hazard of Bucharest. Romania Nat Hazards 63(2):1129–1158

Armas I, Gavris A (2013) Social vulnerability assessment using spatial multi-criteria analysis (SEVI model) and the social vulnerability index (SoVI model)-a case study for Bucharest. Romania Nat Hazards Earth Syst Sci 13:1481–1499

Cencus of India (2011) Cencus of India, Government of India, 2011.

Champatiray PK, Perumal RJ, Thakur VC, Bhat MI, Malik MA, Singh VK et al (2005) A quick appraisal of ground deformation in Indian region due to the October 8, 2005 earthquake, Muzaffarabad, Pakistan. J Indian Soc Remote Sens 33(4):465–473
Chang KT (2006), Spatial interpolation, introduction to geographic information systems, 3rd ed., chapter 16, Beijing: Science Press and McGraw-hill Education (Asia) Co., China
Chen LC, Liu YC, han KC (2006) Integrated community-based disaster management program in Taiwan: a case study of Shang-An village. Nat Hazards 37:209–223
Chen M, Wood MD, Linstead C, Maltby E (2011) Uncertainty analysis in a GIS-based multicriteria analysis tool for river catchment management. Environ Model Softw 26(4):395–405
Chemenda AI, Mattauer M, Malavieille J, Bokun AN (1995) A mechanism for syn-collisional rock exhumation and associated normal faulting: results from physical modelling. Earth Planet Sci Lett 132:225–232
Fekete A (2009) Validation of a social vulnerability index in context to river-floods in Germany. Nat Hazards Earth Syst Sci 9(2):393–403
Fischhoff B, Lichtenstein S, Slovic P, Derby SL, Keeney RL (1981) Acceptable risk. Cambridge University Press, New York
Glennie KW, Boef MGA, Hughes-Clarke MW, Moody-Stuart MN, Pilaar WFH, Reinhart BM (1990) Inter-relationship of Makran–Oman Mountainsbelts of convergence. In: Robertson AHF, Searle MP, Ries AC (eds) The geology and tectonics of the Oman region, vol 49. Geological Society Special Publication, London, pp 773–786
Jianya G (2004) Review of the progress in contemporary GIS. Geomatics Spat Inf Technol 27 (1):5–11
Kaila KL, Narayan H (1976) Evolution of the Himalaya based on seismotectonics and deep seismic soundings. Him Geol Em, Sec II B, New delhi, Geol Surv Ind Misc Pub No 41, Part-IV, pp 1–26
Keeney RL (1995) Understanding life threatening risks. Risk Anal 15:627–637
Lang D, Molina-Palacios S, Lindholm C, Balan SF (2012) Deterministic earthquake damage and loss assessment for the city of Bucharest, Romania. J Seismol 16(1):67–88
Lillesand TM, Kiefer RW, Jonthan WC (2008) Remote sensing and image interpretation, 6th edn. Wiley, New York
Lin CW, Liu SH, Lee SY, Liu CC (2006) Impacts of the chi–chi earthquake on subsequent rainfall-induced landslides in Central Taiwan. Eng Geol 86:87–101
Metternicht G, Hurni l, GOGU R (2005) Remote sensing of landslides: an analysis of the potential contribution to geo-spatial systems for hazard assessment in mountainous environments. Remote Sens Environ 98:284–303
Morrow Betty H (1999) Identifying and mapping community vulnerability. Disasters 23(1):1–18
Okrent D (1980) Comment on societal risk. Science 208:372–375
Onagh M, Kumra VK, Rai PK (2012) Landslide susceptibility mapping in a part of Uttarkashi district (India) by multiple linear regression method. Int J Geol 2(2):102–120
Ouma Y, Kipkorir EC, Tateishi R (2011) MCDA-GIS integrated approach for optimized landfill site selection for growing urban regions: an application of neighborhood-proximity analysis. Ann GIS 17(1):43–62
Pavel F, Vacareanu R (2016) Scenario-based earthquake risk assessment for Bucharest, Romania. Int J Disaster Risk Reduct 20:138–144
Pavel F, Vacareanu R, Douglas J, Radulian M, Cioflan CO, Barbat AH (2016) An updated probabilistic seismic hazard assessment for Romania and comparison with the approach and outcomes of the SHARE project. Pure Appl Geophys 173(6):1881–1905
Roustaei M, Nazi H, Amirmotallebi N (2005) The Seismotectonic and Zonation Map of Salmas Vastness by GIS modeling according to Landsat Satellite Images (ETM+) and Aeromagnetic data. Paper presented at the Map Middle East, 23–25 April, 2005, Dubai, UAE
Rufat S (2009) From mural map to GIS: mapping urban vulnerability in Bucharest. In: Krek A, Rumor M, Zlatanova S, Fendel E (eds) Urban and regional data management: UDMS. Taylor and Francis Group, London, pp 301–313

Sarkar, E and D.R. Nandy (1974) Structure and tectonics of Tripura-Mizoram area, India, sem. Vol. on tectonics and metallogeny of S. and E. Asia, Mis. pub.no.34, geol Surv Ind, Part I, pp. 141–148

Sengar SS, Kumar A, Ghosh SK, Wason HR, Raju PLN, Murthy YVNK. 2013. Earthquake-induced built-up damage identification using fuzzy approach. Geomat Nat Hazards Risk v4: 320-338

Sikkim ISFR (2017) Indian state forest report by forest survey of India, Government of India, 2017

Skarlatoudis AA, Theodulidis N, Papaioannou C, Roumelioti Z (2004) The dependence of peak horizontal acceleration on magnitude and distance for small magnitude earthquakes in Greece. In: 13th world conference on earthquake engineering, Vancouver, Canada, August, 2004

Strasser FO, Bommer JJ, Sesetyan K, Erdik M, Cagnan Z, Irizarry J, Goula X, Lucantoni A, Sabetta F, Bal IE, Crowley H, Lindholm C (2008) A comparative study of European earthquake loss estimation tools for a scenario in Istanbul. J Earthq Eng 12(2):246–256

Sun CG, Chun SH, Ha TG, Chung CK, Kim DS (2008) Development and application of a GIS-based tool for earthquake-induced hazard prediction. Comput Geotech 35:436–449

Toma-Danila D, Cioflan CO, Balan SF, Manea EF (2015) Characteristics and results of the near real-time system for estimating the seismic damage in Romania. Math Modelling Civil Eng 11 (1):33–41

Trifunac MD, Todorovska MI (1997) Northridge, California, earthquake of 1994: density of pipe breaks and surface strains. J Soil Dyn Earthquake Eng 16:193–207

USGS (2014) Reports http://www.sott.net/article/236229-USGS-Reports-Record-Number-Of-Strong-Earthquakes-In-2011. Accessed 10 July 2014

Walker BB, Taylor-Noonan C, Tabbernor A, McKinnon T, Dal H, Bradley D, Schuurman N, Clague JJ (2014) A multicriteria evaluation model of earthquake vulnerability in Victoria, British Columbia. Nat Hazards 74(2):1209–1222

Weiwei W, Jiansi Y, Yirui L, Lingchun F (2009) Applications and development tendencies of WebGIS in earthquake disaster management. Recent Dev World Seismol 2009(3):20–27

White GF, Kates RW, Burton I (2001) Knowing better and losing even more: the use of knowledge in hazards management. Environ Hazards 3:81–92

Wilson TJ (1965) A new class of faults and their bearing on continental drift. Nature 207:343–347

Yungang B, Liqiang C, Chengliang W et al (2004) Application of geographical information system to comprehensive disaster reduction. J Coll Disaster Prev Tech 6(3):16–20

Zhou YX, Liu GJ, Fu EJ, Zhang KF (2009) An object-relational prototype of GIS-based disaster database. In: Proceedings of the 6th international conference on Mining Science & Technology, Procedia Earth and Planetary Science 1:1060–1066

Chapter 9
A Sediment Dynamic Modelling of Landsat OLI Image for Suspended Sediment Drift Along the Southwest Coast of India

Meenu Rani, S. Kaliraj, Raihan Ahmed, Biswajit Tripathy, Bismay Ranjan Tripathy, and Gajendra Singh Pippal

Abstract The movement of suspended sediment along the coastal water is an indicator of erosion and deposition of the coastal landforms. The current study deals with the spatio-temporal movement of suspended sediments in the shallow along the southwest coast of Thiruvananthapuram district, Kerala state in India. The customized model here systematically analyses the spectral properties of multiple bands to mapping the suspended sediments at various concentration and spatial distributions. The study on sediment drift and its impacts on the coast through conventional method are difficult; meanwhile, multi-temporal images may provide effective results for studying sediments concentration and their movement along the coastal water. The geoprocessing modelling of sediment dynamics has executed mathematical algorithm on Landsat OLI image to retrieve SSC from coastal water and demarcate movement of sediments along the coast during pre- and post-monsoon. The study reveals that sediment concentration has estimated at higher rate along the offshore area with depth lower than 30 m, and this is gradually

M. Rani
Department of Geography, DSB Campus, Kumaun University, Nainital, Uttarakhand, India

S. Kaliraj
Central Geomatics Laboratory (CGL), ESSO-National Centre for Earth Science Studies (NCESS), Ministry of Earth Sciences, Government of India, Thiruvananthapuram, India
e-mail: s.kaliraj@ncess.gov.in

R. Ahmed
Department of Geography, Jamia Millia Islamia, New Delhi, India

B. Tripathy
Project Engineer Electrical. Supreme Infrastructure (I) Ltd, Kolkata, India

B. R. Tripathy (✉)
Remote Sensing and GIS, Kumaun University, Almora, Uttarakhand, India

G. S. Pippal
Punjab Remote Sensing centre, Ludhiana, Punjab, India

P. Kumar et al. (eds.), *Applications and Challenges of Geospatial Technology*,
https://doi.org/10.1007/978-3-319-99882-4_9

decreasing towards the sea at the depth beyond 30 m. It is observed that the suspended sediment drift produces depositional landforms at the low wave-energy zone and erosional landforms at high wave-energy zone. This study proves the effectiveness of geospatial technology to estimate sediment concentration and transportation in the shallow coastal water.

Keywords Suspended sediment drifts (SSD) · Seasonal movement of sediment · Landsat OLI image · Single band model algorithm · GIS and remote sensing

9.1 Introduction

Suspended sediment loads in the coastal water generally consist of sand, silt, clay, etc. that are derived from river discharge due to littoral transport to the surf zone by waves and currents. The suspended sediment drift along the coastal area determines shape, size, spatial pattern, and morphology of the coastal landforms based on erosion and accretion processes. The availability of suspended sediment concentration in coastal water influences the coastal dynamics and is being an indicator of erosion and deposition processes in the coastal area. Coastal processes are highly dynamic in nature due to waves, currents, and tides that are interacting with coastal landforms and frequently moving littoral sediments (Kaliraj et al. 2017). Sediment transport changes over time and place depends on wave direction and releasing energy to the coast (Panwar et al. 2017; Kaliraj et al. 2016). The stability of coastal landforms and their environment may strongly relate to sediment deposition and erosion resulting from waves and currents. Understanding suspended sediment drift using multi-temporal satellite images is important for monitoring changes of coastal landforms. Geoprocessing modelling of suspended sediment drift may be useful for measurement and monitoring of suspended sediment load in coastal water transported from the river or other sources; this study provides primary information sources for coastal vulnerability mapping and management (Nechad et al. 2010; Sinha et al. 2004). Seasonal variability of suspended sediment has strong influence on coastal erosion and accretion processes, and they alter pre-morphology of the coastal landforms due to shoaling action induced by wind and waves and littoral currents, and bathymetry may increase or decrease. Suspended sediment is drifting along the coast accordingly based on waves and currents that have certainly not touched the bed by means of turbulence of coastal due to upwelling waves that distinctly differentiate sediment from clear coastal water in the surf zone. Suspended sediment concentration in coastal water reflects unique spectral signature that distinctly differentiate sediment occurrence from clear coastal water and other suspended matter in multispectral remotely sensed images. Suspended sediment drift can be demarcated using multispectral images like ETM+, OLI, ASTER and OCM and other hyperspectral images based on spectral reflectance variability between sediment and coastal water (Kaliraj 2016).

Many scientific studies have been performed worldwide on multispectral satellite image for the estimation of suspended sediment and monitoring spatio-temporal movement suspended sediment concentration (SSC) in the coastal water. Since past decades, much number of studies on suspended sediment concentration using remote sending was observed to estimate and monitor SSC in various parts of the world (Rawat et al. 2011; Whitelock et al. 1981; Curran and Novo 1988; Marcus and Fonstad 2010; Kaliraj et al. 2017). As we know remote sensing offers larger area view for analysing water quality and providing a more efficient and cost-effective method for assessing SSC from the ocean. Remote-sensing image plays vital role in the monitoring of sediments along the shallow coastal water, especially Landsat mission, MODIS, SPOT, OCM etc., which provides a synoptic view as well as the configuration (Ontowirjo, et al. 2013). According to Kaliraj and Chandrasekar (2012), the processing of Landsat ETM+ image using empirical algorithm can be estimated through assemblage of suspended sediments from the coastal water. Advantages of remote sensing have opted for various coastal studies such as sediment transportation, movements and deposition along the shallow water area, which can be effectively helpful for monitoring using multi-temporal images with relative high accuracy (Byers 1992). The spectral reflectance of image at each pixel has explored the characteristics of sediments and its accumulation, which can be calibrated using mathematical algorithms for estimation of sediment load in coastal water (Gerald 1980). The spectral reflectance variability between sediments and clear coastal water may be used to extract the pixels representing sediment occurrence using coefficient values calculated from in situ sampling analysis (Kaliraj et al. 2014).

Experimental analysis of estimation of suspended sediment concentration using Landsat images provides site-specific calculations for sediment load with relative practical accuracy, as it is field derived from field-observed reflectance data (Islam et al. 2001; Kaliraj et al. 2013). Moreover, the multispectral images have distinctly reflects spectral signature from suspended sediments and coastal water based on sediment concentration and depth of coastal water (Kaliraj et al. 2015). The coastal water shows higher level of spectral reflectance at the wavelength of 0.5–0.8 μm (Landsat ETM/OLI bands of visible and infrared) due to occurrence of sediment, and the band combination related to this portion may be used to demarcate suspended sediments. Suspended sediment accumulated with coastal water interacts significantly with spectral properties and thermal emissivity in the electromagnetic spectrum. The energy flux varies between coastal water accumulated with suspended sediments and clear coastal water; whereas the energy flux is significantly decreased in sediment mixed coastal water than clear coastal water; this is due to absorption of energy by clear water than turbidity. The empirical algorithm is used to calculate the energy flux variability using Landsat OLI image to estimate suspended sediment load with relatively high accuracy (Islam et al. 2001; Wang and Lu 2010; Qu 2014). Remotely sensed data only can give information in one- to two-dimensional studies because this cannot give evidence on the vertical circulation of sediments in water as we get the reflection data from only two metre surface water (Wang et al. 2009; Katlane et al. 2013; Warrick et al. 2004a, b). Estimation of suspended sediment concentration using Landsat images is a complex process that involves systematic geoprocessing analysis and is necessary to modelling for monitoring suspended sediment drift along the coast (Yanjiao et al. 2007; Tassan 1998; Zhang et al. 2003). This study is attempted to execute mathematical algorithm for estimation of suspended sediment load and its seasonal transportation using multi-temporal Landsat 8, OLI images (30 m) along the southwest coast of Thiruvananthapuram district, Kerala state, for periods of pre-monsoon and post-monsoon. The result of this study may provide primary information for planning the sustainable coastal monitoring and management activities.

9.2 Study Area

Southwest coast along the Thiruvananthapuram district covers both erosional and accretional landforms, and the study area boundary covers with the stretch of about 72 km along the shoreline and extends up to 10 km width across the sea with the geographical coverage of 8° 17' N – 8° 54' N latitudes and 76° 41′ E – 77° 17′.E longitudes. Major parts of the coastal stretches cover dense human settlements, infrastructures, active business centres, tourist attractions and very densely populated zones. The coastal stretch is highly active due to the hydrodynamic forces such as waves, currents and tides, etc. The coastal area also undergoes variability in terms of nearshore bathymetry and landforms of different types, such as sandy berms, beach ridges, beach cusps, beach scarps, etc. The coastal zones frequently experience erosion of larger quantity of sediments that causes severe vulnerability to human settlements and infrastructures. The landforms along the coast undergo morphological instability due to both natural and anthropogenic factors. The sub-tropical climate prevails along the southwest coast that leads high energy waves and currents interaction to the seashore throughout the year. The annual average rainfall along the study area has recorded with the range of 826–1456 mm; meanwhile the optimum temperature is estimated at minimum range of 23.78 °C and maximum range of 33.95 °C. The coastal tract has experienced dynamic changes due to coastal process fluctuating sediment load to the landforms. The sediment dynamic modelling of Landsat OLI image is performed for estimation of suspended sediment drift during pre-monsoon and post-monsoon (Fig. 9.1).

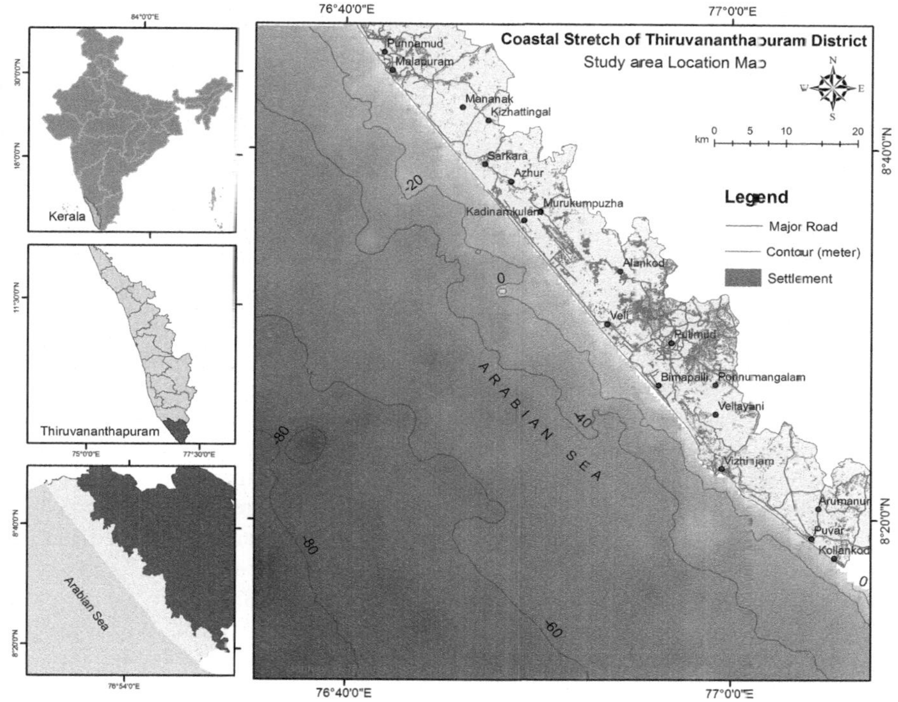

Fig. 9.1 Site-specific map of study area

9.3 Materials and Methodology

9.3.1 Data Used

The data used to carry out the work includes the satellite data and other spatial dataset. Landsat 8 scans the earth surface and measures different ranges of electromagnetic spectrum. Operational Land Imager (OLI) gives the multispectral data for visible as well as infrared range. It covers the entire earth in 16 days. The OLI uses the push broom sensor with 115 mile cross track field of view. We have used OLI data for determining pre-monsoon (March 2017) and post-monsoon (September 2017) suspended sediments from USGS EarthExplorer. Wind direction is derived from the portal European Reanalysis Interim (ERA), provided by European Centre for Medium-Range Weather Forecasts (ECMWF), based on a four-dimensional IFS (Cy31r2) system. Shuttle Radar Topography Mission is a research effort that obtained digital elevation model for generating high-resolution digital topography database for the earth. For extracting the bathymetry, we have used a global relief model which is a composite model of Digital Bathymetry Model (DBM). SRTM30_PLUS consists of both topography and bathymetry model that provides

30 arc seconds (900 m) resolution data for land as well as for ocean, which is derived from depth sounding (SONAR) and satellite altimetry.

9.3.2 Conversion of DN to TOA Reflectance

TOA (top of the atmosphere) reflectance denotes to a most sophisticated method of removing the aids by surface sparkle and atmospheric scattering from the assessed total reflectance to determine the water-leaving reflectance. The total reflectance is the amount of the Rayleigh's reflectance, aerosol reflectance, specular reflection of the sun and the reflectance of foam and whitecaps. All of these parameters can be estimated from the metadata and field measurement using empirical relationship (Gordon and Morel 1994). Water-leaving reflectance is computed as upwelling solar radiance just above the surface of the water divided by downwelling solar irradiance. In NIR band, solar radiation is totally absorbed by the water. As a result, water-leaving radiance is negligible or very small and, therefore, can be eliminated, allowing estimation of aerosol radiance directly. As per many established studies and test, the red (R) and near-infrared (NIR) bands prove to be superior for mapping and modelling medium to relatively high concentrations of SPM.

These raw data can be converted to ToA reflectance from DN (Eq. 9.1), using the required parameters from the metadata file (MTL.txt):

$$\rho_t' = M_p * Q_{\text{cal}} + A_r \tag{9.1}$$

where ρ_t'= TOA planetary reflectance (without solar angle correction), M_p= band-specific multiplicative rescaling factor, A_r =band-specific additive rescaling factor and Q_{cal}= quantized and calibrated standard product pixel values (DN). Next, we have corrected the ρ_t'using Eq. 9.2 for the solar angle:

$$\rho_t = \frac{\rho_t'}{\cos(\theta_{\text{SZ}})} = \frac{\rho_t'}{\sin(\theta_{\text{SE}})} \tag{9.2}$$

where ρ_t = TOA planetary reflectance, θ_{SE}= local sun elevation angle, θ_{SZ}= local solar zenith angle and θ_{SZ}= 90° – θ_{SE}.

9.3.3 Rayleigh's Reflectance Correction

Most of the TOA radiance is light reflected by aerosols and air molecules in the atmosphere, and it is a must to compute these contributions very precisely and subtract from the observed signal. At-sensor total reflectance (Eq. 9.3) can be calculated using the following equation.

$$\rho_{\text{TOA}} = L1B\_ \frac{\text{Radiance}}{\text{Extra Terrestrial Sun Radiation}} \tag{9.3}$$

$$\rho_{0.5} = \frac{\frac{\rho_{\text{TOA}} - \rho_R}{T_R}}{1 + S_R * \frac{\rho_{\text{TOA}} - \rho_R}{T_R}} \tag{9.4}$$

where:

ρ_r = Rayleigh reflectance
Extra-terrestrial sun radiation$= \pi * Es * \cos(\theta s) * e^{-m * OzoneTau} * dsol$

$$m = \frac{1}{\cos(\theta s)} + \frac{1}{\cos(\theta v)}$$

θs and θv are the angle of solar zenith and sensor zenith, respectively
dsol, the eccentricity correction factor of the Earth's orbit
T_R = Rayleigh atmospheric transmittance

$$T_{\text{total}} = Tyu * Tyd * e^{-m * OzoneTau} * e^{-m * WaterVapoTau}$$

Tyu=Rayleigh transmittance for sensor

$$\frac{Tyu = 2/3 + \cos(\theta v) + (2/3 - \cos(\theta v)) * e^{-\frac{RayleighTau}{\cos(\theta v)}}}{4/3 + RayleighTau}$$

Tyd=Rayleigh transmittance for sun

$$\frac{Tyd = 2/3 + \cos(\theta s) + (2/3 - \cos(\theta s)) * e^{-\frac{RayleighTau}{\cos(\theta s)}}}{4/3 + RayleighTau}$$

$RayleighTau$ = Rayleigh optical depth
O_2Tau= Oxygen optical depth
Tau = Optical depth
S_R= Rayleigh spherical albedo

$$\begin{aligned} RayPolynomi = {} & -0.58 + RayleighTau - 0.25 * RayleighTau^2 \\ & + 0.055 * RayleighTau^3 - 0.0098 * RayleighTau^4 \\ & + 0.0011 * RayleighTau^5 \end{aligned}$$

Rayleigh reflectance (ρ_r) is calculated below:

$$\rho_R = (\mu_5, \mu_v, \varnothing_v - \varnothing_5)$$
$$= \rho_{R1}(\mu_5, \mu_v, \varnothing_v - \varnothing_5) + \left(1 - e^{-\frac{\tau}{\mu_v}}\right)\left(1 - e^{-\frac{\tau}{\mu_v}}\right)\Delta(\tau) \quad (9.5)$$

where m $_s$ = cos of sun zenith, m $_v$ = cos of sensor zenith, j $_s$ = sun azimuth, j $_v$ = sensor azimuth, ρ_{R1}= the single-scattering contribution and t = atmospheric optical depth.

Most of the terminologies used in Eq. 9.5 are derived either through precomputed radiative transfer simulations or from models that depended upon the spectral response, solar angle, viewing geometry of sensor and ancillary information. Rayleigh correction can be made by subtracting the value of Rayleigh reflectance from TOA as stated in Eq. 9.6.

$$\rho_C = \rho_R - \rho_{\text{TOA}} \quad (9.6)$$

9.3.4 Marine Reflectance Calculation

Aerosol (ε) is calculated from the reflectance ratio in the band couple over water pixels so that the marine effect in those bands can be minimized and expected to be zero. "ε", the ratio of multiple-scattering aerosol reflectance, is assumed as constant over scene and, therefore, considered 1 in R and NIR band for more standard processing (Vanhellemont et al. 2014). The aerosol reflectance in R-NIR (4, 5) is predicted by assuming a linear relationship between bands 4–5 marine reflectance and constant aerosol types (ε) over the scene. ε can be derived through regression line slope (Neukermans et al. 2009) or the median ratio of Rayleigh corrected reflectance over clear water pixels in band R and NIR (ρc^4, ρc^5). The ratio of oceanic reflectance, "∝", in both bands is constant and can be derived using the mean resemblance spectrum for central wavelengths of the band (Eq. 9.7).

$$\propto = \frac{\rho_w^{(4)}}{\rho_w^{(5)}} = \frac{\overline{\rho}_{wn780}^{(655nm)}}{\overline{\rho}_{wn780}^{(865nm)}} = \frac{4.734}{0.544} = 8.702 \quad (9.7)$$

"γ" is the fraction of diffused atmospheric transmittances in the two bands, which is calculated using the following equation (Eq. 9.8).

$$\gamma = \frac{t_0^{(4)} t_v^{(4)}}{t_0^{(5)} t_v^{(5)}} \quad (9.8)$$

Then, the oceanic reflectance is calculated using $\rho_w^{(3)}$ and $\rho_w^{(4)}$ as noted in Eq. 9.9.

$$\rho_w{}^{(3)} = \frac{\alpha}{t_0{}^{(4)} t_v{}^{(4)}} \left[\frac{\rho_c{}^{(3)} - \rho_c{}^{(4)}}{\alpha\gamma - \varepsilon} \right] \quad (9.9)$$

9.3.5 Physiography and Drainage System

In the study area we will get huge numbers of landforms and several water bodies spatially distributed. Information of those landforms and drain system were extracted using high-resolution satellite images like Landsat 8 OLI (30 m) and district resource map published by GSI, 2010. The study area has also various geomorphic features, and all of these are mapped using Landsat 8 OLI image of 30 metres spatial resolution and SOI published topographic map (1:25,000 Scale), where we have extracted nine to ten features for the study area.

9.3.6 Mean Significant Wave Height

According to the physical oceanography, the mean significant wave height is one third of the highest wave, where the wind blowing across the sea pushes the water surface into waves. Effective coastal sediment management depends on understanding the processes that affect sediment transport at various spatial and temporal scales. So in a short time scale, huge waves can cause large amount of sediment movement and thus potential to cause damages to the coastal environment.

9.3.7 Coastal Erosion and Accretion

As we know shoreline is the most dynamic feature in the coastal landforms, which reflects erosion and accretions of coast in short-term and long-term periods. Shoreline change detection is the simplest way to show the erosional and depositional scenario of the shore using multi-temporal satellite data. So in the present study, we have used Landsat images for delineating the accretion and erosion area. We have taken Landsat 5 TM of Feb, 1997, and Landsat 8 OLI of Feb, 2017, for analysing the changes of shoreline during the past 20 years. Both the data have the same resolution, 30 metre resolution. We have extracted the erosion and depositional area in the study area using analysing tool “Intersect” in ArcGIS 10.4.1.

9.3.8 Relative Sea Level Rise

Sea level rise is a major issue for the shallow coastal communities all over the world. The surface of the ocean is not uniform, and the sea level is changing in response to its chemistry and temperature. Over time, global warming-induced sea level rise may inundate the shallow area of the coast throughout the world that changes the position of the coastline and sinks natural locale and coastal structures. It directly results in a corresponding higher shift to the zone of wave action, which reflects in a larger shoreline recession on milder slopes. Sea-level rise along the coast is the enduring inundation of coastal areas and produces serious impacts on the coastal habitats as well as socio-economic status of the people and causes worsening of coastal erosion by transporting inundated sediments offshore and spreading the effects of coastal flooding by allowing waves to act further, and hence, the coastal landforms along with the human-made construction gradually get affected badly. Sea level data is available for many station near the study area and can be found at PSMSL portal as described at the data used section. Ten stations' location and their sea-level data are listed below (Table 9.1).

9.3.9 Extraction of Suspended Sediments

Suspended sediments in pre-monsoon and post-monsoon were calculated after correcting data. The SSC was analysed using the reflectance of blue (0.45–0.51 μm), red (0.64–0.67 μm) and near-infrared (0.85–0.88 μm) wavelengths. The increase of the red reflectance in the turbid water indicates the presence of sediment or shallow water. SSC was derived using the single band algorithm (Eq. 9.10) used by *Nechad* et al.:

Table 9.1 PSMSL station and their corresponding rate of MSL

Latitude	Longitude	Location	Relative sea level (m)
4.181036	73.50818	Maldives	1.20
3.921824	72.7445	Himandhoo	1.32
4.221673	73.84193	Hulhule	1.35
8.27908	73.05076	Minicoy	1.48
8.460289	78.09659	Tuticorin	1.14
6.940934	79.85113	Colombo	1.19
11.17359	75.79905	Beypore	0.98
15.37592	73.81338	Marmagoa	1.40
12.8816	74.81591	Mangalore	0.82
9.934021	76.2456	Cochin	0.90

$$\mathrm{SPM} = \frac{A\rho_w}{1 - \rho_w/C} \quad (9.10)$$

Where A = 327.84 g m − 3 and C = 0.1708. So in this way we have mapped the solid particles suspended near the shore (within 10 km) and get the idea about transportation of suspended particle along the coast in pre-monsoon as well as in post-monsoon.

9.4 Results and Discussion

9.4.1 Landforms and Drainage System

Varkala, Vizhinjam and Edavai places present in the coastal city of Thiruvananthapuram district have very rough and undulated topography (Fig. 9.2). Three distinguished topographic units identified from west to east in the district are (1) lowland (coastal plains), (2) midlands and (3) highlands. The lowland topography was observed mainly between Thiruvananthapuram and Anjego and between Vizhinjam and Poovar. Marine and fluvial activities are responsible for lowland topography and can be seen all along the coastal plain (younger and older coastal plain), characterized by gentle sloping landscape. The crystalline at Kovalam, Veli and Vizhinjam and laterite at Poovar and Varkala are noticeable landforms that lie at coastal plains and are quite narrow with maximum width of 5 km. The most important landforms are sandy and coastal cliff, rocky beaches and sand ridges. Denudational process formed the major part of the district, including both midlands and highlands. The midland part comprises of valleys occupying 60% part of the district as well as hillocks creating an undulated topography.

Neyyar (56 km or 35 mi), the southernmost river of the Kerala state, one among the three major rivers (Fig. 9.3) in the district, originated from Agasthyamala (second highest peak in the Western Ghats). Karamana (67 km) originated from Vayuvanthol, another mountain in the Western Ghats. The Vamanapuram River originated from Chemunji Mottai in the Western Ghats. Veli, Anchuthengu, Kadinamkulam, Kappil, Edava Nadayara and Akathumuri are major lakes in the district. Apart from this, a freshwater lake that lies at Vellayani in Thiruvananthapuram taluk is a potential lake which can be a major source of water supply for Thiruvananthapuram in the future.

9.4.2 Hydrodynamic Processes Along the Coast

The sediments near the shore are transported instead of being stabled at one place due to the various hydrodynamic action that influences the ocean, and the mean

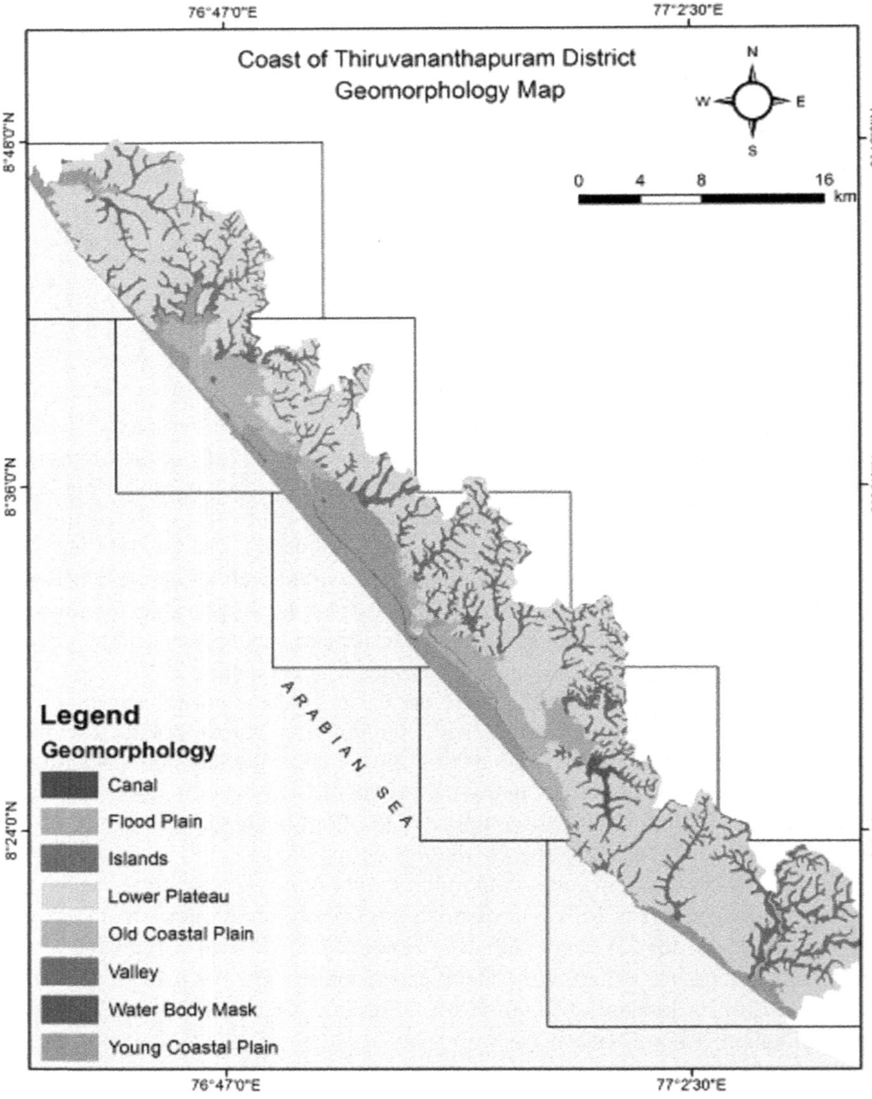

Fig. 9.2 Landform map along the coast

significant wave height (Fig. 9.4) is one of them. The study found that the mean significant wave height for the study area ranges from 0.35 to 1.50 m, where high MSWH was found along southern part, like Pachalur, Vizhinjam, Mullur, Karumkulam, Poovar and Kollankod, which decreases towards the north. The area experience high wave energy can mingle more sediments compared to less energy influential area.

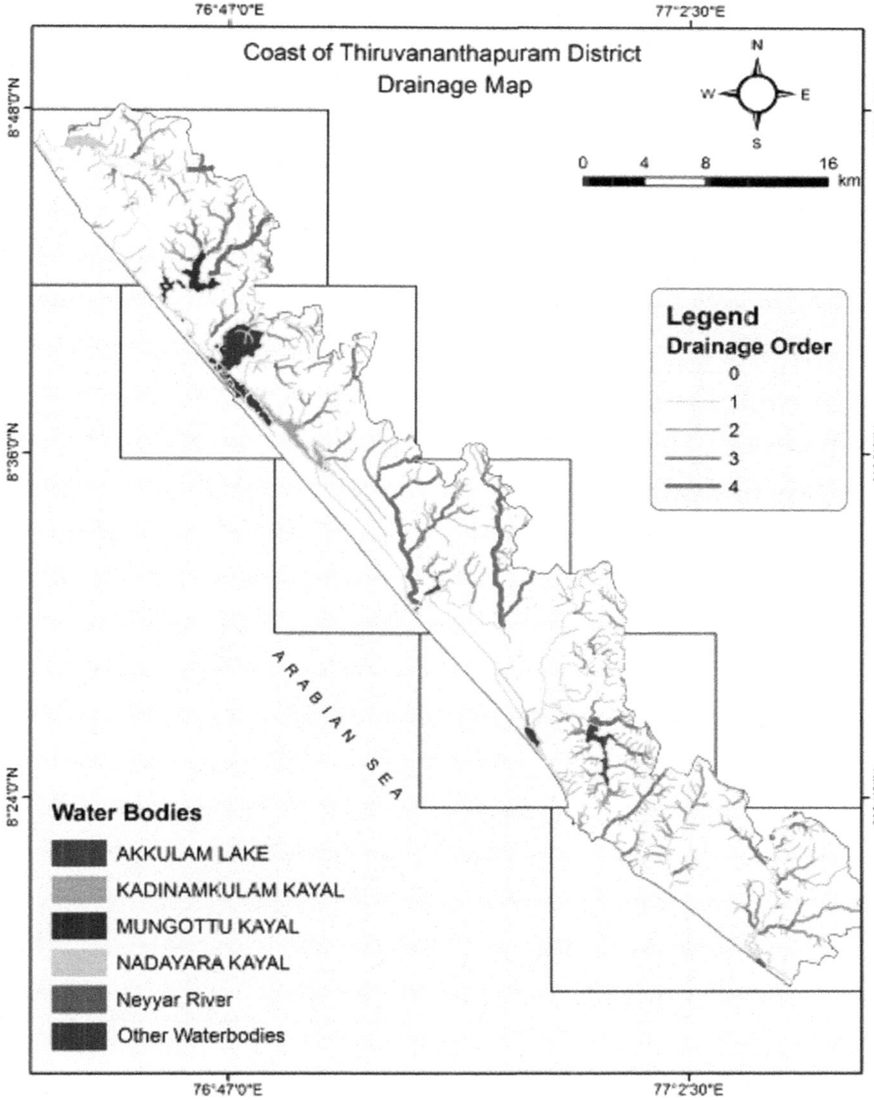

Fig. 9.3 Drainage order map along the coast

The change in the rate of sea-level rise is estimated (Fig. 9.3) by gridding the coastal zones using a linear interpolation method. In the south of the study area near Poovar, Vizhinjam and Kollankod, the sea-level rise is about 0.05 to 0.91 metre above the MSL, and the impact of the wave intensity is more, and other areas like Panathura, Muttathara and Kovalam possess the sea-level rise that approximately ranges from 0.7 to 0.9 m (Table 9.2).

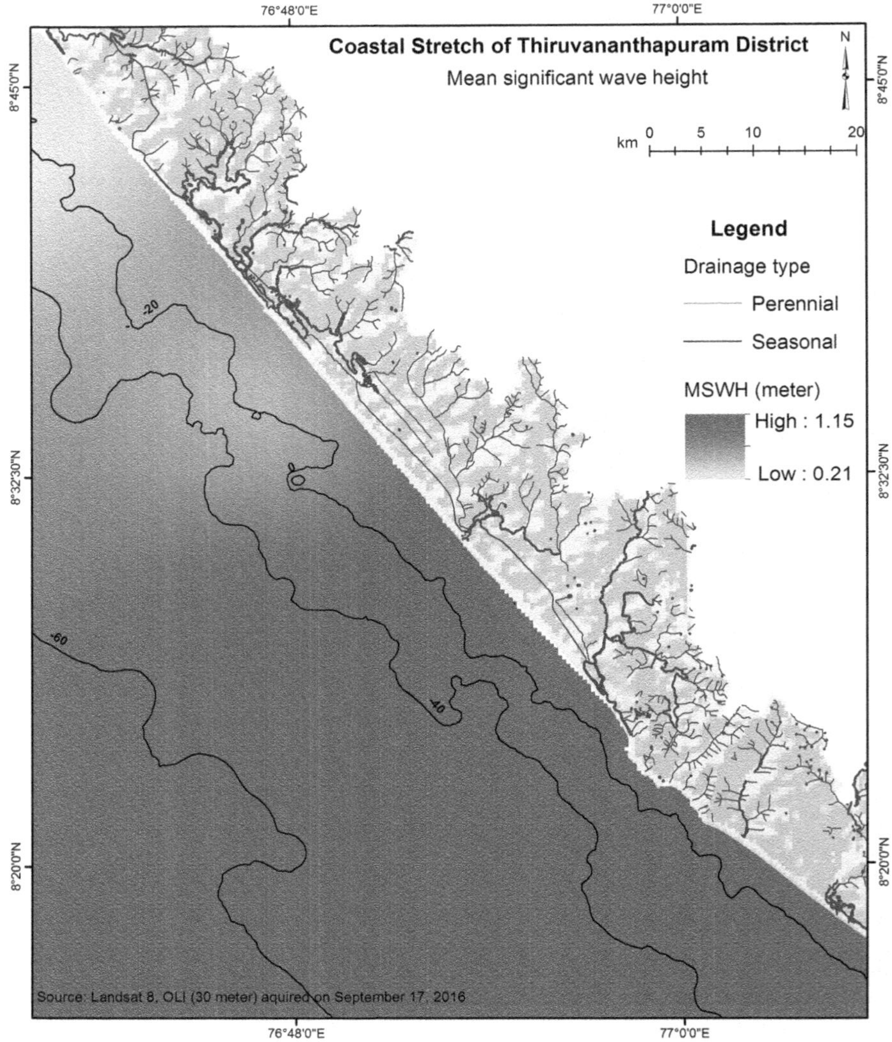

Fig. 9.4 Mean significant wave height along the coast

Areas like Kappil, Palachira, Thumba, Beemapally, Poovar, Vatаturuthu, Vettukad, Pallithura, Kollankod and Kovalam are noticed having high rate of erosion as these area have retreated up to more than 500 m^2 Mullur, Panathura, Puntura, Veli, Vettur and Tiruvampadi regions possess seasonal erosion as well as deposition of about 1–250 m^2 accretion and 1–100 m^2 erosion at a particular interval, whereas Vettur, Vilabhagam, Rathikkal, Vettukad, Puntura and Panathura regions experience 250–500 m^2 of accretion during the last 20 years. The main reason

Table 9.2 Characteristic of Thiruvananthapuram coastal stretch

Coastal stretch	Geomorphology	LULC	Shoreline change (m^2/year)	Coastal slope (m)	RSL (mm/year)
Zone 5	Young coastal plain, flood plain, natural levee, river channel, back swamps, rocky coast and plateau (lateric)	Mixed crop, beaches, perennial, mining, residential, coconut, water bodies, banana, mixed built-up and coastal sand	9–400 (A) More than 500 (E)	**40– (−65)**	**0.69–0.81**
Zone 4	Paleo-beach ridge, canal, swale (old coastal), waterbody mask, lateric plateau, valley, channel bar and river	Beaches, coastal sand, airport, mixed built-up, mixed crop, coconut, double crop, perennial, residential and commercial	11–500 (A) 100 more than 500 (E)	**100– (−48)**	**0.60–0.72**
Zone 3	Estuaries, canal, beach ridge, swale (young and old coast plain), canal, island and young coastal plain	Beaches, coastal sand, coconut dominant mixed crop, mixed built-up, perennial, commercial and dense mixed forest	1–400 (A) More than 500 (E)	**78– (−31)**	**0.56–0.65**
Zone 2	Paleo-beach ridge, tidal flat (younger mud), brackish water creeks, Kayals (estuaries), river canal (flood plain) and old coastal plain	Town/cities, beaches, coastal sand, river/stream, deciduous, villages, mixed crop, coconut, agricultural and built-up land	More than 500 (A) 90–470 (E)	**88– (−40)**	**0.62–0.70**
Zone 1	Beach (young), young coastal plain, canal, valley, lower plateau, river canal, flood plain and Kayals	Banana, teak, beaches, plantation, coconut dominant mixed crop, commercial, mixed built-up, paddy land and villages	More than 500 (A) 0–350 (E)	**110– (−36)**	**0.68–0.77**

behind it is the extending of depositional landforms due to swashing of sediments on the gentle slope.

9.4.3 Monitoring of Suspended Sediments

The northern part of the study area is observed with heavy deposition during the post-monsoon as the swashing of suspended particles by fewer energy waves along with the dominant wave direction in this session (Fig. 9.4). Whereas, in the pre-monsoon, the circulation of the sediments found to be at the lower part of the study area as due to the dominate wave and current action along the shore (Fig. 9.5). Basically, a sediment particle will be suspended when the vertical velocity of the

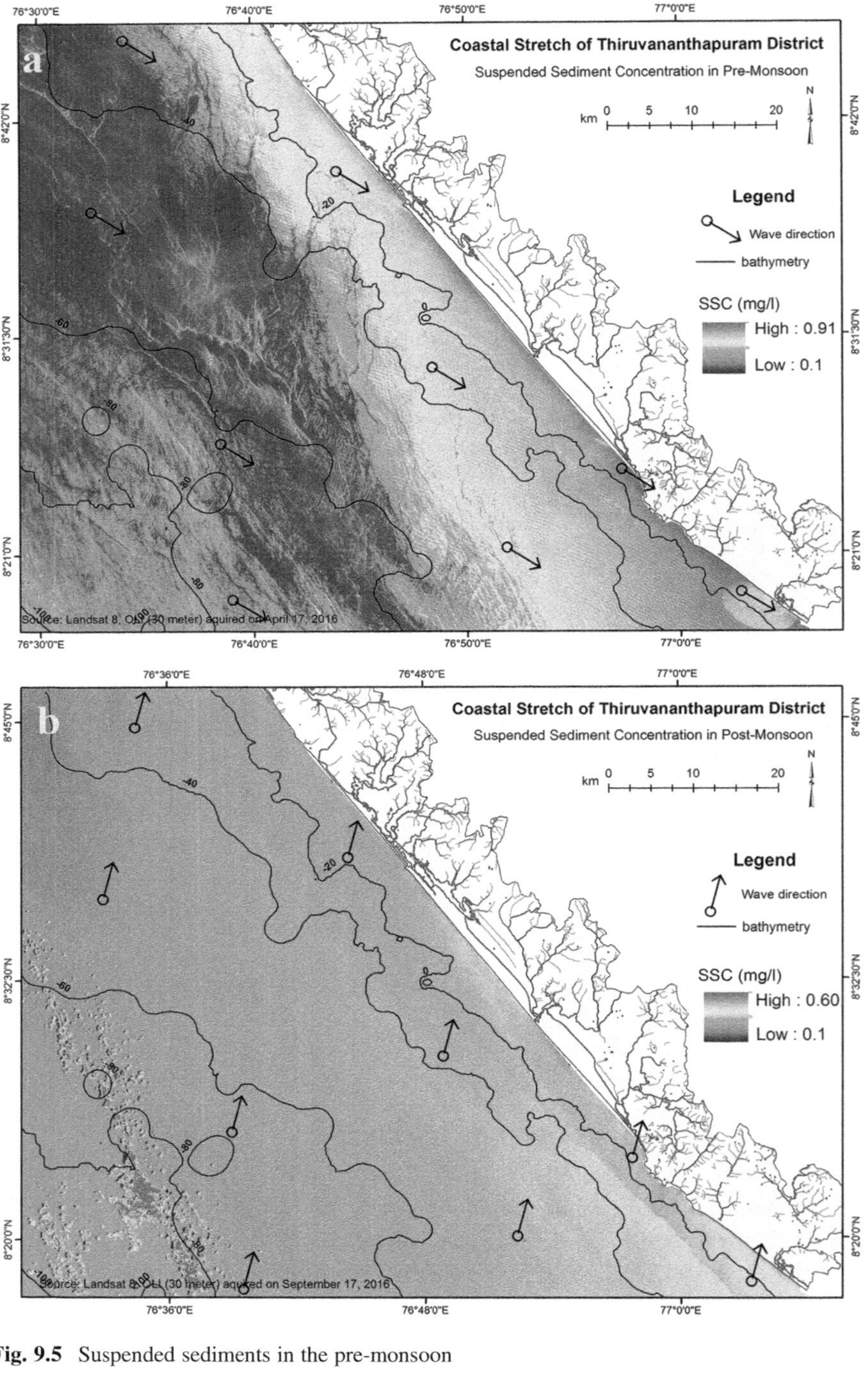

Fig. 9.5 Suspended sediments in the pre-monsoon

fluid motion becomes greater than the settling velocity of the particle. So, in the case of wave field, the vertical motion must result from the combined effects of turbulence and wave orbital motion. Here another phenomenon also acts on the sediments, i.e. gravity of the sediments suspended in the ocean, as the heavy sediments get deposited nearshore, whereas the lighter sediments float. The wave direction in the post-monsoon and pre-monsoon is shown in Figs. 9.4 and 9.5. It is found that SSC is decreased rapidly with the increases of distance (0–10 km) from the shoreline to bathymetry level of 5–10 m. Also, the effect of the wave shoaling in the deeper water is significantly less, which causes sparely distribution sediment concentration. The SSC, beyond 5000 metres from the shoreline and below 10–20 m depth, is rarely observed as they cannot be measured at that depth, which is one of the limitations of the optical data. The movement of SSC indicates that the SSC has a positive correlation with wave direction and littoral current, and according to the deposition along the shore, we have assigned the weight to the layer. Mostly the southern part of the study area is found to be in high concentration of suspended matter as these regions experience high suspended sediments in both periods, whereas the middle part is found to be in cyclic concentration as it experiences seasonal erosion of the shore and deposition sediments near the beach. The northernmost part of the study area is found to have less concentration of the suspended sediments throughout the year.

9.5 Conclusion

It is found that the sediment concentration decreased rapidly with the increase of distance to the beach and depth to the seabed. As the bathymetry increases, the low amount of sediments available moves towards the shore and so it's observed as low concentrations in the surface water. Another factor, wave, which has a frequency at comparatively a large distance in deeper water, causes sparely distribution of sediments. So the sediments remain at a lower depth in high bathymetry (>= 10 m) and distance more than 2 km from the shoreline. The dissimilarity of sediments reveals that the suspended sediments are indirectly proportional to bathymetry and distance from the shoreline and are directly proportional to the wave direction and littoral current at offshore. Since the 1970s, the skill to retrieve the concentration from remotely sensed imagery has opened up the possibility for rebuilding past transport settings. Unfortunately, there is still a great deal of hesitation in the relation between the water-leaving reflectance and these concentrations in turbid, as different particles are having different sizes, and mineralogy properties affect the scattering properties. However, the present study provides interesting visions to monitoring the suspended sediments at nearshore, after radiometric and geometric correction. Another prospectus of this study also revealed the factors affecting the particles to be suspended in the nearshore as well as offshore. OLI (30 m), having spectral range from 0.45 to 0.88 μm for visible as well as NIR, demonstrated the best detail to conduct the analysis on the basis of the spectral response. Its spectral analysis for the suspended sediments may conclude that the high reflection in near IR expresses the high concentration of SSC, and, in contrast, the high reflection in the visible range states the lower concentration of suspended sediments. The monitoring in both periods (pre-monsoon and post-monsoon) revealed that suspended sediments are transported from south to north in post-monsoon, and the contrary situation occurs in the pre-monsoon. Mapping the spatial distribution of suspended materials with remotely sensed data would be supportive for the management of water bodies as well as the seasonal variation in the water quality, and the study can further extend for determining the point and non-point source of water bodies, which discharge in the ocean. Finally, we have concluded that geospatial technology can potentially be used as a tool for monitoring the sediments in the ocean in the pre-and post-monsoon.

Acknowledgement The authors declare that there is no conflict of interests regarding the publication of this paper. BT and KR designed and proposed the research; BT processed data; PK, KR and BT implemented techniques; PK and BT analysed results; and PK and JSR drafted and edited the article.

Reference

Byers AC (1992) Soil loss and sediment transport during the storms and landslides of May 1988 in Ruhengeri prefecture, Rwanda. Nat Hazards 5(3):279–292

Curran PJ, Novo EMM (1988) The relationship between suspended sediment concentration and remotely sensed spectral radiance, a review. J Coast Res 4(3):351–368

Gerald KM (1980) Satellite remote sensing of water turbidity/Sonde de télémesure par satellite de la turbidité de l'eau. Hydrol Sci Bull 25(4):407–421

Gordon HR, Morel AY (1994) Remote assessment of ocean colour for interpretation of satellite visible imagery, a review. Lecture Notes Coast Estuar Stud 4:114

Islam MR, Yamaguchi Y, Ogawa K (2001) Suspended sediment in the Ganges and Brahmaputra River in Bangladesh, observation from TM and AVHRR data. Hydrol Process 15(3):493–509

Kaliraj S (2016) Geospatial analysis of coastal geomorphological vulnerability on South West coast of Kanyakumari using Remote sensing Technology. Unpublished Ph.D. Thesis, Manonmaniam Sundaranar University, Tirunelveli, pp 83–110

Kaliraj S, Chandrasekar N (2012) Spectral recognition techniques and MLC of IRS P6 LISS III image for coastal landforms extraction along south west coast of Tamil Nadu, India. BonfringInt J Adv Image Process 2(3):01–07

Kaliraj S, Chandrasekar N, Magesh NS (2013) Impacts of wave energy and littoral currents on shoreline erosion/accretion along the south-west coast of Kanyakumari, Tamil Nadu using DSAS and geospatial technology. Environ Earth Sci. https://doi.org/10.1007/s12665-013-2845-6

Kaliraj S, Chandrasekar N, Magesh NS (2015) Evaluation of coastal erosion and accretion processes along the south-west coast of Kanyakumari, Tamil Nadu using geospatial techniques. Arab J Geosci 8(1):239–253

Kaliraj S, Chandrasekar N, Ramachandran KK (2016) Mapping of coastal landforms and volumetric change analysis in the south west coast of Kanyakumari, South India using remote sensing and GIS techniques. Egypt J Remote Sens Space Sci. https://doi.org/10.1016/j.ejrs.2016.12.006

Kaliraj S, Chandrasekar N, Ramachandran KK, Srinivas Y, Saravanan S (2017) Coastal landuse and land cover change and transformations of Kanyakumari coast, India using remote sensing and GIS. Egypt J Remote Sens Space Sci. https://doi.org/10.1016/j.ejrs.2017.04.003

Katlane R, Nechad B, Ruddick K, Zargouni F (2013) Optical remote sensing of turbidity and total suspended matter in the Gulf of Gabes. Arab J Geosci 6:1527–1535

Marcus WA, Fonstad MA (2010) Remote sensing of rivers, the emergence of a sub discipline in the river sciences. Earth Surf Process Landf 35(15):1867–1872

Nechad B, Ruddick K, Park Y (2010) Calibration and validation of a generic multi-sensor algorithm for mapping of total suspended matter in turbid waters. Remote Sens Environ 114:854–866

Neukermans G, Ruddick K, Bernard E, Ramon D, Nechad B, Deschamps PY (2009) Mapping total suspended matter from geostationary satellites, a feasibility study with SEVIRI in the southern North Sea. Opt Express 17:14029–14052

Ontowirjo B, Paris R, Mano A (2013) Modeling of coastal erosion and sediment deposition during the 2004 Indian Ocean tsunami in Lhok Nga, Sumatra, Indonesia. Nat Hazards 65 (3):1967–1979

Panwar S, Agarwal V, Chakrapani GJ (2017) Morphometric and sediment source characterization of the Alaknanda river basin, headwaters of river ganga, India. Nat Hazards:1–23

Qu L (2014) Remote sensing suspended sediment concentration in the Yellow River. Ph.D. dissertation paper 383, University of Connecticut, website accessed Dec 2014 at http,// digitalcommons.uconn.edu/dissertations/383/

Rawat PK, Tiwari PC, Pant CC, Sharama AK, Pant PD (2011) Modelling of stream run-off and sediment output for erosion hazard assessment in lesser Himalaya, need for sustainable land use plan using remote sensing and GIS, a case study. Nat Hazards 59(3):1277–1297

Sinha PC, Guliani P, Jena GK, Rao AD, Dube SK, Chatterjee AK, Murty T (2004) A breadth averaged numerical model for suspended sediment transport in Hooghly estuary, East Coast of India. Nat Hazards 32(2):239–255

Tassan S (1998) A procedure to determine the particulate content of shallow water from thematic mapper data. Int J Remote Sens 19:557–562

Vanhellemont Q, Neukermans G, Ruddick K (2014) Synergy between polar-orbiting and geostationary sensors, remote sensing of the ocean at high spatial and high temporal resolution. Remote Sens Environ 146:49–62

Wang JJ, Lu XX (2010) Estimation of suspended sediment concentrations using Terra MODIS, an example from the lower Yangtze River, China. Sci Total Environ 408(5):1131–1138

Wang JJ, Lu XX, Liew SC, Zhou Y (2009) Retrieval of suspended sediment concentrations in large turbid rivers using Landsat ETM plus, an example from the Yangtze River, China. Earth Surf Process Landf 34:1082–1092

Warrick JA, Merters LAK et al (2004a) Estimating suspended sediment concentrations in turbid coastal waters of the Santa Barbara Channel with SeaWiFS. Int J Remote Sens 25:1995–2002

Warrick JA, Merters LAK, Siegel DA, Mackenzie C (2004b) Estimating suspended sediment concentrations in turbid coastal waters of the Santa Barbara Channel with SeaWiFS. Int J Remote Sens 25:1995–2002

Whitelock CH, Witte WG, Taly TA, Morris WD, Usry JW, Poole LR (1981) Research for reliable quantification of water sediment concentrations from multispectral scanner remote sensing data, US National Aeronautics and Space Administration, Langley Research Center, Hampton, VA, NASA-TM-82372, 243–255

Yanjiao W, Feng Y, Peiqun Z, Wenjie D (2007) Experimental research on quantitative inversion models of suspended sediment concentration using remote sensing technology Chin Geogr Sci 17(3):243–249

Zhang Y, Pulliainen J, Koponen S, Hallikainen M (2003) Water quality retrievals from combined Landsat TM data and ERS-2 SAR data in the Gulf of Finland. IEEE Trans Geosci Remote Sens 41:622–629

Part IV
Progress and Perspective Scenario of Urban Growth Models

Chapter 10
Inter-calibration and Urban Light Index of DMSP-OLS Night-Time Data for Evaluating the Urbanization Process in Australian Capital Territory

Christopher D. Elvidge, Himangshu Kalita, Upasana Choudhury, Sufia Rehman, Bismay Ranjan Tripathy, and Pavan Kumar

Abstract The magnification mechanism of human settlement is called urbanization, involving various other activities such as population transition, resource consumption, etc. leading to various growing patterns of urban augmentation. The assessment of such spatial pattern is crucial in developing a sustainable urban agglomeration. Night-time light (NTL) data is an important machination for such assessment and detailed monitoring. In this paper, DMSP-OLS (the Defence Meteorological Satellite Program/Operational Linescan Program) datasets have been used to assess the urban straggle of Australian Capital Territory (ACT) and to delineate the urban extent from 1992 to 2012, at an interval of 3 years. DMSP-OLS has the unique capability to detect synthetic lights from cities, towns, industrial sites, ports, etc. Moreover, using ARC GIS calligraphies, 20 random points were selected from the extracted area of interest (AOI). Pixels values of those 20 random points are derived from the given time series dataset (1992–2012). A regression value was extracted from each year by using a second-order polynomial equation. A polynomial regression model is also constructed by taking the regression values and the time series as the two variables, respectively. Urban light index (ULI) is also constructed for analysing the progression of urbanization in ACT from 1992 to 2012 with the help of a derived formula. Furthermore, a unit circle buffer zone having a radius of 20 km is established by taking the centre of each built-up zone as the focal point of the constructed buffer, to compare easily the rate of expansion of urbanization. The

C. D. Elvidge
Applied Earth Sciences, National Oceanic and Atmospheric Administration, Stanford University, Boulder, CO, USA
e-mail: chris.elvidge@noaa.gov

H. Kalita · U. Choudhury · B. R. Tripathy
Remote Sensing and GIS, Kumaun University, Almora, Uttarakhand, India

S. Rehman · P. Kumar (✉)
Department of Geography, Jamia Millia Islamia, New Delhi, India

P. Kumar et al. (eds.), *Applications and Challenges of Geospatial Technology*,
https://doi.org/10.1007/978-3-319-99882-4_10

present paper indicates the growing potential of the DMSP-OLS night-time satellite data to define the urban light space information, which truly describes the attributes of urban sprawl, and to delineate the evolution of urban morphology and urban extension.

Keywords Australian Capital Territory · DMSP-OLS night-time data · Inter-calibration · Urban light index · Urbanization

10.1 Introduction

The congenital nature of man as a societal visceral directly or indirectly has led to the development of urbanization. With the plodding increase in the concreteness of people living in urban zones than in rural, this increase of people leads to progression of towns and cities making urban to sprawl. Urban sprawling causes an outward migration from urban to low-density areas. Metamorphosis of agrarian economy to an industrial and transition of natural ecological land cover simulated land use environment (Han et al. 2009; Feng et al. 2002; Elvidge et al. 1999; Li 2004). The synchronized process associated with population statistical transition inducing force on nature and biological cycle, aiding to anthropogenic activities, resource consumption and economic dynamics (Milesi et al. 2003; Ma et al. 2012). Therefore, the measurement of urbanization progress can be deliberated from only with the values of demographic transition (Zhang et al. 2013; Pandey et al. 2013; Letu et al. 2012; He et al. 2006; Lo 2010). Exploration of the potencies influencing urban enlargement and its impact on environment is crucial for scrutinizing ecological urban augmentation (He et al. 2012). Estimation and monitoring of the transition of urban agglomeration over the years is crucial for the development of a sustainable metropolitan planning (Cinzano et al. 2001; Fujita et al. 1999; Small 2003). Constituting of a very small segment of the facade of the earth, urban escalation exerts colossal influence on the other interrelated activities such as mass movements, mass compactness and exploitation of recourses (Small 2005; Sutton et al. 2001). The structural formation of the spatial allocation and the occurrence of the large-scale urban netting have imperative allegation ranging from socio-economic development to ecological urban augmentation to evolution of urban fabrication (Elvidge et al. 2007; Imhoff et al. 1997). Assessment of worldwide urban extent through corporeal supervision is challenging. Currently, night-time dataset for the measurement of the large-scale lighted extension is being employed (Liu et al. 2012; Shu et al. 2011).

The night-time satellite imagery is an important contrivance in analysing the urban straggle and delineating urban extent. Synthetic lights of towns, cities, manufacturing locations and anthropological actions during night-time can be detected using the satellite images of night-time data from the Defence Meteorological Satellite Program/Operational Linescan Program (DMSP/OLS) (Schneider et al. 2010; Elvidge et al. 1999), and VNIR (visible near infrared, 0.4–1.1 μm) and TIR (thermal infrared, 10.5–12.6 μm) are two spectral bands of DMSP with radiometric resolution of 6 bit and 8 bit, respectively (Owen 1998; Elvidge et al. 1997). In a spatial resolution of 2.7 km and a swath width of 3000 km, OLS covers a wider extent of the surface of the earth (Miller et al. 2009). Although previously used for cloud observation (Henderson et al. 2003; Huang et al. 2016), DMSP-OLS images are currently being used extensively for urban morphology assessment due to its ability to detect low light during night without moonlight (Letu et al. 2012; Ma et al. 2012). Delineate global dispersal of metropolitan extents through stable satellite imaging, with the help of DMSP/OLS in a precise and economical manner. The spatial resolution of 30 arcsecond of the DMSP/OLS continues to provide night-time data on urban morphology at a national and global measure, but the coarse resolution debilitates the accurate extraction of the built-up areas (Liu et al. 2015; Li et al. 2016; Lu et al. 2008). The dataset of DMSP-OLS can be arranged into three types: the mutated radiance, the stable NTL and the standardized DN (digital number) values (Welch and Zupko 1980). Among them, the stable NTL dataset is used for assessment of urban morphology (Lo 2002; Ma et al. 2012).The acquired dataset from DMSP-OLS utilized to map the proportion of urban extension, urban nucleus and the progression of urbanization. The major challenged faced while using DMSP-OLS dataset is to acquire quantitative association between urbanization factors and the pixel brightness of the dataset in different types of urban magnification mechanism (Sutton 2003; Ting Ma et al. 2012).

However, the saturated pixel values restrict the possibilities to differentiate between interurban discrepancies (Mertes et al. 2015). Previous studies have revealed inconsistency between the actual lighted zones and the geographical periphery of the urban areas (Su et al. 2015). Sometimes the global synthetic lighted zones aligned to as the advanced built-up area allude due to blooming effecting (Small et al. 2005). The abrasive radiometric and spatial resolution of the DMSP/OLS also detecting scattered low-lighted zones, large amount formation of mixels in OLS satellite imagery and geo-positional errors are introduced during the processing stage; these are some of the constraints of DMSP-OLS which results in blooming effects, yielding erroneous information of synthetic lights of site which in actual does not restrain any energy resource. So, the frequently identified lighted area of threshold 89% was proposed by (Imhoff et al. 1997) to reduce the effects of blooming and to establish a consistent relationship between substantial lighted zones and the perimeter of the urban extent, thus shrinking the lighted zones (Small et al. 2005). But fixation of a threshold evades other minor frequency lights, thus reducing the information content in the dataset (Small et al. 2005). Moreover, DMSP-OLS dataset cannot be applied directly due to unavailability of on-broad calibration producing erroneous DN (digital number) values caused by varied orientation and displacement of sensor and atmospheric and radiometric condition (Li et al. 2014).

NPP-VIIRS, another regional night-time amalgamated data, is also used to the exact statistics of urban morphology chronicling synthetic light at night, having a spatial resolution of 15 arcsecond which is equal to about 500 m and a radiometric resolution of 14 bit (Elvidge et al. 2007; Kaifang Shi et al. 2014). Such high spatial and radiometric resolutions of NPP-VIIRS dataset have the potential to stipulate better product of information. To evaluate urban agglomeration, urban extension and socio-demographic pattern at a variety of global scales such as demographic transition, power consumption and conservational modification, the NPP-VIIRS dataset is a powerful tool for such assessment. However, there are still some constraints in NPP-VIIRS dataset such as area dependency on referenced data affecting the result of urban area extractions; the inability to remove the noise created from fires, gas flares and volcanic eruptions from recent NPP-VIIRS dataset and the eminence of the dataset however requires to be enhanced (Kaifang Shi et al. 2014). A debatable competition is arising between the application of the dataset of DMSP-OLS and the NPP-VIIRS. So, a comparative analysis of the accuracy quality of the dataset of NPP-VIIRS with that of DMSP-OLS will provide better result in scrutinizing the urban sprawl and extraction of the built-up area and their application in future research but being in the toddler stage, VIIRS night-time images are not able to fulfil the demands of the long duration time sequence requirements of analyst and also involves a sophisticated data processing (Dewan and Yamaguchi 2009). However other remote sensing data such as QuickBird, IKONOS and SPOT images are expensive and need intricate data processing along with miscellaneous spectral differentiation, and inadequate spatial and radiometric resolution causes satellite like Landsat TM less capable to detect urban expansion and urban dynamics.

This paper aims to appraise the capabilities of NTL dataset to identify and delineate the urban morphology, urban augmentation and growth of suburban zones over the years of the Australian Capital Territory, using night-time satellite imagery from DMSP-OLS dataset from 1992 to 2012 at an interval of 3 years, for evaluation of the Australian Capital Territory growing pattern of urban sprawl and urban extension which is critical for sustainable development relative to ecological urban planning, management and decision-making for future urban development projects.

10.2 Study Area

The Australian Capital Territory (ACT) embraces a population of 410,300 out of 24 billion and a population density of 179.96 per km^2 (ABS 2017). ACT is located on the south-western side of the country, with geographical extension of 35°18′29″S latitude and 149°7′28″E longitude. Surrounded by the Australia's capital city Canberra on the north, the Namadgi National Park on the south constitutes about 40% of the territory (Linsie Tan 2017), the Goulburn-Cooma railway line in the east and in the west by the Cotton River. Canberra is a sprawling city with heavily concentrated economic activities of the Australian Capital Territory allied with rapidly increasing population associated with steady employment. The urban morphology of Canberra as a planned city was initially designed by a major American architect Walter Burley Griffin (Wigmore 1971). The transportation network follows a prototype of wheel and spoke. The nucleus of the city's morphology is arranged as a 90° angle, with two

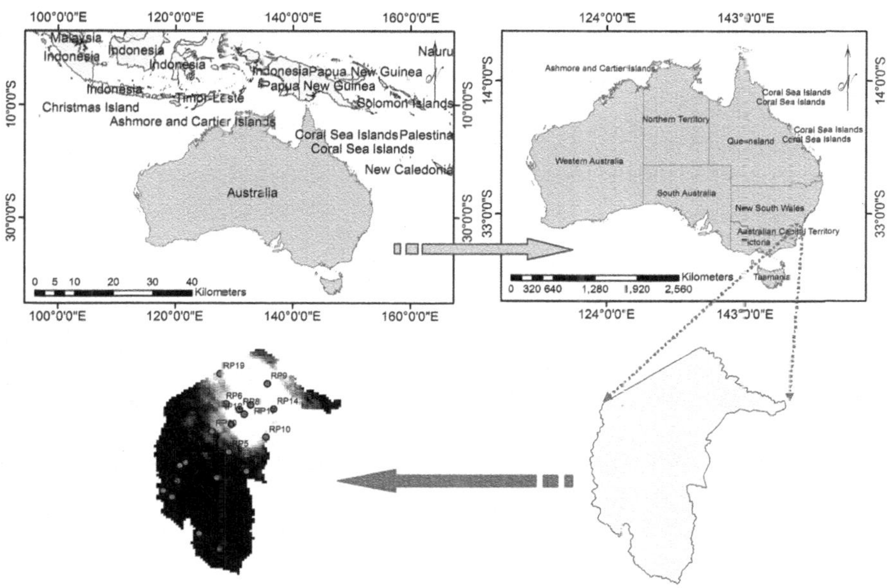

Fig. 10.1 Study area

directional axes, one stretching along Lake Burley Griffin and the other stretching from north-eastern Anzac Parade to the foot of the Mount Ainslie. Settlement hierarchy of the urban areas of Canberra is organized in a chronological order of districts, city hubs, assembly hubs, native fringes and rural community (Fig. 10.1).

10.3 Materials and Methodology

10.3.1 Spatial Analysis: Estimation of Detected Area Under Threshold Frequency

The lighted zone and their frequency distributions size are detected through frequency threshold to quantify the light detected area and the adjoining light polygons and the regularity of centroid for every dataset, and the frequency of dataset is from 1992 to 2012, which is booked at diverse edge. Using ArcGIS calligraphies, we planned the numeral of sunlit polygons and their extent and outside for both onsets at 10% breaks among 0% and 90% symmetry. The longitude and latitude tones were then bestowed to the centroids of each concerning lighted magnitude. For enumerating the communication amid size and regularity of recognition, we calculated the finding incidence at the centroid of each polygon at a different interval. This was finished by coalescing the inventive light incidence datasets with the case comprehending the longitude and latitude directs of the centroids of the attached bright polygons. By the system, every uncovering regularity value agreeing with connecting lighted area is dispensed to the area centroid. In several cases, people find undeviating distance which is more instinctual than expanse, and we generally embody the size of each irregularly fashioned neighbouring lighted area as comparable circular span defined.

10.3.2 Urban Built-Up Area Extraction

ConstrutION of artificially lighted urban areas provides the radiance values for corresponding DN values which surrounds the different values. In other different revisions, a non-set assessment was regularly assistance to division of urban areas on DMSP-OLS documents (Imhoff et al. 1997; He et al. 2006; Liu et al. 2012; Small et al. 2005). Larger than the different threshold values which is considered as the building part of the different urban areas and the corresponding pixel values is the inception value for that particular areas. With the otherwise use of DMSP-OLS data overhead and in lieu of this, the NPP-VIIRS data can be used which is the best suited method as used to extract the onset value of urban modelling. The optimization of different threshold values for urban mapping causing extraction of night-time light data is also a provident of government statistical data (He et al. 2006). The dawn rate that created slightest difference among image-derived assessment and arithmetic figures was scheduled as the foundation for assessment of night-time light-based urban mapping. The high-accuracy implementations based on this method are the more accurate based on the DMSP-OLS data night-time light data-based urban mapping (Shu et al. 2011). This revision includes the adoption of a comparable tactic to decide the peak beginning threshold value of residential urban area abstraction for every one different city from two types of nocturnal graceful NTL-based data images. However, the main objectives to find different diplomatic process are the following: Firstly, the extraction of different types of night-time light images for every municipal (city) was mined from the different international data cliques by using a façade polygon of the managerial periphery. Secondly, an onset of the tiniest or radiance value was used to divide the images hooked on built-up non-urban area.

The extracted area and the different statistical data which gives an absolute value grouping of the differences between the threshold values of different statistical data were chronicled. The recapitulation of different threshold values such a process increases the optimum value of night-time light data and reaches the extreme pixel value of the image. The difference in the maximum and minimum value of the NTL-based data provides the threshold value for the built-up extraction of the urban areas.

10.3.3 Inter-calibration of the Night-Time Light Dataset

For the different dramatic mapping picture of urbanization through large city light monitoring, the use of night-time images of the earth shows visible light secretions which have become a great interest in recent times (Small et al. 2011). The night-time light data do not quantify land shelter unswervingly, but most of the non-urbanized places which have light during night-time, considering various features, has been shown a many interrelated with the populace density (Lo 2001). Nevertheless, but the DMSP-OLS night-time light series dataset is not being able to rightly be used for studying suburbanization due to the absenteeism of on-board setting in the DMSP-OLS dataset (Miller et al. 2009; Liu et al. 2012). The data used in this paper (1992–2012), which is obtained from different NTL satellites contains no firm inter-calibration. Direct utilization of NTL data cannot be done due to quotation the undercurrents of worldwide and local built-up urban increase because of the deficiency of comparability and endurance revenue of NTL data. To dominate the inconsistencies and instruct confusability to the NTL dataset, the NTL data needs to be inter-calibrated which is indispensable to strict use. The stability and various comparisons to these data of NTL time series data, which we take from 1992 to 2012 for DMSP-OLS data, are monitored by a statistical data formulation of second-order regression algometry that improves data solidity (Elvidge et al. 2009).

In this study, we have created 20 random points in each year data of regional area within a unit loop so that every dataset can cover a unique and equally distributed data. The total random points taken are 20, and a sample version of this pictorial view is given below in the figure (Fig. 10.2).

The DN (digital number) values from each random point are extracted using extract values from point calibration. And the different sets of a DN value from every year data (1992–2012) are extracted. The random points which extracted the values were then sampled, and by using the second-order polynomial equation, the regression value from each set is extracted. These illustration data of Australian Capital Territory (ACT) were extracted from each year data (1992–2012) (ACT, urban area and semiurban). In the process of swotting the night-time light series dataset, it has initiated that the data from the satellite −2003 (F152003) apprehended with extreme numeral of stricken pixels in the region of Australian Capital Territory. From the year 2003, the main urban area begins to encompass in the form of different sides. The extracted various DN values of the raster image are shown in Table 10.1.

The data inter-calibration of DMSP-OLS satellite was steered using the regression formula of second-order polynomial regression (In Eq. 10.1), relaying on capricious as the reference image and self-regulating inconstant as images to be calibrated.

$$\mathrm{DN}_{\mathrm{calibrated}} = a\,X\,\mathrm{DN}^2 + b\,X\,\mathrm{DN} + c \qquad (10.1)$$

The DN values which were extracted from 20 random points were inter-calibrated by using Eq. 10.1, where the reliant flexible values are used as the reference data to

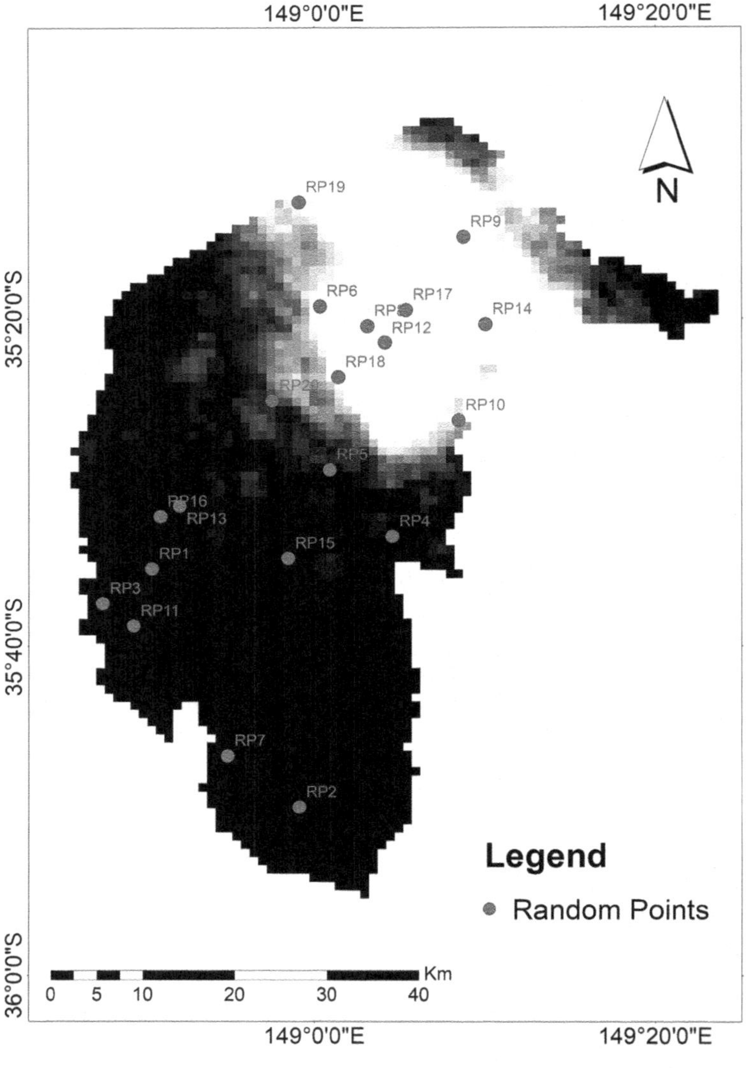

Fig. 10.2 Extracted Defence Meteorological Satellite Program (DMSP)/Operational Linescan System (OLS) night-time satellite imagery of Australian Capital Territory (ACT)

use the regression equation. The typical restrictions *a*, *b* and *c* were appraised with conventional slightest tetragonal relapse equation for specific NTL image dataset (Pandey et al. 2013) (Fig. 10.3).

The core-centred model of urban indices in two different circles shows that the centred area is how much expanded accordingly with the evaluation of time. The second image correlates the random points to extract the DN values, where DN 53–63 belongs in core mid area, 41–52 belongs to second circle and 0–40 belongs to

Table 10.1 The extracted pixel value from the random points

	DN values of randomly selected point							
Sample point	1992	1995	1998	2000	2003	2006	2009	2012
RP1	0.00	0.00	3.70	0.00	2.44	2.33	2.44	2.33
RP2	0.00	0.00	0.00	0.00	0.00	0.00	0.00	0.00
RP3	0.00	0.00	0.00	1.52	1.20	0.00	1.20	2.22
RP4	0.00	6.56	1.79	2.90	1.15	2.27	1.15	3.57
RP5	18.75	9.84	11.11	11.76	12.36	2.33	12.36	4.82
RP6	63.00	55.67	52.00	61.43	48.86	43.17	48.86	46.84
RP7	0.00	1.69	2.04	1.47	1.22	0.00	1.22	1.18
RP8	63.00	63.00	63.00	63.00	63.00	63.00	63.00	63.00
RP9	63.00	63.00	63.00	63.00	63.00	63.00	63.00	63.00
RP10	5.36	6.35	7.51	7.15	9.51	8.54	9.37	9.65
RP11	0.00	0.00	2.08	1.54	1.19	0.00	1.19	1.10
RP12	63.00	63.00	63.00	63.00	63.00	63.00	63.00	63.00
RP13	0.00	1.69	1.96	0.00	2.33	0.00	2.33	1.20
RP14	63.00	63.00	63.00	63.00	63.00	63.00	63.00	63.00
RP15	6.25	1.56	0.00	1.49	0.00	4.76	0.00	2.33
RP16	0.00	1.64	3.70	1.47	2.30	0.00	2.30	0.00
RP17	63.00	63.00	63.00	63.00	63.00	63.00	63.00	63.00
RP18	63.00	63.00	35.96	36.76	29.31	57.67	59.31	55.29
RP19	53.75	44.85	52.55	59.12	64.37	61.11	64.37	51.43
RP20	37.50	36.07	13.21	5.71	19.77	12.50	19.77	10.59

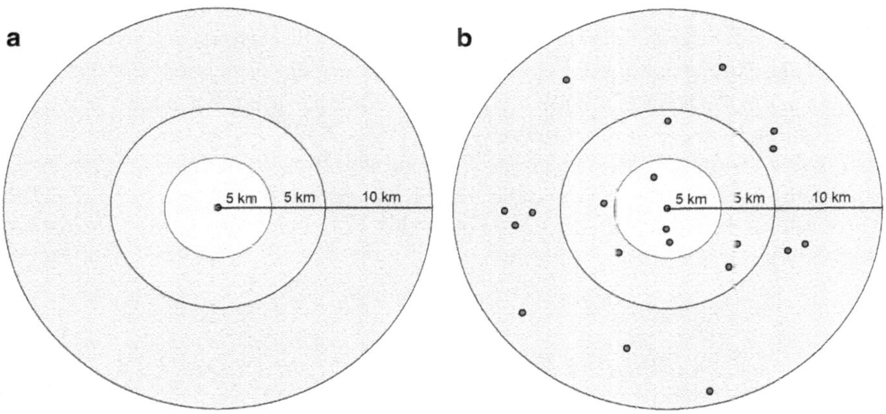

Fig. 10.3 A unit circle buffer zone of radius 20 km established by taking the centre of each built-up zone as the focal point of the constructed buffer, to compare easily the rate of expansion of urbanization

last circle. This is how about the random points show their significance in this core research.

10.3.4 Statistical Data for Measuring ULI

In this research, NTL time series study period (1992–2012) was separated into eight periods (1992–2012) with a hiatus of 3 years. F10 for 1992, F12 for 1995 and F16 for 2006 were particularly selected due to elevated coefficients than the erstwhile substitute satellite of F14 and F15 which was used for the year 1998, 2000 and 2003, respectively (Table 10.2). Demographic transition indicator like urbanization, economical population, economic growth, integration of industrial progression in the national gross domestic product (GDP) and built-up area are obtained from the Australian Capital Territory (ACT) (1992, 1995, 1998, 2003) (ABS 2017).

10.3.5 Frequency of Calibration Data

The change in lighted space is the number of adjoining lights for changed recognition regularity inceptions. Increasing threshold from 14% to 20%, the total lighted area monotonically regulates, but the number of adjoining lights primarily intensifies as large cities splinter from the data of DMSP-OLS. As inception value of light increases supplementary on the number of lights moderates as greater quantities of slighter or less frequently area are detected, the data of night-time lights are hampered. The following figures of different years which comprised two graphical representations—before calibrated and after calibrated—give the intensification of urbanization increase in the different years (1992–2012). A frequency threshold of 14% value resulting in the detection of real urbanization from the NTL time series data gives the individual pixel value for the DN ranges in the dataset 1992–2012. Threshold of 10% exploits the numerical value of lights in the 1994–2012 dataset (Fig. 10.4).

10.3.6 Inter-calibrated Resulted Urban Light Index (ULI)

In this study paper, the different versions of DMSP/OLS NTL data series were used where version 4 data of different satellites, viz., F10, F12, F14, F15, F16 and F18, is used as the coverage of the data is 149°E and 35°S. The bit of DMSP/OLS data is 6 bit, and the range of digital number for each pixel is 0–63. The concentration of the area's light unswervingly signposts assessment of this value. The zero value of DN in each pixel is from the non-light area. Pixel with 63 DN value is an inundation pixel, and much blizzards are instituted in the fundamental built-up urban area

Table 10.2 The DN inter-calibrating night-time lights regression evaluation

Satellite	Year	c	b	a	R^2	Second-order regression
F10	1992	0.0000497	0.0496	7.2711	0.795	Var1 = 7.2711 + 0.0496*x + 4.9752E-6*x^2
F12	1995	0.0000558	0.0557	7.1325	0.812	Var1 = 7.1325 + 0.0557*x + 5.5869E-6*x^2
F14	1998	0.0000491	0.049	7.5702	0.724	Var1 = 7.5702 + 0.049*x + 4.9136E-6*x^2
F15	2000	0.0000445	0.0444	7.7692	0.702	Var1 = 7.7692 + 0.0444*x + 4.4533E-6*x^2
F15	2003	0.0000497	0.0496	7.587	0.718	Var1 = 7.587 + 0.0496*x + 4.9749E-6*x^2
F16	2006	0.0000485	0.0484	7.5053	0.773	Var1 = 7.5053 + 0.0484*x + 4.8545E-6*x^2
F16	2009	0.0000465	0.0464	7.7683	0.702	Var1 = 7.7683 + 0.0464*x + 4.65E-6*x^2
F18	2012	0.0000477	0.0476	7.6998	0.713	Var1 = 7.6998 + 0.0476*x + 4.7726E-6*x^2

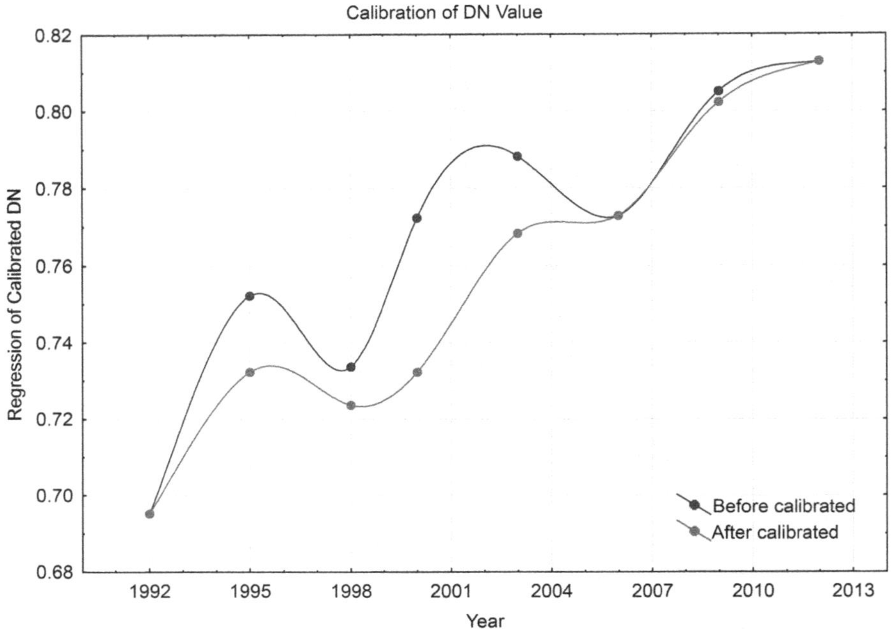

Fig. 10.4 The statistical model showing the variations in regression of calibrated DN values from 1992 to 2012 and a comparison between before and after calibrated DN values

(Elvidge et al. 1997; Letu et al. 2012). The grid data of the night-time light data images accounts the intensity value of the DN and latitudinal magnitude evidence of ACT. The different light areas which are growing or increasing are redirected in association with the marginal areas that are promising (Imhoff et al. 1997). This partakes in the consequence for the broadening of small reimbursements and intensifying the borders of large cities which are expanding and can be validated by NTL data. Therefore, these attributes are accustomed to paradigm expanse sunlit catalogues. The shrewdness formula for urban light index is (Yi et al. 2014):

$$\text{Urban Light Index}(\text{ULI}) = 100\,X \sum_{i=10}^{\text{maxDN}} \frac{\text{DN}_i}{\text{maxDN}} X \frac{C_i}{\text{sum}C} \tag{10.2}$$

The ULI indicates the urban light index, whereas DN_irefers to the calibrated DN value from Eq. 10.1. sum$C$ is the count of total DN values in a unit area. By Eq. 10.2, the value of urban light index for the different data taken from 1992 to 2012 is calculated, and finally the main outcome of urban sprawl map is determined with a final generalized value.

10.4 Results and Discussion

10.4.1 Light Detection Threshold Information Retrieval

The light data series of DMSP/OLS NTL have several important properties with some percentage threshold value and their discrepancy in size occurrence disseminations. The minimum inception seen in the 1992 dataset is the value of 14% and which increases as on 2000-year dataset that provides a useful blue-pencilled to exploit the number of the individual lights thus corresponding to diminution of insignificant different lights with the lessening of latitudinal data completed magnitude over around large cities of ACT. The size occurrence dispersals of the threshold value increase from the 14% value to the 20% over the sharp intensification in the value of threshold where a tiny value of light can distribute the DN value which perseveres to intensification of the threshold value. Another appearance of the tributary mode suggests more areas observed at lower than 14% to 20% of the peak value of threshold value of NTL data; interstitial prospering among reimbursements of the ACT moderately grows than individual small light areas. The maintenance of this allegation of the centroid of NTL data occurrences dissemination. As they show that the dissemination of specific decorations, which mountaintops around 14% beginning and dewdrops accelerating at lower thresholds (Welch 1980). The data observation of different years (1992–2012) recommends that furthermost of the lighted DN value (pixel value) within the threshold range of 14–20% belongs to light blooming on the sideline of the grander reimbursements. The areas which are distinguished in the regression value $R^2 < 50\%$ corresponds the total area which threshold value belongs within the 14–20% for the dataset 1992–2012 correspondingly (Small 2002). The stratum of 14% threshold value which significantly reduces the up-and-coming threshold DN ranges in many different tempering small settlements also. To retrieve the extreme information from the NTL time series data, a minimum of 14% threshold value for data collecting is used for significant guesstimate of ACT size. Comparable threshold value during 1992–2012 dataset occurs at 17% exposure in this study research for total lighted area and 14% for the non-lighted area.

10.4.2 Urban Light Indexation (ULI) of NTL Time Series Data

The hardness of urban area level extraction from any data is not easy based on an administrative area because of the great transformations whichever in spatial extent or populace of the dataset. The minimization of this problem levels the study of need of DMSP/OLS dataset which is thoroughly discussed in this paper. In the comparison of the different scale expansions of the urban areas with the total data, a process of fixing the whole data by a filtering through the ULI (urban light index) gives the distinct urbanization boundary from the entire data. Here the study area which comprises of 20 random points and the extracted DN values enhances the urban centre, and to show the entire urbanization, we use a perfect buffer zone for the whole random points that can retreat the entire DN value extraction easily. The given following figures show the increase in ULI value from the different datasets (1992–2012) in Australian Capital Territory (Fig. 10.5).

As the most developed region in the northern Australian Capital Territory (ACT), the Canberra region shows a centre of urbanization which is increasing in time extension. The increasing evolution of the urbanization, the population and economy is the main factor for the growing activities of the ACT in the Canberra region. The system for the northern Australian Capital Territory is integrating towards a wide range in the year from 2000 to 2012 in comparison with the year from 1992 to 2000. The main axis of growing in the southern ACT region is Canberra of ACT. The other cities including Canberra are Acton, Parkes, Melba, Hall, etc. The demography of the ACT shows the actual integration of urbanization in comparison to the population increase with different respective years.

10.5 Discussion

The suburbanization development parades palpable terrestrial altitudinal powers. DMSP/OLS dark light data was converted for efficient expansion inquiry than the rest daytime satellite data for the succeeding two explanations: the first one is the altitudinal tenacity of DMSP/OLS NTL data is 1–1.2 km, and anthropological domestic expanses are essential for provincial suburbanization inquiry; on the other hand, NTL data is alongside a dusky family, and the night light data metaphors confiscate the troubles of terminated evidence on city surface settings in afternoon metaphors so they can thoughtfully internment the framework of the conurbations, which is advantageous for confiscating urban astrophysical suggestion. The present paper utilizes ULI which is instituted to scrutinize along with quantitative estimation of the speedy and method of suburbanization by incorporating DMSP/OLS NTL Data from the period from 1992 to 2012. The municipal well-lit concentration is incapable to replicate the agglomeration flat of a city's inhabitants and pecuniary doings, whereas dissimilarity in Metropolitan Dark Space records the trace of municipal space enlargement. For associating levels of suburbanization in unlike cities under the identical average development of unit circle assessment were recognized in this broadside to painstakingly investigate the suburbanization process in ACT in gently years (1992–2012). Substantial variations are found in reverted cities and populous without slanting comparability predisposed by influences as antiquity, position and policy implementation. Hence, suburbanization research is of vital significance in order to ascertain the progression and advancement of urbanizations. Unit circle model up to certain hefty magnitude eradicates the deficiency of urbanization exploration with the directorial area as the appraisal unit. This is prototypically relevant for estimating research of expansion levels midst cities

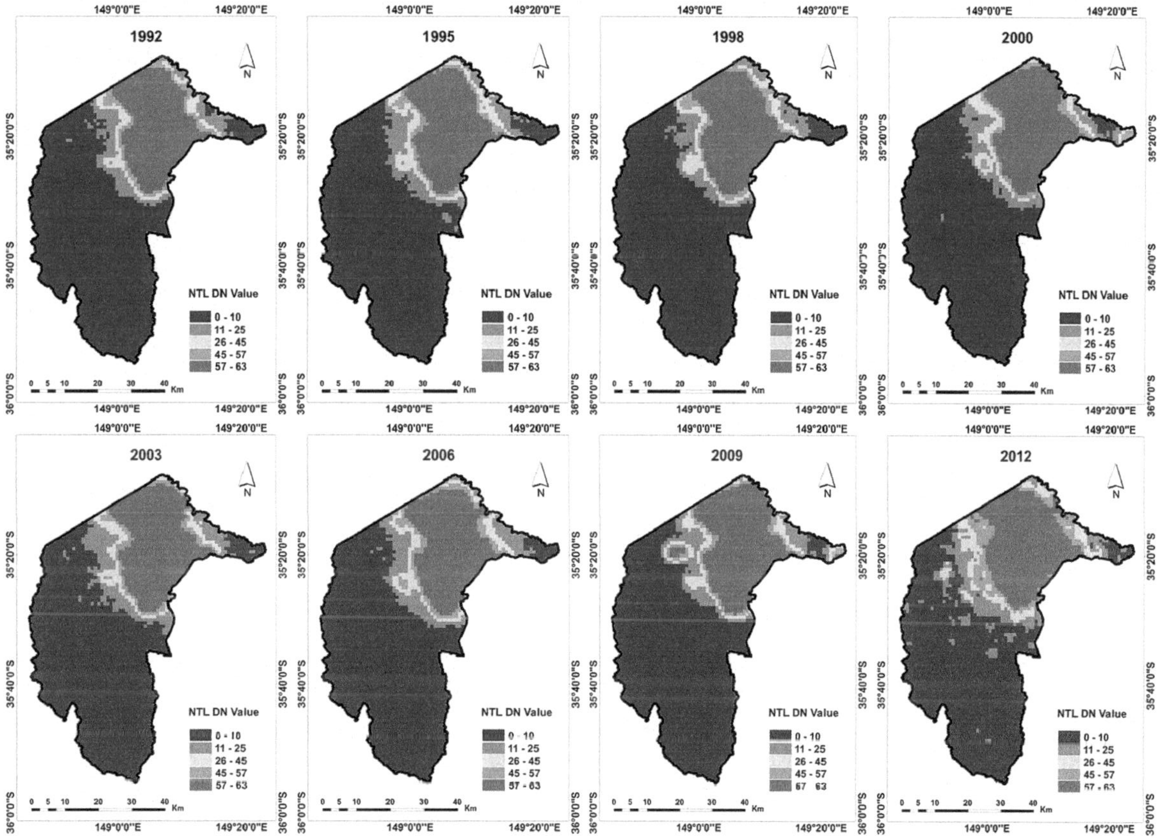

Fig. 10.5 Showing the trend expansion of built-up zone through Urban Light Indexation (ULI) of NTL DMSP/OLS time series dataset in different year variations

indoors as a province. This urban light index (ULI) is gauged to study the recompenses of weaknesses and the attainability of studies on urbanization. Results showed wider application of ULI in monetary needles, extent of built-up areas in cities and urban inhabitants.

Arrangement of urban light space as times gone by of expansion which effectively imitate the detached landscapes of town space fruition. There is rectilinear bearing among the urban inhabitants, monetary movement, built-up cover and ULI linearity (the correlation coefficients with 70–75%). This study embryonically utilized urban light space for the first time. In agreement with diverse supplies in urban light concentration, the urban light space is classified into central urban area (CUA) (DN 57–63), urban fluorescence space (DN 15–25), suburban changeover zone (SCZ) (DN 45–57) and outlying area (OA) (DN 25–45). From the four light spaces observed built-up types, time-based three-dimensional growth patterns from the time-based and longitudinal dissimilarity of the ACT are given due to consideration of the present. Outcomes derived by analysing built-up cover and ULI are closely associated in course of urbanization with spatial pattern of cities and their geographical distribution.

References

Australian Bureau of Statistics (ABS) (2017) Population projections, Australia, 2016 (base) to 2101. ABS cat. no. 3222.0. Canberra: ABS

Cinzano P, Falchi F, Elvidge CD, Baugh KE (2001) The artificial sky brightness in Europe derived from DMSP satellite data. Preserving the astronomical sky, pp 95–102

Dewan AM, Yamaguchi Y (2009) Land use and land cover change in Greater Dhaka, Bangladesh: using remote sensing to promote sustainable urbanization. Appl Geogr 29:390–401

Elvidge CD, Baugh KE, Kihn EA, Kroehl HW, Davis ER (1997) Mapping city lights with nighttime data from the DMSP operational linescan system. Photogramm Eng Remote Sens 63(6):727–734

Elvidge CD, Baugh KE, Dietz JB, Bland T, Sutton PC, Kroehl HW (1999) Radiance calibration of Dmsp-Ols low-light imaging data of human settlements. Remote Sens Environ 68(1):77–88

Elvidge CD, Tuttle BT, Sutton PC, Baugh KE, Howard AT, Milesi C, Bhaduri B, Nemani R (2007) Global distribution and density of constructed impervious surfaces. Sensors 7(9):1962–1979

Elvidge C, Ziskin D, Baugh K, Tuttle B, Ghosh T, Pack D, Erwin E, Zhizhin M (2009) A fifteen year record of global natural gas flaring derived from satellite data. Energies 2:595–622

Feng Y, Kugler J, Zak PJ (2002) Population growth, urbanisation and the role of government in China: a political economic model of demographic change. Urban Stud 39:2329–2343

Fujita M, Krugman P, Mori T (1999) On the evolution of hierarchical urban systems. Eur Econ Rev 43(2):209–251

Han J, Hayashi Y, Cao X, Imura H (2009) Application of an integrated system dynamics and cellular automata model for urban growth assessment: a case study of Shanghai, China. Landsc Urban Plan 91:133–141

He CY, Shi PJ, Li JG, Chen J, Pan YZ, Li J, Zhuo L, Toshiaki I (2006) Restoring urbanization process in China in the 1990s by using non-radiance calibrated DMSP/OLS nighttime light imagery and statistical data. Chin Sci Bull 51:1614–1620

He CY, Ma Q, Li T, Yang Y, Liu ZF (2012) Spatiotemporal dynamics of electric power consumption in Chinese Mainland from 1995 to 2008 modeled using DMSP/OLS stable nighttime lights data. J Geogr Sci 22:125–136

Henderson M, Yeh ET, Gong P, Elvidge C, Baugh K (2003) Validation of urban boundaries derived from global night-time satellite imagery. Int J Remote Sens 24(3):595–609

Huang X, Schneider A, Friedl MA (2016) Mapping sub-pixel urban expansion in China using Modis and Dmsp/Ols nighttime lights. Remote Sens Environ 175:92–108

Imhoff ML, Lawrence WT, Stutzer DC, Elvidge CD (1997) A technique for using composite DMSP/OLS 'City Lights' satellite data to map urban area. Remote Sens Environ 61:361–370

Letu H, Hara M, Tana G, Nishio F (2012) A saturated light correction method for Dmsp/Ols nighttime satellite imagery. IEEE Trans Geosci Remote Sens 50(2):389–396

Li SM (2004) Population migration and urbanization in China: a comparative analysis of the 1990 population census and the 1995 national one percent sample population survey. Int Migr Rev 38:655–685

Li Q, Linlin L, Weng Q, Xie Y, Guo H (2016) Monitoring urban dynamics in the Southeast U.S.A. using time-series Dmsp/Ols nightlight imagery. Remote Sens 8(7):578

Liu ZF, He CY, Zhang QF, Huang QX, Yang Y (2012) Extracting the dynamics of urban expansion in China using DMSP-OLS nighttime light data from 1992 to 2008. Landsc Urban Plan 106:62–72

Liu X, Guohua H, Bin A, Xia L, Shi Q (2015) A normalized urban areas composite index (Nuaci) based on combination of Dmsp-Ols and Modis for mapping impervious surface area. Remote Sens 7(12):17168–17189

Lo CP (2001) Modelling the population of China using DMSP operational line scan system nighttime data. Photogramm Eng Remote Sens 67:1037–1047

Lo CP (2002) Urban indicators of China from radiance-calibrated digital DMSP-OLS nighttime images. Ann Assoc Am Geogr 92:225–240

Lo CP (2010) Urban indicators of China from radiance-calibrated digital DMSP-OLS nighttime images. Ann Assoc Am Geogr 92(2):225–240

Lu D, Tian H, Zhou G, Ge H (2008) Regional mapping of human settlements in southeastern China with multisensor remotely sensed data. Remote Sens Environ 112(9):3668–3679

Ma T, Zhou C, Pei T, Haynie S, Fan J (2012) Quantitative estimation of urbanization dynamics using time series of Dmsp/Ols nighttime light data: a comparative case study from China's cities. Remote Sens Environ 124:99–107

Mertes CM, Schneider A, Sulla-Menashe D, Tatem AJ, Tan B (2015) Detecting change in urban areas at continental scales with Modis data. Remote Sens Environ 158(0):331–347

Milesi C, Elvidge CD, Nemani RR, Running SW (2003) Assessing the impact of urban land development on net primary productivity in the southeastern United States. Remote Sens Environ 86:401–410

Miller JD, Knapp EE, Key CH, Skinner CN, Isbell CJ, Creasy RM, Sherlock JW (2009) Calibration and validation of the Relative differenced Normalized Burn Ratio (RdNBR) to three measures of fire severity in the Sierra Nevada and Klamath Mountains, California, USA. Remote Sens Environ 113:645–656

Owen TW (1998) Using DMSP-OLS light frequency data to categorize urban environments associated with US climate observing stations. Int J Remote Sens 19:3451–3456

Pandey B, Joshi PK, Seto KC (2013) Monitoring urbanization dynamics in India using DMSP/OLS night time lights and SPOT-VGT data. Int J Appl Earth Obs Geoinf 23:49–61

Runkui L, Zhipeng L, Wenju G, Wenjun D, Qun X, Xianfeng S (2014) Diurnal, seasonal, and spatial variation of PM2.5 in Beijing. Sci Bull 60(3):387–395

Schneider A, Friedl MA, Potere D (2010) Mapping global urban areas using Modis 500-M data: new methods and datasets based on 'urban ecoregions'. Remote Sens Environ 114 (8):1733–1746

Shi K, Huang C, Yu B, Yin B, Huang Y, Wu J (2014) Evaluation of NPP-VIIRS night-time light composite data for extracting built-up urban areas. Remote Sens Lett 5(4):358–366

Shu S, Yu B, Wu J, Liu H (2011) Methods for deriving urban built-up area using night- light data: assessment and application. Remote Sens Technol Appl 26:169–176

Small C (2002) A global analysis of urban reflectance. In: D.M.a.F.S.-E.C. Jurgens (ed) Proceedings of the third international symposium on remote sensing of urban areas. Istanbul, Turkey

Small, C. (2003). High spatial resolution spectral mixture analysis of urban reflectance. Remote Sens Environ, 88(1–2), 170–186

Small C (2005) A global analysis of urban reflectance. Int J Remote Sens 26(4):661–681

Small C, Pozzi F, Elvidge CD (2005) Spatial analysis of global urban extent from DMSPOLS night lights. Remote Sens Environ 96:277–291

Small C, Elvidge CD, Balk D, Montgomery M (2011) Spatial scaling of stable night lights. Remote Sens Environ 115:269–280

Su Y, Chen X, Wang C, Zhang H, Liao J, Yuyao Y, Wang C (2015) A new method for extracting built-up urban areas using Dmsp-Ols nighttime stable lights: a case study in the Pearl River Delta, Southern China. Gisci Remote Sens 52(2):218–238

Sutton PC (2003) A scale-adjusted measure of "Urban sprawl" using nighttime satellite imagery. Remote Sens Environ 86:353–369

Sutton P, Roberts D, Elvidge C, Baugh K (2001) Census from Heaven: an estimate of the global human population using night-time satellite imagery. Int J Remote Sens 22(16):3061–3076

Tan L (2017) Australian states and territories: New South Wales, 2017, Geography & Environment/ Geo/Env: Countries/Continents/Geo/Env: CC – Australia/Oceani. ISBN 9780994624710

Welch R (1980) Monitoring urban-population and energy-utilization patterns from satellite data. Remote Sens Environ 9(1):1–9

Welch R, Zupko S (1980) Urbanized area energy-utilization patterns from Dmsp data. Photogramm Eng Remote Sens 46(2):201–207

Wigmore L (1971) Canberra: history of Australia's national capital. Dalton Publishing Company. ISBN 0-909906-06-8

Yi K, Tani H, Li Q, Zhang J, Guo M, Bao Y, Wang X, Li J (2014) Mapping and evaluating the urbanization process in Northeast China using DMSP/OLS nighttime light data. Sensors 14(2):3207–3226

Zhang J, Zhu T, Kipen H, Wang G, Huang W, Rich D et al (2013) Cardiorespiratory biomarker responses in healthy young adults to drastic air quality changes surrounding the 2008 Beijing Olympics. Res Rep Health Eff Inst 174:5–174

Chapter 11
Modelling Spatial Patterns of Urban Growth in Pune Metropolitan Region, India

Bhartendu Pandey, P. K. Joshi, T. P. Singh, and A. Joshi

Abstract Explaining urban growth patterns is a fundamental need to understand the recent rapid urbanization globally. This study identifies geographic features explaining the spatial patterns of urban land expansion (ULE) in the rapidly urbanizing Pune metropolitan region (India). ULE maps were derived from Landsat Thematic Mapper and Operational Land Imager images using support vector machine (SVM) classification. Relation between geographic features and spatial patterns of ULE was analyzed using statistical modelling including ordinary least squares (OLS) regression, spatial lag model (SLM), spatial error model (SEM), and geographically weighted regression (GWR). SEM specification best modeled ULE patterns. High density of existing urban areas is identified to negatively affect ULE, suggesting dominant dispersed urban growth. In addition, proximity to special economic zones and transportation infrastructure explains multicentric growth in the region. GWR model was identified inappropriate due to the presence of high local collinearity. Models accounting for spatial dependencies are recommended while studying ULE patterns.

Keywords Urban growth · Spatial lag model · Spatial error model · Geographically weighted regression · Special economic zone

---

B. Pandey
School of Forestry & Environmental Studies, Yale University, New Haven, USA

P. K. Joshi (✉)
School of Environmental Sciences, Jawaharlal Nehru University, New Delhi, India

T. P. Singh
Symbiosis Institute of Geo-informatics, Symbiosis International University (SIU), Pune, India

A. Joshi
Department of Statistics, Kumaun University, Almora, India

P. Kumar et al. (eds.), *Applications and Challenges of Geospatial Technology*,
https://doi.org/10.1007/978-3-319-99882-4_11

11.1 Introduction

Globally the rate of urban land-use change is faster than urban population growth rate (Angel et al. 2005). China and India are the two countries where recent rapid urban land-use change has shown similar trends (Zhang and Seto 2011). Contemporary urbanization is unprecedented in the rate and scale of demographic, economic, and urban land-use changes (Taubenböck et al. 2009; Zhang and Seto 2011). In India, retrospective studies of urban land-use and land-cover (LULC) changes—that essentially measured changes in built-up or impervious surface area from satellite image analysis—have shown that the rate of urban land expansion (ULE) is higher than that of demographic transitions (Sudhira et al. 2004; Farooq and Ahmad 2008; Jat et al. 2008). ULE has direct effects on regional climate (Pathirana et al. 2014), agriculture (Fazal 2001), biological diversity (Concepción et al. 2015), and other systems. Therefore, considering these effects, there is an urgency to understand the magnitude, rate, and pattern of urban land-use changes in cities with rapid growth. Hereafter, the terms ULE and urban growth are used interchangeably to denote urban land-use changes measured as changes in built-up or impervious surface area.

Several studies have examined the magnitude, rate, and pattern of ULE using remote sensing (RS) and geographic information system (GIS). A conventional approach involves temporal comparison of landscape ecology metrics for discrete administrative units (Sudhira et al. 2004; Schneider and Woodcock 2008; Taubenböck et al. 2009; Kowe et al. 2015; Triantakonstantis and Stathakis 2015). The relative influences of different causal variables on these metrics are then examined using multivariate regression (Sudhira et al. 2004; Jat et al. 2008). Such an approach may have three limitations. First, regression models estimated for a spatially relevant process such as ULE using conventional statistical methods are prone to bias in error-variance and may have modified significance levels or overestimated R^2 (Anselin and Griffith 1988). Second, aggregated measurements of ULE based on discrete administrative units result in information loss and may add to statistical bias (Openshaw 1984; Fotheringham and Wong 1991). Third, ULE process is assumed to be spatially independent and non-stationary (Luo and Wei 2009; Luo et al. 2008). Recently, a gradient analysis approach has been proposed in which the study area is divided into concentric circles with radius varying with respect to the city center (Ramachandra et al. 2012). The algorithm then computes spatial metrics for each concentric circle to reveal broad spatial patterns of ULE. Although this approach is prone to the modifiable areal unit problem (MAUP) when aggregating LULC data into radial geometries (Openshaw 1983), it may be used to study the spatial patterns of ULE relative to other geographic features, such as transportation corridors, industrial centers, and others. More recently, statistics-based spatial clustering approaches have been utilized to study urban land-use change hotspots (Ganguly et al. 2016). In summary, the conventional framework to study ULE embraces its cogent measurement but has ignored explaining its relation with geographic features.

In addition to measuring ULE, scholars have modeled ULE using different techniques, including cellular automata-based models (Moghadam and Helbich 2013; Al-sharif and Pradhan 2015) and machine learning algorithms (Linard et al. 2013). While these techniques are robust and important for simulation and prediction, these do not directly determine how different geographic features explain the spatial patterns of ULE relative to each other. Local to regional geography plays a significant role in shaping urban growth. In fact, geographic features have been conventionally used as proximate driver variables in several dynamic models of urban growth to estimate spatially explicit urban growth probabilities (Linard et al. 2013; Moghadam and Helbich 2013; Al-sharif and Pradhan 2015; Mondal et al. 2016), which are pertinent for urban land-use change science and policy studies. Therefore, identifying geographic determinants of urban growth patterns is highly relevant.

Multivariate statistical techniques can study the relationships between ULE and geographic features (Cheng and Masser 2003; Hu and Lo 2007; Luo et al. 2008; Luo and Wei 2009; Shafizadeh-Moghadam and Helbich 2015). Classical multivariate statistical models, however, assume spatial independence and stationarity (Yu 2006; Luo and Wei 2009; Lafazani and Lagarias 2016). Due to decentralized urbanization in India and elsewhere (Luo and Wei 2009; Mondal et al. 2015; Shafizadeh-Moghadam and Helbich 2015), geographic features explaining spatial patterns of ULE are likely to show spatial dependency and non-stationarity. In this context, spatial regression tools may be used to study the spatial patterns of ULE. Previous studies have considered global and local multivariate regression models to model urban growth patterns. A majority of these studies concluded that local multivariate regression models or geographically weighted regression (GWR) models outperform local models (Luo and Wei 2009; Mondal et al. 2015; Shafizadeh-Moghadam and Helbich 2015). However, in local regression, dealing with local collinearity between explanatory variables remains a challenge (Wheeler and Tiefelsdorf 2005; Lu et al. 2014). More recently spatial econometric models such as spatial lag models (SLM) and spatial error models (SEM) have been used to model ULE (Zhang et al. 2013; Zeng et al. 2015). These studies suggest that SLM and SEM outperform global regression models estimated using ordinary least squares (OLS) regression. Given that the aforementioned regression models deal with spatial effects and spatial heterogeneity differently, it is worth exploring how each of these modelling tools identifies geographic determinants of ULE patterns.

The overarching goal of this paper is to propose a statistical regression modeling-based framework to analyze the spatial patterns of urban growth with a focus on geographic determinants. Here, the emphasis is on variables influencing spatial patterns of ULE in rapidly urbanizing regions considering decentralized urbanization and importance for place-specific planning strategies. Specific research objectives of this case study are, first, to understand the capability of spatial regression to study the spatial patterns of ULE and, second, to identify key contemporary geographic determinants of ULE within the study area.

11.2 Study Area and Data

Pune metropolitan region (PMR) with a total area of 1870 km^2 was selected as the study site. Pune district is located in the State of Maharashtra in India (Fig. 11.1). The district encompasses Pune metropolitan region or Pune urban agglomeration which is currently the eighth largest—in terms of population—of all metropolitan regions in India (Census of India 2011). There is significant urbanization and economic activity in the metropolitan region. This region is an industrial hub known for manufacturing, especially automotive, and more recently service industries (e.g., information technology (IT) and biotechnology). High job and economic opportunities in the region have caused significant population influx (Pimpri-Chinchwad Municipal Corporation 2008; Pune Municipal Corporation 2008).

The study site has a heterogeneous landscape characterized by an undulating terrain (minimum, maximum, and mean elevations are 450, 1150, and 625, respectively). Furthermore, it has a characteristic polycentric urban growth pattern (Taubenböck et al. 2009; Kantakumar et al. 2016); however, it is not clear which underlying geographic features explain the polycentricism.

The datasets used in this study are a combination of official and open-source resources. This study largely relies on satellite images to identify ULE and Survey of India (SOI) toposheets for information on geographic features such as transportation networks, streams, water bodies, and protected forests (Fig. 11.1). Geographic features digitized from SOI toposheets were edited using Google Earth™ images and Google Maps™ data. Addresses of universities and special economic zones (SEZs) were determined from government websites and later geocoded using Google Maps™ Geocoding service. Slope parameter was calculated from ASTER

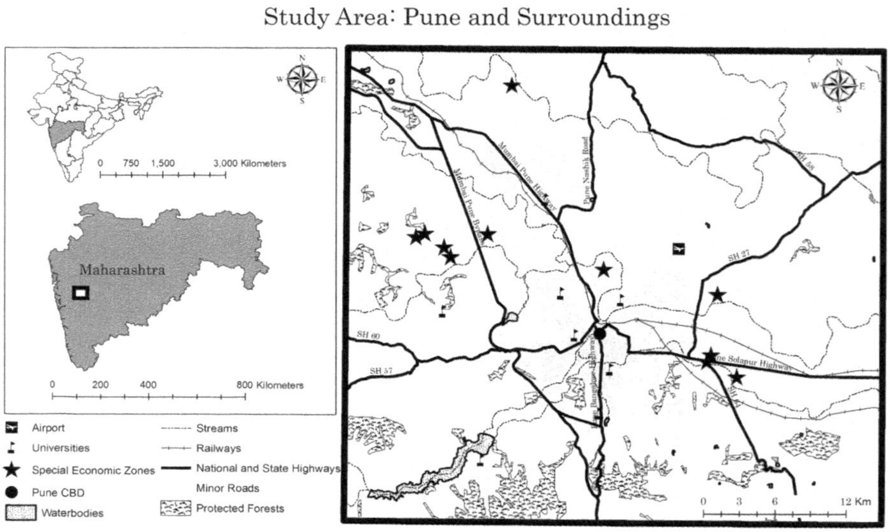

Fig. 11.1 Location map of the study area and the vector database generated

Global Digital Elevation Model (GDEM) Version 2. Due to known inaccuracies in the GDEM product, the digital elevation model (DEM) was first processed by identifying abrupt depressions in elevation values. Later the values for identified depressions were interpolated by calculating the median value in a 5 × 5 window. Slope parameter was calculated from this depression-less DEM and was used during satellite image classification and in subsequent modeling exercise.

11.3 Methods

The approach undertaken in this study comprised of two stages: (a) satellite image classification to identify urban LULC and ULE and (b) model development and interpretation. Figure 11.2 summarizes the methodology used in the case study.

11.3.1 Satellite Image Classification

Two Landsat 5 TM and one Landsat 8 OLI images with ~0% cloud cover were downloaded, each for March months of the year 2000, 2010, and 2015. For each year, all the bands, except thermal, were converted to top-of-atmosphere reflectance

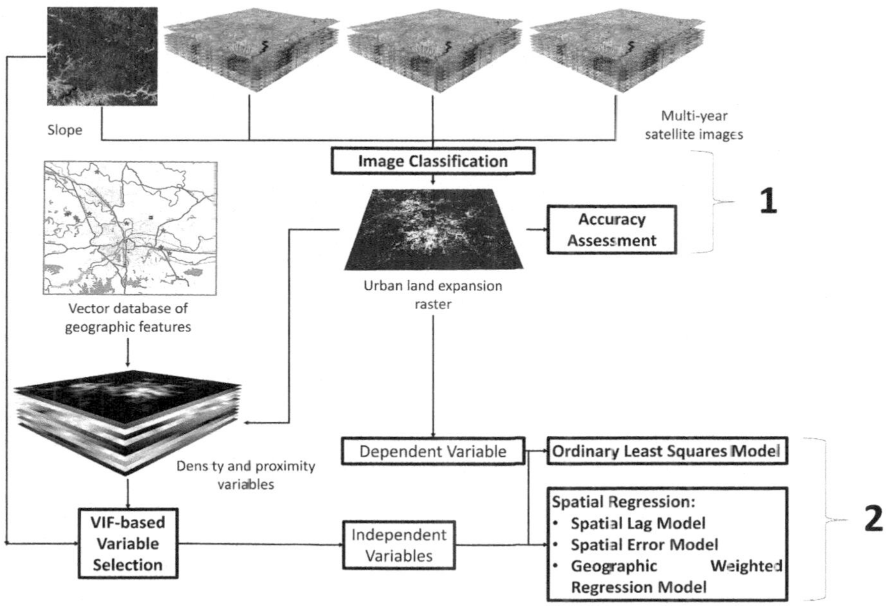

Fig. 11.2 Overall methodology to study spatial patterns of urban growth and identify geographic determinants. Note the two stages involved in our analysis: (1) satellite image classification to identify urban land expansion and (2) model development and interpretation

images that were used along with the slope image to classify LULC into four classes: water, vegetation, bare land, and urban. Training sites for each of these classes were randomly collected from Google Earth™ and raw Landsat images. Next, a support vector machine (SVM) classifier with radial basis function (RBF) kernel was used to perform image classification using the e1071 package in R statistical software v3.0.0 (Dimitriadou et al. 2006). Here, the γ parameter of the RBF kernel was optimized using a ten-fold cross-validation procedure, and the optimized classifier was used to predict LULC classes for the entire study area.

Accuracies of classified images were assessed using 200 random points generated with equalized random sampling. These random points were labeled with reference from Google Earth™ and raw satellite images. These random points along with the class distributions were used to compute an unbiased confusion matrix from which overall accuracies were computed (Pontius Jr and Millones 2011). The classified images showed overall accuracies of 88.90%, 94.19%, and 93.62% for the year 2000, 2010, and 2015, respectively. The classified images were then used to compute ULE raster with four classes: nonurban, urban (2000), urban growth (2000–2010), and urban growth (2010–2015). During ULE raster generation, a pixel identified as urban was also considered urban in the subsequent years. Again for accuracy assessment, 400 random points were generated with equalized random sampling and labeled using temporal Google Earth™ images. Here, an overall accuracy of 67.44% was obtained from the unbiased confusion matrix.

11.3.2 Model Development

11.3.2.1 Dependent and Explanatory Variables

Dependent and explanatory variables were computed at two resolutions, i.e., 300 m and 750 m. Here, the dependent variable represented urban growth in a continuous scale ranging from 0 to 100. Specifically, urban growth class from the ULE raster was aggregated to coarser spatial resolutions, and total pixel counts were recorded for each aggregated pixel. Thirteen explanatory variables were used in regression modeling (Table 11.1). These variables, including the dependent variable, will be referenced in the remaining text with abbreviations listed in Table 11.1. Previous studies have indicated topography as one of the important factors that limit urban growth; relatively flat regions are conducive to ULE (Jantz et al. 2004; Li et al. 2013). The slope parameter was therefore computed at 300 and 750 m spatial resolutions and used to model topographic constraints to urban growth.

Socioeconomic factors typically drive ULE. For instance, population and economic growth are usually positively associated with urban expansion. Often data for these variables is collected for administrative divisions and is too coarse compared to per-pixel representation of geographic features. Thus, due to the lack of socioeconomic data at the required spatial and temporal scale, proximity to CBD (Prox_CBD), SEZ (Prox_SEZ), and the university (Prox_Univ) were included as

Table 11.1 Dependent and explanatory variables used in regression modelling

Variable	Variable type	Abbreviation
Urban growth (2000–2010)	Dependent	UG_00_10
Urban growth (2010–2015)	Dependent	UG_10_15
Slope	Explanatory	Slope
Proximity to Pune CBD	Explanatory	Prox_CBD
Proximity to SEZs	Explanatory	Prox_SEZ
Proximity to universities	Explanatory	Prox_Univ
Proximity to minor roads	Explanatory	Prox_Road
Proximity to national and state highways	Explanatory	Prox_High
Proximity to railways	Explanatory	Prox_Rail
Proximity to airport	Explanatory	Prox_Airp
Proximity to protected forests	Explanatory	Prox_Forest
Proximity to streams	Explanatory	Prox_Str
Proximity to water bodies	Explanatory	Prox_Water
Density of existing urban areas (2000)	Explanatory	Den_Urb_00
Density of existing urban areas (2010)	Explanatory	Den_Urb_10
Density of vegetated areas	Explanatory	Den_Veg

explanatory variables. Clearly, these variables will not capture the actual socioeconomic dynamics but will reveal the spatial relationships between socioeconomic centers and urban growth.

Physical contiguity and connectivity of urban areas through transportation networks are usually the major determinants of ULE (Jantz et al. 2004; Hu and Lo 2007; Li et al. 2013; Linard et al. 2013). Euclidean distance raster for transportation features such as road (Prox_Road), highway (Prox_High), railway (Prox_Rail), and airport (Prox_Airp) were computed to represent connectivity levels. In addition, an urban density raster (Den_Urb) was computed to represent contiguities to urban areas.

Aesthetics and recreation is also an important factor that promotes urban growth (Clark et al. 2002). Here proximity to forests (Prox_Forest), water bodies (Prox_Water), and streams (Prox_Str) was used to evaluate whether these geographic features affect urban growth positively or negatively in the study area. Furthermore, a density of vegetation raster (Den_Veg) was computed to assess whether contiguity to vegetated areas is a determinant of ULE in the study area.

All the proximity and density variables were computed in Quantum GIS v1.8.0 at spatial resolutions of 300 m and 750 m. The euclidean distance rasters were transformed such that a location (i) closer to the geographic features has a higher value and finally were rescaled to a range of 0 to 100 (Eq. 11.1). Den_Urb raster was computed by calculating stable urban pixel count in the ULE image at aggregated spatial resolutions (i.e., 300 m and 750 m) and scaling the output to range from 0 to 100. Similarly, Den_Veg raster was computed by calculating stable vegetation pixel count in the classified image at the aforementioned resolutions. Figure 11.3 shows all

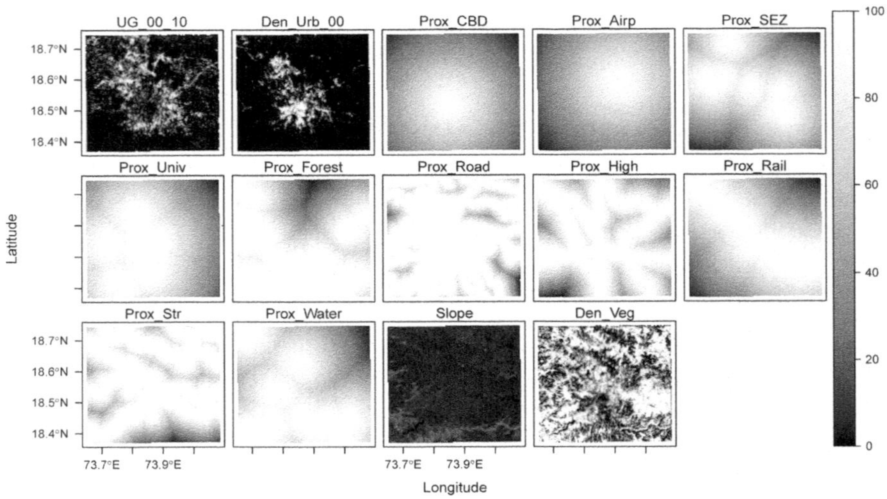

Fig. 11.3 Dependent and independent variables computed at 300 m resolution for the 2000–2010 period

the dependent and independent variables computed at 300 m resolution for ULE model for the 2000–2010 period.

$$\text{Proximity}_i = \left(1 - \left(\frac{\text{Euclidean Nearest Distance}_i}{\text{Maximum Euclidean Distance}}\right)\right) * 100 \tag{11.1}$$

11.3.2.2 Variable Selection

Inclusion of too many explanatory variables in the model could cause multicollinearity effects; strong correlation between variables may bias the parameter estimates. On the contrary, choosing too few variables could cause omitted variable bias. Therefore, dealing with the multicollinearity problem and simultaneously avoiding omitted variable bias necessitated a three-stage selection of explanatory variables. First, a matrix of pair-wise correlations was examined to identify variables with high collinearity. Figure 11.4 shows the correlation matrix and the pair-wise density scatterplots. Here, Prox_CBD, Prox_SEZ, Prox_Univ, and Prox_Rail showed high correlations (>0.6). Second, a variable inflation factor (VIF)-based backward variable elimination procedure was used such that variables were dropped until all the VIF values were below 5. Here, Prox_CBD and Prox_Rail were identified as collinear. Using outputs of the two stages, Prox_CBD and Prox_Rail were eliminated from subsequent modeling. Third, step-wise Akaike information criterion (AIC)-based model selection was carried out with the

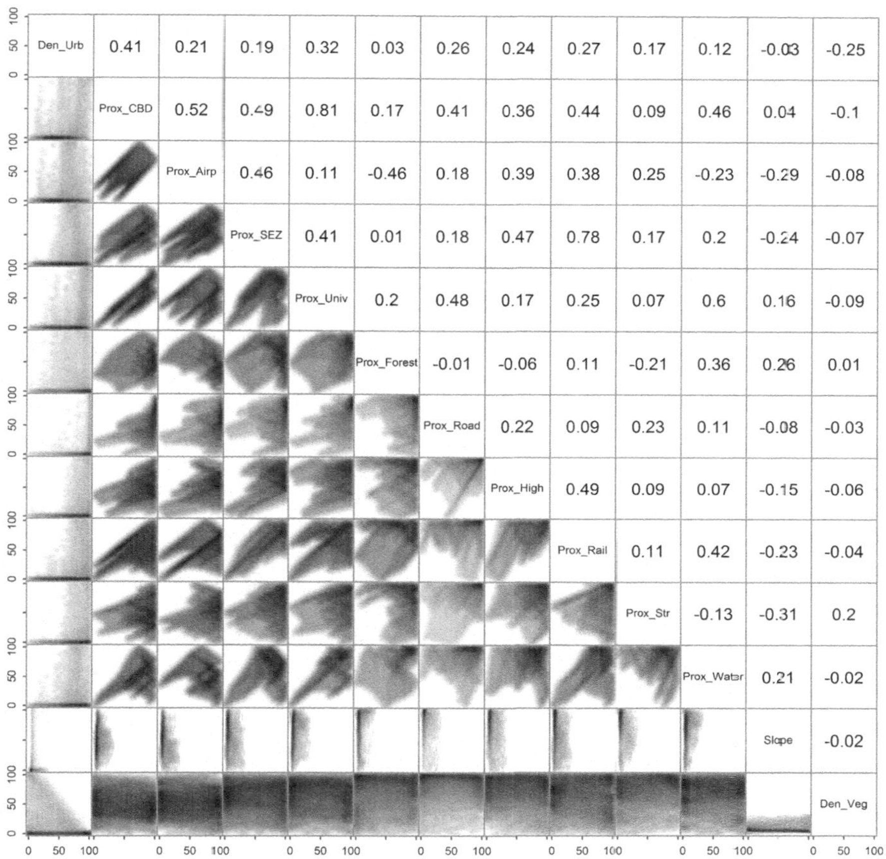

Fig. 11.4 Density scatterplots and pair-wise correlation matrix of all the explanatory variables used in this study

remaining variables to select the final model. Here, the final model included all the variables, and none of the variables were dropped. Therefore, after excluding Prox_CBD and Prox_Rail, 11 explanatory variables were used in subsequent modeling.

11.3.2.3 Ordinary Least Squares Regression

Two nonspatial regression models were estimated using OLS regression technique. In order to check parameter stability across spatial resolutions and ensure that the choice of coarser resolution does not cause any bias, the model specified in Eq. 11.2, with UG_00_10 as the dependent variable, was estimated with variables computed at 300 m and 750 m resolutions, whereas the model specified in Eq. 11.2 with

UG_10_15 was estimated only with variables computed at 300 m, with a purpose of evaluating the parameter stability across time. After estimation, residuals were tested for the presence of spatial autocorrelation using Moran's I statistic. Here, a first-order contiguity-based spatial weights matrix was used (Anselin 1988). Furthermore, spatial residual autocorrelation and spatial lag dependence were tested using the Lagrange Multiplier tests.

$$\begin{aligned}\text{Dependvar} = {} & \beta_0 + \beta_1 * (\text{Den\_Urb\_00}) + \beta_2 * (\text{Prox\_Airp}) \\ & + \beta_3 * (\text{Prox\_SEZ}) + \beta_4 * (\text{Prox\_Univ}) + \beta_5 * (\text{Prox\_Forest}) \\ & + \beta_6 * (\text{Prox\_Road}) + \beta_7 * (\text{Prox\_High}) + \beta_8 * (\text{Prox\_Str}) \\ & + \beta_9 * (\text{Prox\_Water}) + \beta_{10} * (\text{Slope}) + \beta_{11} * (\text{Den\_Veg}) \\ & + \varepsilon \end{aligned} \tag{11.2}$$

where Dependvar∈ (UG_00_10, UG_10_15).

11.3.2.4 Spatial Regression

Besides OLS-based regression, models that account for spatial dependencies were also estimated. Specifically, the following SLM and SEM were estimated using the maximum likelihood estimation technique (Anselin 1988):

$$\begin{aligned}\text{Dependvar} = {} & (\rho * W * \text{UG\_00\_10}) + \beta_0 + \beta_1 * (\text{Den\_Urb\_00}) \\ & + \beta_2 * (\text{Prox\_Airp}) + \beta_3 * (\text{Prox\_SEZ}) + \beta_4 * (\text{Prox\_Univ}) \\ & + \beta_5 * (\text{Prox\_Forest}) + \beta_6 * (\text{Prox\_Road}) \\ & + \beta_7 * (\text{Prox\_High}) + \beta_8 * (\text{Prox\_Str}) + \beta_9 * (\text{Prox\_Water}) \\ & + \beta_{10} * (\text{Slope}) + \beta_{11} * (\text{Den\_Veg}) + \varepsilon \end{aligned} \tag{11.3}$$

$$\begin{aligned}\text{Dependvar} = {} & \beta_0 + \beta_1 * (\text{Den\_Urb\_00}) + \beta_2 * (\text{Prox\_Airp}) \\ & + \beta_3 * (\text{Prox\_SEZ}) + \beta_4 * (\text{Prox\_Univ}) + \beta_5 * (\text{Prox\_Forest}) \\ & + \beta_6 * (\text{Prox\_Road}) + \beta_7 * (\text{Prox\_High}) + \beta_8 * (\text{Prox\_Str}) \\ & + \beta_9 * (\text{Prox\_Water}) + \beta_{10} * (\text{Slope}) + \beta_{11} * (\text{Den\_Veg}) \\ & + (\lambda * W * \varepsilon) + \upsilon \end{aligned} \tag{11.4}$$

where Dependvar∈ (UG_00_10, UG_10_15), ρ, and λ are the spatial autoregressive coefficients, W is the first-order contiguity weights matrix, and υ is the idiosyncratic disturbance. In order to check parameter stability across spatial resolutions, models with UG_00_10 as the dependent variable specified in Eqs. 11.3 and 11.4 were estimated with variables computed at 300 and 750 m resolutions.

A number of studies have used GWR to take spatial non-stationarity of parameter estimates into account. Intrinsic differences in social, economic, and environmental conditions are conducive to different responses to the same stimuli but in different locations (Fotheringham et al. 2003). In this study, the capability of GWR to study non-stationary relationships between urban growth and geographic determinants was evaluated.

In GWR, the algorithm assigns weights for a given location (i) to its neighboring values. GWR requires specification of a kernel function that further requires a bandwidth specification that defines distance decay in the weighting scheme. The bandwidth could be specified as fixed or adaptive across space, and its optimum value can be estimated by analyzing how well the model fits the data while accommodating model complexity (Fotheringham et al. 2003). In this study, the bi-square distance decay function was used with an adaptive weighting scheme. The optimum value of the bandwidth parameter was determined by minimizing the corrected AIC. The following GWR model was estimated using GW model package in R (Gollini et al. 2013; Lu et al. 2014):

$$\begin{aligned} \mathrm{UG\_00\_10_i} = {} & \beta_{0i} + \beta_{1i} * (\mathrm{Den\_Urb\_00_i}) + \beta_{2i} * (\mathrm{Prox\_Airp_i}) \\ & + \beta_{3i} * (\mathrm{Prox\_SEZ_i}) + \beta_{4i} * (\mathrm{Prox\_Univ_i}) \\ & + \beta_{5i} * (\mathrm{Prox\_Forest_i}) + \beta_{6i} * (\mathrm{Prox\_Road_i}) \\ & + \beta_{7i} * (\mathrm{Prox\_High_i}) + \beta_{8i} * (\mathrm{Prox\_Str_i}) \\ & + \beta_{9i} * (\mathrm{Prox\_Water_i}) + \beta_{10i} * (\mathrm{Slope_i}) + \beta_{11i} * (\mathrm{Den\_Veg_i}) \\ & + \varepsilon_i \end{aligned} \tag{11.5}$$

where *i* is a given location. The optimum model was identified using the model. selection.gwr function from the GW model package. This function selects the appropriate model through a step-wise forward selection procedure. Additionally, multicollinearity in the model was explored using the gwr.collin.diagno function.

11.4 Results

11.4.1 Ordinary Least Squares Regression

Table 11.2 shows the OLS regression outputs for urban growth from 2000 to 2010 using data at 300 and 750 m. The OLS regression models obtained from data at 300 m ($R^2 = 0.18$) and 750 m ($R^2 = 0.26$) have low explanatory power. Additionally, the parameter estimates are not stable across spatial scales. When comparing the estimates across spatial scales, the results show that Prox_Forest, Prox_Water, and Den_Urb_00 variables have opposite relationships with the dependent variable. Significant change in parameter values, however, is noted for Prox_Water and Slope. Furthermore, two variables, i.e., Prox_Forest and Prox_Airp, became statistically insignificant at 700 m compared to 300 m. Clearly, these results indicate MAUP in multivariate statistical models. Nonetheless, results also indicate that OLS models cannot be trusted due to spatially autocorrelated residuals (Table 11.2). Table 11.3 shows OLS regression outputs for urban growth from 2010 to 2015 using data at 300 m. Again, low explanatory power and spatially autocorrelated residuals were observed (Table 11.3). Results from the Lagrange Multiplier tests showed statistically significant spatial residual autocorrelation and spatial lag

Table 11.2 OLS, SLM, and SEM regression for urban growth in the 2000–2010 time-period

	Dependent variable: UG_00_10					
	Model 1	Model 2	Model 3	Model 4	Model 5	Model 6
	(300 m)	(300 m)	(300 m)	(750 m)	(750 m)	(750 m)
	OLS	SLM	SEM	OLS	SLM	SEM
Constant	−30.498***	−6.842***	−55.618***	−25.589***	−6.018***	−35.030***
	(1.20)	(0.802)	(5.005)	(2.500)	(1.745)	(7.757)
Den_Urb_00	−0.019**	−0.056***	−0.233***	0.046**	−0.016	−0.082***
	(0.007)	(0.005)	(0.009)	(0.019)	(0.013)	(0.023)
Prox_Airp	0.049***	0.014***	0.076*	0.026	0.001	−0.012
	(0.009)	(0.006)	(0.042)	(0.019)	(0.013)	(0.078)
Prox_SEZ	0.149***	0.014***	0.151***	0.151***	0.025*	0.161**
	(0.008)	(0.005)	(0.036)	(0.019)	(0.013)	(0.071)
Prox_Univ	0.067***	0.019***	0.004	0.050***	0.005	0.049
	(0.009)	(0.006)	(0.040)	(0.018)	(0.012)	(0.067)
Prox_Forest	−0.027***	0.002	−0.014	0.038	0.005	−0.047
	(0.006)	(0.004)	(0.030)	(0.025)	(0.017)	(0.086)
Prox_Road	0.187***	0.040***	0.472***	0.182***	0.061***	0.314***
	(0.009)	(0.006)	(0.034)	(0.020)	(0.014)	(0.040)
Prox_High	0.074***	0.019***	0.117***	0.072***	0.017	0.175***
	(0.008)	(0.005)	(0.034)	(0.016)	(0.011)	(0.054)
Prox_Str	0.090***	0.020***	0.044	0.101***	0.021	0.033
	(0.007)	(0.005)	(0.031)	(0.016)	(0.011)	(0.049)
Prox_Water	0.016**	0.005	0.041	−0.093***	−0.009	−0.046
	(0.008)	(0.005)	(0.040)	(0.023)	(0.016)	(0.085)
Slope	−0.235***	−0.075***	−0.102***	−0.318***	−0.103*	−0.130
	(0.030)	(0.020)	(0.035)	(0.080)	(0.054)	(0.085)
Den_Veg	−0.043***	−0.016***	−0.027***	−0.047***	−0.029***	−0.053***
	(0.004)	(0.002)	(0.004)	(0.010)	(0.007)	(0.010)

ρ	–	0.852***	–	–	0.835***	–
		(0.005)			(0.012)	
λ	–	–	0.872***	–	–	0.853***
			(0.004)			(0.012)
Observations	20,976	20,976	20,976	3355	3355	3355
Moran's I	0.575	−0.012	−0.020	0.524	0.006	−0.002
R^2	**0.179**	**0.658**	**0.670**	**0.264**	**0.658**	**0.668**
AIC	178,498	163,204	162,568	27,647	25,548	25,479

OLS Ordinary least squares model, *SLM* Spatial lag model, *SEM* Spatial error model
$*p < 0.1$; $**p < 0.05$; $***p < 0.01$

Table 11.3 OLS, SLM, and SEM regression for urban growth in the 2010–2015 time-period

	Dependent variable: UG_10_15		
	Model 1	Model 2	Model 3
	(300 m)	(300 m)	(300 m)
	OLS	SLM	SEM
Constant	−1.584*	−1.387**	−16.418***
	(0.928)	(0.675)	(2.917)
Den_Urb_10	−0.018***	−0.022***	−0.081***
	(0.004)	(0.003)	(0.005)
Prox_Airp	−0.131***	−0.024***	−0.112***
	(0.007)	(0.005)	(0.023)
Prox_SEZ	0.093***	0.019***	0.122***
	(0.006)	(0.004)	(0.020)
Prox_Univ	0.073***	0.029***	0.030
	(0.007)	(0.005)	(0.023)
Prox_Forest	−0.015**	−0.001	−0.020
	(0.005)	(0.003)	(0.017)
Prox_Road	0.004	0.007	0.119***
	(0.007)	(0.005)	(0.022)
Prox_High	0.066***	0.020***	0.089***
	(0.006)	(0.004)	(0.020)
Prox_Str	0.074***	0.009**	0.082***
	(0.006)	(0.004)	(0.018)
Prox_Water	−0.001	−0.001	0.014
	(0.006)	(0.005)	(0.023)
Slope	−0.701***	−0.249***	−0.336***
	(0.023)	(0.017)	(0.029)
Den_Veg	0.037***	0.021***	0.044***
	(0.003)	(0.002)	(0.004)
ρ	–	0.782***	–
		(0.006)	
λ	–	–	0.800***
			(0.006)
Observations	20,976	20,976	20,976
Moran's I	0.493	−0.025	−0.033
R^2	**0.123**	**0.537**	**0.547**
AIC	166,950	155,931	155,644

OLS Ordinary least squares model, *SLM* Spatial lag model, *SEM* Spatial error model
*p < 0.1; **p < 0.05; ***p < 0.01

dependence in all the OLS models. These results emphasize that OLS regression is unable to capture spatial dependencies in the data. Consequently, it is unsuitable for identifying the geographic determinants of ULE patterns.

11.4.2 Spatial Regression

Table 11.2 shows the SLM and SEM results for urban growth from 2000 to 2010 using data at 300 m and 750 m. Models that accounted for spatial dependencies, i.e., SLM and SEM, showed much higher explanatory power compared to the OLS models and negligible spatial autocorrelation in residuals. Although there is only a slight difference in AIC statistic and R^2 values, when comparing SLM and SEM, SEM models consistently have higher R^2 values and lower AIC values (Table 11.2). This consistency is also observed in models estimated for urban growth from 2010 to 2015 at 300 m (Table 11.3). This suggests that, for this study, SEM specification is better than the OLS and SLM specifications and that SEM is suitable for identifying the geographic determinants of spatial patterns of ULE in this study. Therefore, SEMs were interpreted to determine geographic determinants of ULE in the study area.

Figure 11.5a shows the comparison between model parameters estimated from SEM at 300 and 750 m. SEM estimated using data at 300 m and 750 m largely showed consistency in parameter estimates. Two parameters, i.e., Den_Urb_00 and Prox_Road, however, showed maximum changes in absolute values, i.e., 0.151 and 0.158, respectively. Nonetheless, Prox_Road continued to be a major and statistically significant variable positively affecting ULE (Fig. 11.5a). Similarly, Den_Urb_00 is found to affect ULE negatively at both the scales but with a higher magnitude at 300 m (Fig. 11.5a). Overall, while there may be minor differences in parameter estimates across spatial scales, these results emphasize that one can identify geographic determinants of ULE by interpreting a model estimated at a given scale. In this study, SEM estimated at 300 m resolution is selected for further interpretation given relatively higher R^2 and sample size.

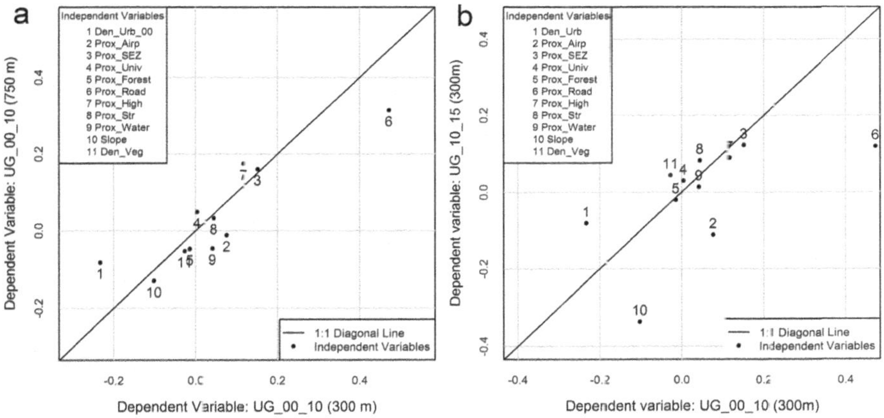

Fig. 11.5 Comparison between parameter estimates of spatial error models estimated for (**a**) urban growth during 2000–2010 time-period at 300 m and 750 m spatial resolutions and (**b**) urban growth during 2000–2010 and 2010–2015 time-periods

SEM parameters underline Prox_Road as the major determinant of urban growth during the 2000–2010 time-period (Table 11.2). This is not surprising as proximity to transportation infrastructure is a typical precursor for ULE, which is further emphasized by the model results that show Prox_High as a major determinant of urban growth during the 2000–2010 time-period. Additionally, Prox_SEZ is another major determinant of urban growth during the 2000–2010 time-period (Table 11.2). In the study area, recently approved and operational SEZs have increased employment opportunities, and based on a recent fieldwork, better infrastructure was observed in areas contiguous to SEZs than elsewhere in the PMR. In fact, establishing SEZs and a lack of urban jurisdiction have caused the surrounding areas to be more attractive, and realizing the attractiveness and paucity of housing infrastructure, several private developers initiated housing projects near SEZs providing better amenities and promoting better living standards. The SEM results corroborate these observations from the fieldwork and findings from a recent case study (Kantakumar et al. 2016) and thus are realistic. Den_Urb_00 is identified to be the greatest barrier to ULE indicating dominant suburbanization or dispersed form of ULE in the study area. In addition to Den_Urb_00, SEM parameters highlight topography as another major ULE barrier (Table 11.2).

Next, SEM estimates for urban growth during the 2000–2010 time-period were contrasted with SEM estimates for urban growth during the 2010–2015 time-period. Figure 11.5b shows the comparison between parameter estimates from SEM estimated for urban growth during the 2000–2010 time-period and SEM estimated for urban growth during the 2010–2015 time-period. Here, the results showed no broad changes in the geographic determinants of ULE; Prox_Road, Prox_SEZ, and Prox_High remain the major determinants, and Slope and Den_Urb remain the major barriers to urban growth across the two time-periods. Nonetheless, there are significant differences in some of the parameter estimates. Specifically, Prox_SEZ in the recent time-period is the strongest positive determinant of urban growth as opposed to Prox_Road in the former time-period (Fig. 11.5b). Similarly, Slope in the recent time-period is the strongest barrier to urban growth as opposed to Den_Urb in the former time-period. These results suggest that the relationship of a geographic feature with urban growth may vary across time, but the underlying reasons for these temporal dynamics remain unclear.

As opposed to OLS, SLM, and SEM, GWR necessitated additional model selection procedure to identify the best model while taking into account spatial non-stationarities. Here, the model represented in Eq. 8 was selected in the forward selection procedure, i.e., no variables were dropped. Table 11.4 shows the results from GWR model estimation. The estimated GWR model has higher explanatory power ($R^2 = 0.50$) and lower AIC value compared to the OLS model. Conversely, the GWR model has relatively low explanatory power and higher AIC value compared to SLM and SEM. This suggests that even though GWR is able to take into account spatial non-stationarity of parameter estimates, overall SLM and SEM are superior. Next, GWR residuals were examined, and a statistically significant

Table 11.4 Summary of GWR model for urban growth in the 2000–2010 time-period

	Dependent variable: UG_00_10				
	Minimum	Maximum	Mean	% Positive	% Negative
Constant	−1321.339	1145.715	−39.777	42.758	57.242
Den_Urb_00	−0.547	10.818	0.492	70.900	29.100
Prox_Airp	−32.083	14.026	−0.217	53.423	46.577
Prox_SEZ	−6.356	18.346	−0.036	41.710	58.290
Prox_Univ	−41.011	29.281	−0.071	54.381	45.619
Prox_Forest	−3.927	5.098	−0.086	38.701	61.299
Prox_Road	−0.539	3.595	0.576	85.097	14.903
Prox_High	−1.757	1.660	0.181	62.419	37.581
Prox_Str	−1.228	2.289	0.088	66.047	33.953
Prox_Water	−9.623	42.617	0.325	47.707	52.293
Slope	−2.717	3.503	−0.166	27.169	72.831
Den_Veg	−0.354	0.162	−0.042	30.683	69.317
Observations	20,976				
Moran's I	0.356				
AIC	168,511				
R^2	0.506				

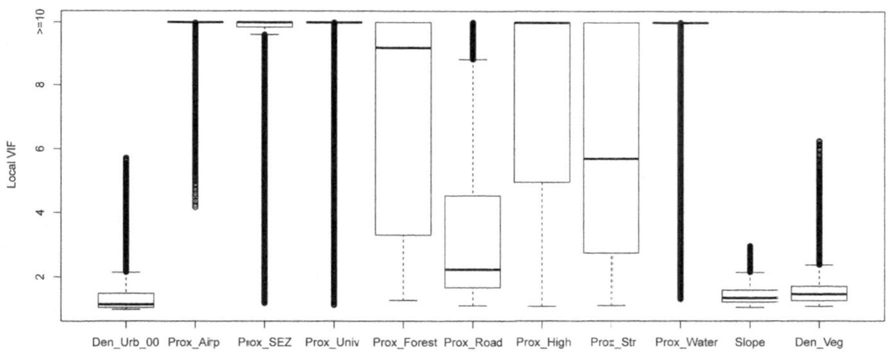

Fig. 11.6 Local collinearity in parameter estimates of GWR model estimated for urban growth during 2000–2010 time-period

spatial autocorrelation is identified suggesting presence of bias in model parameters. Furthermore, the examination of local collinearity diagnostics results suggests that there is significant local collinearity between variables that will aver: unbiased estimation of local model parameters (Fig. 11.6). Overall, these results suggest that GWR is not an appropriate tool for identifying geographic determinants of ULE in the study area.

11.5 Discussion

11.5.1 Key Findings

From a methodological perspective, this study highlights three key findings. First, variable selection using a combination of different methods is more robust compared to a single method. Previous research focused on a single method such as pair-wise comparison (Luo and Wei 2009) or VIF-based selection (Shafizadeh-Moghadam and Helbich 2015). Analysis in this study, however, showed differences in outputs from different techniques and that a combination of these techniques can address the multicollinearity problem. Second, global regression models, in which spatial dependencies are adequately accounted, can actually perform better than local regression models. Previous research suggests that local regression using GWR outperforms global regression that considers parameter stationarity. On the contrary, results from this study suggest that global models that consider parameter stationarity with error or spatial lag terms can have higher strength and robustness when compared to a GWR model. Third, results from this study suggest that analyzing local collinearity is important and cannot be neglected when attempting to estimate a GWR model. Several previous research studies that estimated GWR models only considered collinearity problem in the entire spatial dataset. However, this research showed that collinearity could exist locally, which could bias the spatially varying parameter estimates. While the current study found GWR application unsuitable, one may explore the prospects of using GWR models. In the presence of local collinearity, one may use locally compensated ridge term (Gollini et al. 2013; Lu et al. 2014). Note that due to very high local collinearity (Fig. 11.6), GWR with locally compensated ridge term was not explored in the present study.

Our results highlight two key findings about the urban growth patterns. First, locations that are sparsely urbanized, are near SEZs, and have higher transportation accessibility are more likely to urbanize compared to other locations. This study showed that transportation infrastructure is a key determinant, that existing urban area density is negatively associated with ULE, and that a location near a SEZ will experience greater ULE in the study area. These findings point toward suburbanization or diffused patterns of ULE since these conditions hold generally at locations farther from the core urban area. Second, in the recent time-period, i.e., 2010–2015, this study found that SEZ is the most important determinant of ULE compared to transportation infrastructure in the 2000–2010 time-period. This suggests that recent ULE in the study area is associated with the establishment of SEZs during the 2000–2010 period.

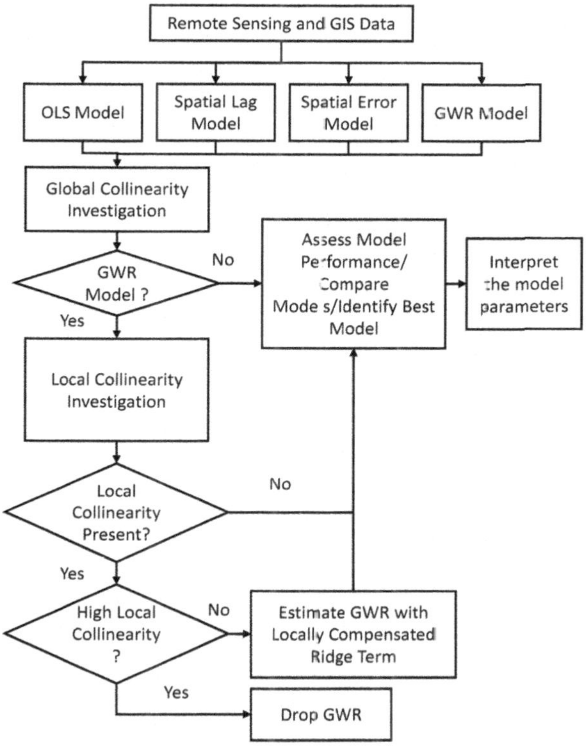

Fig. 11.7 A simplified representation of the proposed statistical analysis-based framework to identify geographic determinants of urban growth

11.5.2 Proposed Framework

The analysis and results put forward a statistical regression modeling-based analytical framework to study the spatial patterns of urban growth. As shown in this study, comparing different ULE models, albeit for same time-periods, based on different factors, such as model performance, multicollinearity, spatial-scale considerations, parameter non-stationarity, and spatial dependency, is critical. This is so because model selection based on one of these criteria may be biased and may lead to the selection of an inappropriately specified model. For example, even though a GWR model might show higher performance compared to OLS model, local multicollinearity may bias the parameter estimates, thus making GWR an inappropriate model. Nonetheless, after identifying a stable unbiased model, one can study the geographic determinants of ULE. A simplified representation of the proposed framework is given in Fig. 11.7.

The second important facet of the proposed framework is that one can easily examine the geographic determinants of specific land-use changes while estimating separate models each with the specific land-use change type as the dependent variable, which is continuous in nature. With this approach one can also examine the differences or similarities between the geographic determinants of different land-use change types by comparing between different models. However, when the dependent variable is categorical in nature, a single multinomial model can study the relationship between different land-use changes and the geographic features.

11.5.3 Methodological Limitations

This analysis highlights two methodological problems in using RS and GIS data in a statistical regression modeling-based framework. First, RS-derived information about ULE accompanies some measurement error. Similarly, digitizing detailed geographic features are also subject to multiple error sources. Consequently, one must not expect high model fit. In the present study, the best spatial regression models explained ~67% of the variance. Second, identifying appropriate spatial scales to model ULE could be challenging. While previous studies identified an optimum scale by referring to goodness-of-fit measures (Shafizadeh-Moghadam and Helbich 2015) or fractal dimension (Hu and Lo 2007), Fotheringham and Wong (1991) suggested that the MAUP effects in multivariate models are unpredictable and that one must analyze models estimated at different aggregation levels. The present study emphasizes that one must look for consistent information from parameter estimates when identifying geographic determinants of ULE, in the absence of which an optimum scale may be identified. Despite the aforementioned problems, this study suggests that multivariate statistical tools and spatial regression, in particular, are appropriate for identifying geographic determinants of ULE.

11.6 Conclusion

A first step toward understanding urban-environment linkages is to develop a scientific understanding of how urban landscapes are changing. Much of the current research using RS and GIS tools has focused on quantifying the spatial patterns of growth in cities. However, examining the relevance of different geographic features toward observed changes can further our understanding of underlying reasons for similarities and differences in the morphologies of different urban areas. In the present study, first, the capability of multivariate statistics, spatial regression in particular, to understand the spatial patterns of ULE is examined. Second, key contemporary geographic determinants of ULE in the Pune metropolitan region are identified.

This study uses state-of-the-art regression techniques, including spatial econometric tools, and presents a novel investigation of how these tools can be used to understand the geographic determinants of ULE. The results show that comparing models estimated from different techniques based on model performance, multicollinearity, spatial-scale considerations, parameter non-stationarity, and spatial dependency is critical. Most previous studies on statistics-based ULE modeling have carried out model selection following one of these criteria. This study shows, in such cases, an inappropriate model may be selected and interpreted. Therefore, based on the analysis and results, this study proposes a robust multivariate statistics-based framework to understand the geographic determinants of ULE. The proposed framework does not invalidate using conventional methods, such as landscape analysis using spatial metrics, but is intended to supplement these methods for detailed examinations. The framework can also be used with spatially explicit socioeconomic datasets to explain ULE patterns. However, developing countries often have a paucity of high-quality data on urban landscapes due to several political and economic reasons. In such cases, empirical model estimations from RS and GIS data with appropriate model selection can study the relationships between land-use changes and corresponding spatial drivers. Here, the proposed framework will be a useful tool in decision-making, planning, and assessing land-use policy outcomes.

This study also identified several geographic determinants of the spatial patterns of urban growth in the urbanizing region of Pune. The results show that contemporary urban growth in one of India's major metropolitan areas is concentrated in outlying areas with low existing urban density and that the SEZs are causing immediate surrounding areas to be attractive for urbanization and are acting as growth nuclei for ULE. Furthermore, results highlight that SEZs became a significant determinant of urban growth during the 2010–2015 time-period, i.e., after most SEZs were notified or established in the 2000–2010 time-period. This finding also highlights SEZs as one significant factor explaining characteristic polycentric growth pattern observed in the Pune metropolitan region.

In summary, in data-sparse conditions, analyzing RS images and GIS data of urbanizing regions in an appropriate analytical framework can provide insights into urban growth dynamics, from both urban and regional perspectives. This study suggests that multivariate statistical analysis tools that explore spatial dependencies and spatial non-stationarity in influencing features (such as SLM, SEM, and GWR) can identify the geographic determinants but with some limitations.

References

Al-sharif AAA, Pradhan B (2015) A novel approach for predicting the spatial patterns of urban expansion by combining the chi-squared automatic integration detection decision tree, Markov chain and cellular automata models in GIS. Geocarto Int 30:858–881

Angel S, Sheppard S, Civco DL, Buckley R, Chabaeva A, Gitlin L, Kraley A, Parent J, Perlin M. (2005 The dynamics of global urban expansion [Internet]. [place unknown]: Citeseer; [cited

2016 Jul 26]. Available from: http://citeseerx.ist.psu.edu/viewdoc/download?doi=10.1.1.309.2715&rep=rep1&type=pdf

Anselin L (1988) Spatial econometrics: methods and models. Dordrecht. Kluwer Academic Publishers, Boston

Anselin L, Griffith DA (1988) Do spatial effects really matter in regression analysis? Pap Reg Sci 65:11–34

Census of India (2011 Census of India, 2011. India Provisional Popul Totals Pap. 1

Cheng J, Masser I (2003) Urban growth pattern modeling: a case study of Wuhan city, PR China. Landsc Urban Plan 62:199–217

Clark TN, Lloyd R, Wong KK, Jain P (2002) Amenities drive urban growth. J Urban Aff 24:493–515

Concepci6n ED, Moretti M, Altermatt F, Nobis MP, Obrist MK (2015) Impacts of urbanisation on biodiversity: the role of species mobility, degree of specialisation and spatial scale. Oikos 124:1571–1582

Dimitriadou E, Hornik K, Leisch F, Meyer D, Weingessel A, Leisch MF (2006) The e1071 package. Misc Funct Dep Stat E1071 TU Wien [Internet]. [cited 2016 Jul 26]. Available from: http://ftp.auckland.ac.nz/software/CRAN/doc/packages/e1071.pdf

Farooq S, Ahmad S (2008) Urban sprawl development around Aligarh city: a study aided by satellite remote sensing and GIS. J Indian Soc Remote Sens 36:77–88

Fazal S (2001) The need for preserving farmland: a case study from a predominantly agrarian economy (India). Landsc Urban Plan 55:1–13

Fotheringham AS, Wong DW (1991) The modifiable areal unit problem in multivariate statistical analysis. Environ Plan A 23:1025–1044

Fotheringham AS, Brunsdon C, Charlton M (2003) Geographically weighted regression: the analysis of spatially varying relationships [Internet]. [place unknown]: John Wiley & Sons; [cited 2016 July 26]. Available from: https://books.google.com/books?hl=en&lr=&id=9DZgV1vXOuMC&oi=fnd&pg=PR7&dq=Geographically+weighted+regression:+Wiley+New+York+book&ots=64FJNgo9KG&sig=rvfajcybZupWNRMGTen6intGITc

Ganguly K, Kumar R, Reddy KM, Rao PJ, Saxena MR, Shankar GR (2016) Optimization of spatial statistical approaches to identify land use/land cover change hot spots of Pune region of Maharashtra using remote sensing and GIS techniques. Geocarto Int 0:1–20

Gollini I, Lu B, Charlton M, Brunsdon C, Harris P (2013) GWmodel: an R package for exploring spatial heterogeneity using geographically weighted models. ArXiv Prepr ArXiv13060413 [Internet]. [cited 2016 July 26]. Available from: http://arxiv.org/abs/1306.0413

Hu Z, Lo CP (2007) Modeling urban growth in Atlanta using logistic regression. Comput Environ Urban Syst 31:667–688

Jantz CA, Goetz SJ, Shelley MK (2004) Using the SLEUTH urban growth model to simulate the impacts of future policy scenarios on urban land use in the Baltimore-Washington metropolitan area. Environ Plan B Plan Des 31:251–271

Jat MK, Garg PK, Khare D (2008) Monitoring and modelling of urban sprawl using remote sensing and GIS techniques. Int J Appl Earth Obs Geoinf 10:26–43

Kantakumar LN, Kumar S, Schneider K (2016) Spatiotemporal urban expansion in Pune metropolis, India using remote sensing. Habitat Int 51:11–22

Kowe P, Pedzisai E, Gumindoga W, Rwasoka DT (2015) An analysis of changes in the urban landscape composition and configuration in the Sancaktepe District of Istanbul Metropolitan City, Turkey using landscape metrics and satellite data. Geocarto Int 30:506–519

Lafazani P, Lagarias A (2016) Applying multiple and logistic regression models to investigate periurban processes in Thessaloniki, Greece. Geocarto Int 31:927–942

Li X, Zhou W, Ouyang Z (2013) Forty years of urban expansion in Beijing: what is the relative importance of physical, socioeconomic, and neighborhood factors? Appl Geogr 38:1–10

Linard C, Tatem AJ, Gilbert M (2013) Modelling spatial patterns of urban growth in Africa. Appl Geogr 44:23–32

Lu B, Harris P, Charlton M, Brunsdon C (2014) The GWmodel R package: further topics for exploring spatial heterogeneity using geographically weighted models. Geo-Spat Inf Sci 17:85–101
Luo J, Wei YD (2009) Modeling spatial variations of urban growth patterns in Chinese cities: the case of Nanjing. Landsc Urban Plan 91:51–64
Luo J, Yu D, Xin M (2008) Modeling urban growth using GIS and remote sensing. GISci Remote Sens 45:426–442
Moghadam HS, Helbich M (2013) Spatiotemporal urbanization processes in the megacity of Mumbai, India: a Markov chains-cellular automata urban growth model. Appl Geogr 40:140–149
Mondal B, Das DN, Dolui G (2015) Modeling spatial variation of explanatory factors of urban expansion of Kolkata: a geographically weighted regression approach. Model Earth Syst Environ 1:29
Mondal B, Das DN, Bhatta B (2016) Integrating cellular automata and Markov techniques to generate urban development potential surface: a study on Kolkata agglomeration. Geocarto Int 32:1–19
Openshaw S (1983) The modifiable areal unit problem. GeoBooks, Norwich
Openshaw S (1984) The modifiable areal unit problem. In: [place unknown]: Geo Abstracts University of East Anglia
Pathirana A, Denekew HB, Veerbeek W, Zevenbergen C, Banda AT (2014) Impact of urban growth-driven landuse change on microclimate and extreme precipitation—a sensitivity study. Atmos Res 138:59–72
Pimpri-Chinchwad Municipal Corporation (2008) Comprehensive mobility plan (CMP) for PCMC. [place unknown]
Pontius RG Jr, Millones M (2011) Death to Kappa: birth of quantity disagreement and allocation disagreement for accuracy assessment. Int J Remote Sens 32:4407–4429
Pune Municipal Corporation (2008) Comprehensive mobility plan for Pune city. Pune
Ramachandra TV, Setturu B, Aithal BA (2012) Per-urban to urban landscape patterns elucidation through spatial metrics. Int J Eng Res Dev 2(12):58–81
Schneider A, Woodcock CE (2008) Compact, dispersed, fragmented, extensive? A comparison of urban growth in twenty-five global cities using remotely sensed data, pattern metrics and census information. Urban Stud 45:659–692
Shafizadeh-Moghadam H, Helbich M (2015) Spatiotemporal variability of urban growth factors: a global and local perspective on the megacity of Mumbai. Int J Appl Earth Obs Geoinf 35:187–198
Sudhira HS, Ramachandra TV, Jagadish KS (2004) Urban sprawl: metrics, dynamics and modelling using GIS. Int J Appl Earth Obs Geoinf 5:29–39
Taubenböck H, Wegmann M, Roth A, Mehl H, Dech S (2009) Urbanization in India–spatiotemporal analysis using remote sensing data. Comput Environ Urban Syst 33:179–188
Triantakonstantis D, Stathakis D (2015) Examining urban sprawl in Europe using spatial metrics. Geocarto Int 30:1092–1112
Wheeler D, Tiefelsdorf M (2005) Multicollinearity and correlation among local regression coefficients in geographically weighted regression. J Geogr Syst 7:161–187
Yu D-L (2006) Spatially varying development mechanisms in the Greater Beijing Area: a geographically weighted regression investigation. Ann Reg Sci 40:173–190
Zeng C, Zhang M, Cui J, He S (2015) Monitoring and modeling urban expansion—a spatially explicit and multi-scale perspective. Cities 43:92–103
Zhang Q, Seto KC (2011) Mapping urbanization dynamics at regional and global scales using multi-temporal DMSP/OLS nighttime light data. Remote Sens Environ 115:2320–2329
Zhang Z, Su S, Xiao R, Jiang D, Wu J (2013) Identifying determinants of urban growth from a multi-scale perspective: a case study of the urban agglomeration around Hangzhou Bay, China. Appl Geogr 45:193–202

Chapter 12
Analysing Urban Sprawl and Spatial Expansion of Kolkata Urban Agglomeration Using Geospatial Approach

Mehebub Rahaman, Shyamal Dutta, Mehebub Sahana, and Dipendra Nath Das

Abstract Since the last quarter of the twentieth century, India has been witnessing predominantly outward expansion of most large megacities in the form of sprawl, and peripheries have been engulfing many small towns and villages rather than accommodating the migrants from rural areas in the city core. Amidst this transformation, the condition of people living in peripheral areas becomes precarious which is explained by 'degenerated periphery'. In this backdrop, the present study aims to assess the spatiotemporal urban expansion of different municipal areas and municipal corporation areas of Kolkata urban agglomeration of West Bengal, India, during 1990–2015. Landsat Thematic Mapper and Landsat 8 OLI satellite data of the years 1990 and 2015 along with Shannon's entropy model and urban built-up index were used to assess the spatial dispersion of and consistency of urbanization. The investigations reveal a rapid increase of built-up areas outside the municipal boundaries during the last two and half decades. Shannon's entropy at local level is computed, which shows dispersed unplanned urban growth, specifically in the outskirts of the city. The study indicates that the core of the city has experienced negative growth. Land use and land cover change analysis revealed that the built-up area has increased drastically over the study periods. The agriculture land and open land have transformed into built-up area, indicating the sprawl growth within the Kolkata urban agglomeration. The overall result shows that urban expansion of Kolkata urban agglomeration is not compact in nature and it is an evidence of concentration of sprawl growth over the municipalities.

M. Rahaman · D. N. Das
Centre for the Study of Regional Development, School of Social Sciences, Jawaharlal Nehru University, New Delhi, India

S. Dutta (✉)
Department of Geography, The University of Burdwan, Barddhaman, West Bengal, India

M. Sahana
Department of Geography, Jamia Millia Islamia, New Delhi, India

P. Kumar et al. (eds.), *Applications and Challenges of Geospatial Technology*,
https://doi.org/10.1007/978-3-319-99882-4_12

Keywords Land use/land cover change · Urban expansion · Shannon's entropy index · Kolkata urban agglomeration

12.1 Introduction

At present, we the global people are living in an 'urban age' (Brenner and Schmid 2014). For the first time in the human history in today's world, the number of inhabitants residing in urban areas had outstripped the figure of those living in rural villages. It is estimated that by 2050, about two-thirds of human population would be living in urban areas (United Nations 2014). Future prediction reveals that by 2030 out of 8.1 billion population in the world, 5 billion people will live in urban areas. Most of this growth will be concentrated in the developing countries, especially sub-Saharan Africa and southern as well as south-eastern Asia. Though the growth of urban population in the countries like China and India has slowed down, the absolute number is astonishing. From the recent Census of India data, it is evident that despite the slow rate of urbanization, the absolute addition of population in urban areas was even more than the total rural population in India (Census of India 1991, 2011). However, urbanization continues in the cities of developing courtiers without adequate planning or expansion of infrastructure. As a result, unlike developed countries, the rapid urbanization in developing countries (Montgomery 2008) leads to unplanned and haphazard urban expansion (Cohen 2006; Grimm et al. 2008). One of the striking features of India's urbanization is the rapid growth of metropolitan suburbs in spatial transformation (United Nations Population Fund 2007; World Bank 2013). Since the last quarter of the twentieth century, India has been witnessing predominantly outward expansion of most of the large megacities in the form of sprawl, and peripheries have been engulfing many small towns and villages rather than accommodating the migrants from rural areas in the city core. After globalization, liberalization and privatization, many Indian large cities like Kolkata, Mumbai, Delhi, etc. had started to spill over their administrative boundary (Shaw and Satish 2007), and this process was fuelled by urban planner's proposed new development areas. Urbanization can fundamentally be defined as the process of transformation of land mostly happening as a consequence of rural to urban migration (Velmurugan and Sajjad 2009; Bhagat and Mohanty 2009; Taubenböck et al. 2009). The concept of evolution of urbanization as a process starts with the configuration of urban areas in terms of towns and cities which later on takes the dimension of metropolitan as well as urban agglomeration (Jokar et al. 2013). Urban growth came under the domain of a multifaceted vibrant progression comprising of dynamism in physical as well as functional constituents of built environment which pick up the pace of transition of landscape to urban forms (Castle and Crooks 2006). Those drivers of landscape change to configure urban identity incorporated several geographical, environmental, as well as sociopolitical factors (Su et al. 2012; Dutta et al. 2017; Sahana et al. 2018). Across the several parts of the world, a new global economy boost up resulted in the process of urbanization that is factually reforming the visage of the globe (Sajjad and Iqbal 2012; Abbas 2016). Metropolitan cities in most of the developing countries across the globe have grown up speedily due to rapid population increase and growth in different economic sectors (Luo and Mountrakis 2010; Al-sharif and Pradhan 2015). So, managing urban growth is a very intricate phenomenon and an immense challenge of the present century (Cohen 2004). Though rapid urbanization has been measured on the scale of expansion of suburban growth or by urban sprawl, there are several temporal phases of urban development (Duany et al. 2001; Sajjad 2014). In the first stage of the process of urbanization, the land use/land cover changes are confined to urban fringes as a result of urban expansion (Nuissl et al. 2009). This process results in changes in landscape function and consequently land degradation (Dewan and Yamaguchi 2009; Jamil et al. 2018).

Remote sensing and geographical information system (GIS) techniques have made it achievable to produce modelling and analysis of urban growth and its predication globally (Awasthi et al. 2011; Al-shalabi et al. 2013). These techniques have been extensively used for mapping urban expansion and prediction of change over the years (Sahana et al. 2018; Haack and Rafter 2006). Many models have been introduced to analyse land use/land cover transformation and urban expansion as these sophisticated techniques have become powerful instrument in monitoring and analysing urban landscape dynamics too (Al-shalabi et al. 2012; Hashem and Balakrishnan 2015; Prenzel 2004; Jat et al. 2008). Many efforts have been made to model urban growth and define urban spatial patterns using Shannon's entropy model over the world (Torrens and Alberti 2000; Angel et al. 2007; Jiang et al. 2007; Ahmed and Bramley 2015). During the last 50 years, the population of India (today 1.2 billion) has become more than double, but the urban population has grown nearly five times. India has been undergoing rapid urbanization over the last three decades (Bhagat and Mohanty 2009). With the passage of time as well as in every decade, the number of urban agglomeration (UA) has grown rapidly as evidenced from the growth scenario of Mumbai, Delhi and Kolkata, Bangalore, Chennai and Hyderabad in India, and it would be expected that India will have the biggest concentration of urban agglomeration in the world by the year 2021 (Chakrabarti 2001, Taubenböck et al. 2009). The Kolkata urban agglomeration is the tenth largest in the world and the third largest in the country (only in Eastern India) (UN 2011). KUA is continuously expanding over the urban agglomeration from the last few decades (Bhatta 2009). Various studies have attempted (Bhatta 2009; Ramachandra et al. 2014; Mondal et al. 2015, 2016) to analyse urban growth as well as land transformation especially urban areas within Kolkata Municipal Corporation and its adjoin regions, but hardly any work has been conducted to analyse urban expansion and spatiotemporal growth of the whole Kolkata urban agglomeration. Apart from the assessment of changing land use/land cover dynamics over the last 25 years (1990–2015), the present study used Shannon's entropy model of urban expansion in Kolkata urban agglomeration. Incorporation of specific administrative area (municipality level) along the river line of urban development for calculation of built-up density change as well as entropy dynamics of sprawled urban area gives the present work a newer dimension. The main focus of the present research encompasses to comprehend the LULC dynamics using multi-temporal satellite data along with spatial outlook of urban growth by computing built-up density and by using Shannon's entropy model.

12.2 Study Area

Kolkata urban agglomeration (KUA), also identified as Kolkata Metropolitan Area (KMA), has been incessantly escalating over the last 100 years. With the areal coverage of approximately 1851 sq.km, KUA is located between 22°0′19″ N to 23°0′01″ N latitude and 88°0′04″ E to 88°0′33″ E longitude (Fig. 12.1). The Kolkata urban agglomeration has communicable linear urban prototype along both east and west bank of the river *Hooghly*, which is one of the lifelines of Southern Bengal. The agglomeration is surrounded by rural hinterland lying as a ring around the metropolitan area and acting as a shielding green belt (KMC 2015). It consists of a complex set of administrative entities of 3 important municipal corporations (MC) of South Bengal (Kolkata, Howrah and Chandernagore), 38 municipalities, 77 Census towns (CT), 16 outgrowths and 445 rural villages.

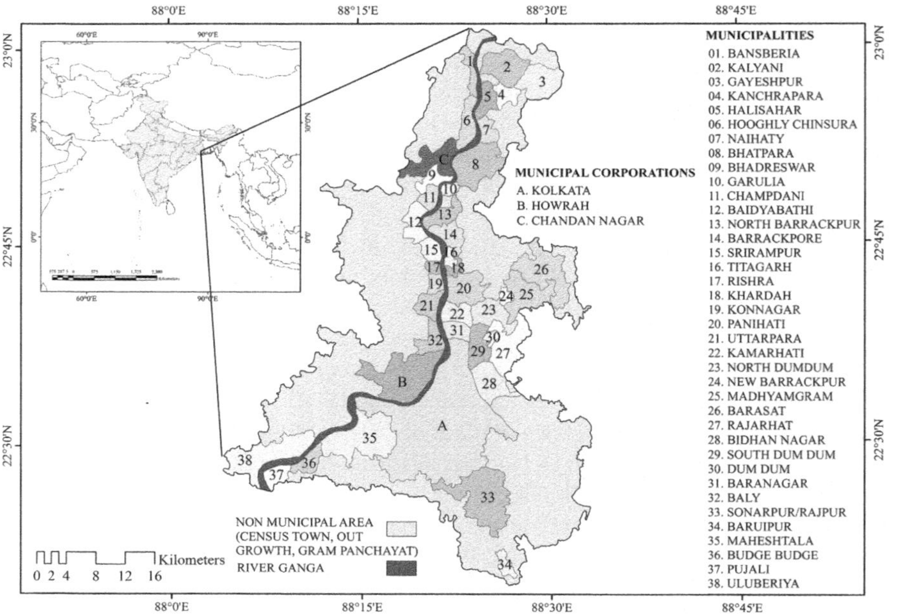

Fig. 12.1 Location of Kolkata urban agglomeration and its spatial identity

According to the 2011 Census, the overall population of the KMA has been recorded as 14.72 million with population density of 7950 persons per square kilometre. Its annual population growth rate is 1.8% by 2011, and its population would be wished-for to rise to 20 million in the year 2021 as well as 21.1 million in the year of 2025 (Census of India 2011; KMDA 2011). Kolkata urban agglomeration features the 30 largest megacities of the world having a population of more than 10 million (UN 2007). Kolkata is the third largest urban agglomeration and the third largest city in India. As a growing metropolitan city (almost likely to other cities around the globe) in a developing country, Kolkata confronts with considerable urban pollution in terms of air-water and noise, traffic congestion, poverty, overpopulation as well as numerous socio-economic nuisances (Bhatta 2009; Mukherjee 2012). Moreover KMA has huge number of slum population which is more than 33% of the total urban population. The mixed nature of urban land uses comprising of residential and commercial as well as industrial character is found in slums of the urban agglomeration (Roy et al. 2014; Sugiyama 2008; Bhatta 2009). The back swamp and marshy lands lie in the eastern corner of the KMA, especially in the Bidhan Nagar, Rajarhat, Maheshtala and Sonarpur. These marshy lands are being encroached by local inhabitants for residential purposes and fitting for urbanized tract mostly without any planning (Ghosh and Sen 1987; Dasgupta et al. 2013).

12.3 Database and Methodology

At the initial stage of study, LULC change and its spatial patterns over time have been analysed by using a set of two specific Landsat images, acquired for two different years, i.e. 1990 (TM) and 2015 (OLI/TIRS). In contemporary research in RS-GIS, Landsat data has been extensively used as satellite data for assessment of urbanization process. The source of data and their characteristics are given in Table 12.1.

Table 12.1 Details of spatial and nonspatial data used for this study

Data	Data types	Source of data	Details about data	Period
Landsat TM	Spatial	USGS satellite images	(30 m resolution) path/raw 138, 45	November 1990
Landsat 8 OLI	Spatial	USGS satellite images	(PAN 15 m, 30 m resolution) path/ raw 138, 45	November, 2015
Base map	Spatial	KMDA	Polygon shape	2011
Population	Nonspatial	Census of India	PCA, 1991, 2001 and 2011 town and village abstract	1991, 2011, 2001

12.3.1 Assessment of the Degree of Urbanization

Landsat satellite images of 1990 (Landsat 5) and 2015 (Landsat 8 OLI), respectively, have been used to generate land use/land cover maps of Kolkata urban agglomeration. In the present study, the supervised classification based on Landsat images of two different time phases has been assessed to identify the spatiotemporal dynamics in different land use parameters. Maximum likelihood method and supervised classification techniques were employed for this classification. The classification error was reduced by reclassifying the generalized images and improved the classification accuracy. Kappa index (Cohen 1968) and KHAT statistic (an estimate of kappa) were used for accuracy assessment. Cohen's kappa index is a multinominal sampling model used to measure the accuracy assessment (Galton 1892; Sahana and Sajjad 2017). In this processing, the whole accuracy of land use/land cover classes was 88.46% for 1990 and 91.92% for 2015, and kappa coefficient values were 0.88 and 0.93, respectively. Areal change discovery for each LULC class over a period of two and half decades has been assessed. The obtained values of individual land use/land cover classes have been used to compute the rate of change in every LULC class using the subsequent rule (Puyravaud 2003; Sahana and Sajjad 2018):

$$R = \left[\frac{1}{t_1 - t_2}\right] \times \left[\ln\left(\frac{C_1}{C_2}\right)\right]$$

where R indicates rate of land use/land cover change and $C1$ and $C2$ are the representatives of built-up area in time periods t1 (2015) and t2 (1990), respectively.

Builtup density (BD) is percentage proportion of built-up area with its total area of a particular spatial unit, which has been calculated by using this formula:

$$BD = \frac{\text{builtup area}}{\text{total area}} \times 100$$

12.3.2 Shannon's Entropy Model of Urban Expansion

Various scholars applied Shannon's entropy model (Shannon 1948) to study the relative urbanization phenomena. This model is competent to weigh up the spatial disparity of urban areas within GIS environment (Li and Yeh 2004). This model is basically used to analyse disparity of urban patterns and assess urban sprawl growth. In our study we used this model for assessing urban expansion in urban subcentre region. Five nonurban land use/land cover classes were considered, water bodies, vegetation, agricultural lands, barren lands and urban areas, which include built-up areas. The entropy value was used to identify the level of urbanization. This value varies between zero and one. The compact urban distribution was indicated by a zero entropy value, and dispersed urban distribution was indicated by nearly one value (Bhatta 2010). So the higher entropy value indicates a high level of sprawl concentration. The below equation was used to assess the relative Shannon's entropy values (Shannon 1948):

$$H_n = \sum\nolimits_{i=1}^{n} p_i \log\left(\frac{1}{p_i}\right)/\log(n)$$

where $Pi =$ the probability (%) of the variable occurring within block i (i.e. urban area (%) in the ith block determined by the urban area in the ith block/block area) and $n =$ total number of block (i.e. 8).

12.4 Result and Discussion

12.4.1 *Spatiotemporal Dynamics of LULC and Urban Built-Up Area*

Advancement in geospatial applications in spatial modelling has significantly enabled the process of assessing spatial mosaic in the analysis of land use dynamics (Sang et al. 2011). Analysis and replication rely on dynamic and wide-ranging satellite databases that portray physical parameters of contemporary urban expansion, incorporation of transportation networks and change in protected areas and topography which controls such land use dynamics. Main characteristics of land use/land cover classification of Kolkata urban agglomeration for the years 1990 and 2015 have been generalized into eight classes, namely, (1) crop land, (2) agricultural fallow, (3) vegetation/plantation, (4) aquatic vegetation, (5) open land, (6) river, (7) wetland and (8) built-up. In this context, KUA practised a severe transformation different in land use/land covers over the last two and half decades. There was a rapid increase in built-up and agricultural fallow whereas a rapid decrease in crop land, vegetation and open land (Table 12.3). Increase in agricultural fallow land has been observed in close proximity to the built-up area. Both the land uses, i.e. agricultural fallow and built-up area, have experienced 62.4 and 45.1% positive change during these 25 years. Mainly the Rajarhat-Sonarpur in south-eastern corner and adjoining region of Bally in south-western part of the study area have experienced enormous increase in agricultural fallow. It is the indication of urban expansion by abolition of primary agricultural land (Fig. 12.2), which has happened tremendously. Open land has also been transformed in built-up area and registered a decline of 67.5 sq. km (3%) during the study period.

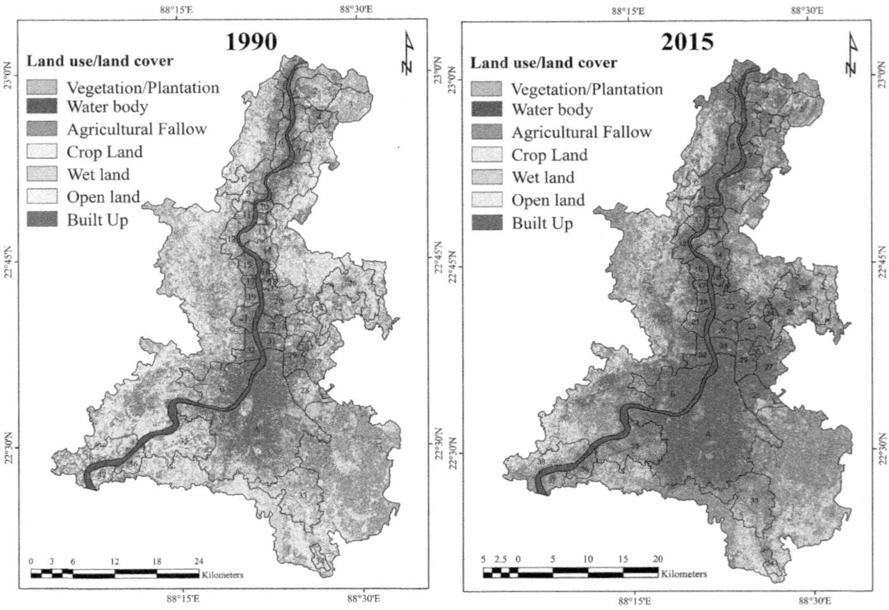

Fig. 12.2 Land use and land cover of Kolkata urban agglomeration in 1990 and 2015

Table 12.2 Change in the area of different land use/land cover classes over the study periods

LULC classes	Area in km^2 (1990)	Area in %	Area in km^2 (2015)	Area in %	Rate of change (%) (1990–2015)
Agricultural land	621.1	34.71	405.1	22.64	−34.8
Agricultural fallow	118.2	6.61	191.9	10.72	62.4
Vegetation/ plantation	154.3	8.62	135.7	7.58	−12.1
Aquatic vegetation	64	3.58	58.1	3.25	−9.2
Open land	177.6	9.92	110.1	6.15	−38
River	55.9	3.12	53.78	3.01	−3.8
Wetland	61.9	3.46	55.9	3.12	−9.7
Built-up	536.5	29.98	778.7	43.52	45.1

Though these particular land uses occur in a very scattered manner over the study area, it has been positively converted into built-up or other land uses. In this conversing scenario, other land use/land cover classes under green space in terms of vegetation as well as plantation, wetlands and aquatic vegetation have been experiencing a drop-off in their respective areas (Table 12.2). In this period (1990–2015), remarkable change has occurred in the percentage of agricultural fallow land which has risen by 62.4 percentage point. Land speculation and losing suitability of the land for agriculture due to the change in physical environment induced by urban expansion could be the main reasons behind such phenomenal increase.

12.4.2 Builtup Density

Proportion of built-up area of a particular unit to total unit area in its percentage figure was indicated as the built-up density of any urban area. In 1990, the highest built-up density is identified in Bally (87.85%) followed by Baranagar (76.55%) and Dumdum (75%) whereas the lowest in Barrackpore (4.4%). During the mentioned 25-year time span, all the 42 units experienced significant positive increase in built-up area. In 2015, the highest built-up density is identified in Dumdum (97.67%) followed by Bally (96.5%) and South Dumdum (95.92%) whereas the lowest in Uluberia (19.20%). So the highest value of built-up density almost reached cent percent, and the minimum threshold also increased significantly. Maximum change in built-up area occurred in case of North Dumdum (46.44%) followed by Rajarhat (38.3%) and North Barrackpore (28.54%). From the spatial output in 1990, area under 60% built-up has been concentrated around the two major urban cores of Kolkata MC and Howrah MC where only Bally Municipality crosses the limit and recorded more than 80% built-up area (Fig. 12.3). After 25 years the scenario has

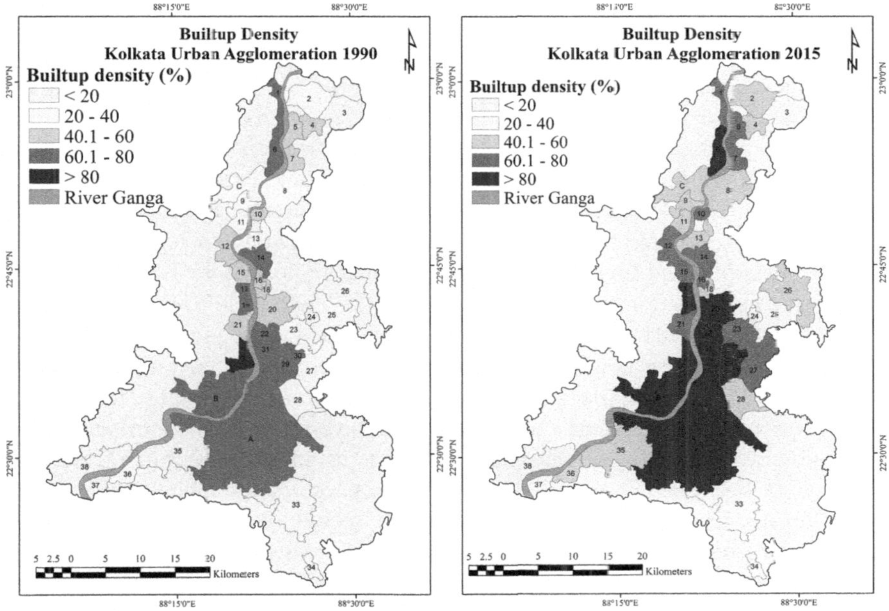

Fig. 12.3 Built-up area density in Kolkata urban agglomeration in 1990 and 2015

changed drastically where the previous occupied area adjoining to Kolkata MC and Howrah MC has crossed the 80% built-up area. In the northern part of the study area, Hooghly-Chinsurah Municipality also recorded 83.65% built-up area as well as area under lowest record (< 20%) of built-up area only limited to Uluberia municipal area (Fig. 12.3).

12.4.3 Shannon's Entropy Model for Urban Expansion

Shannon's entropy which is an indicator of sprawl indicates that the land use is fragmented in all orders due to the coming out of new urban tracts with point in time. Extreme growth is practical at the centre of the whole urban agglomeration as a single unit. The main interest of this segment of study highlights an increase of entropy values during the last two and half decades, demonstrating the sprawl tendency that demands appropriate strategy interventions for the condition of basic facilities. The relative entropy values over the study periods are larger than 0.5 (Table 12.3). Thus, it is clear that urban growth of the study area is not compact in nature, and it is an evidence of concentration of sprawl growth, and it's clearly seen that the urban expansion is occurring mainly in Khardah (0.54), Bidhan Nagar (0.55), Bally (0.56), Kolkata (0.58), Hooghly-Chinsurah as well as Howrah (0.59) and Uttarpara (0.63) (Table 12.3). Moreover, during the 25 years (1990–2015), sprawling trend for the mentioned municipalities is increasing except in the case of Naihati (−0.04), Titagarh (−0.06), Bidhan Nagar (−0.03), Kolkata (−0.14) and Howrah (−0.08). So planned urban policy is needed to this study area to control the disorganized expansion. Table 12.3 shows that higher positive values of entropy change can be seen in Rajarhat, North Dumdum, Pujali, North Barrackpore and New Barrackpore, i.e. 0.57, 0.54, 0.52, 0.51 and 0.47, which shows that the rate of urban expansion is much more in these three areas over the last 25 years (Figs. 12.4 and 12.5). The higher value of overall entropy for the whole Kolkata urban agglomeration represents higher spreading of the built-up area, which is an indication of urban growth. Spatial mosaic of urban sprawl during this time period has happened in major four directions concentrating in the main urban core of Kolkata MC starting from northern (Barrackpore region) to north-eastern (Dumdum region) to south-eastern (Sonarpur) and south-western (Budge Budge region) as the Kolkata-Howrah MC became saturated in their built-up density. The increase in dispersion between 1990 and 2015 is due to new-fangled areas being added to the municipal area limits and some of the latest housing schemes implemented by the government. Beside this, there is a noteworthy boost in the commercial, industrial as well as recreational complexes in the periphery of Kolkata MC. The government also contributed in the expansion by establishing some of its educational institutions (e.g. Aliah University Campus) as well as major recreational facilities (e.g. eco-park) in the Rajarhat, South Dumdum and Dumdum which aggravated the sprawling of built-up area in the north-eastern corner of Kolkata MC. In the southern extension of KUA, Maheshtala, Budge Budge, Pujali and Sonarpur have been added to the high sprawling zone during the 25 years.

Table 12.3 Built-up density and entropy index of 1990 and 2015

Name of the municipality	Total area (sq. km.)	Built-up 1990 (sq. km.)	Built-up density 1990	Built-up 2015 (sq. km.)	Built-up density 2015	Change in built-up density 1990–2015	Entropy 1990	Entropy 2015	Change in entropy value
Baidyabati	12.97	5.71	44.02	8.32	64.15	20.12	0.31	0.57	0.26
Bally	10.86	9.54	87.85	10.48	96.50	8.66	0.56	0.61	0.05
Bansberia	8.7	5.91	67.93	6.53	75.06	7.13	0.41	0.58	0.17
Baranagar	9.17	7.02	76.55	8.52	92.91	16.36	0.27	0.42	0.15
Barasat	36.44	6.29	17.26	16.46	45.17	27.91	0.39	0.81	0.42
Barrackpore	12.33	7.75	62.85	9.3	75.43	12.57	0.44	0.85	0.41
Baruipur	7.01	0.65	9.27	1.76	25.11	15.83	0.13	0.27	0.14
Bhadreswar	8.25	2.61	31.64	4.46	54.06	22.42	0.31	0.57	0.26
Bhatpara	31.06	12.32	39.67	16.67	53.67	14.01	0.45	0.46	0.01
Bidhan Nagar	23.21	7.96	34.30	13.81	59.50	25.20	0.55	0.25	−0.3
Budge Budge	11.06	3.14	28.39	5.32	48.10	19.71	0.29	0.71	0.42
Champdani	9.95	3.61	36.28	4.1	41.21	4.92	0.25	0.37	0.12
Chandan Nagar	21.52	7.1	32.99	10.98	51.02	18.03	0.45	0.52	0.07
Dumdum	3	2.25	75.00	2.93	97.67	22.67	0.41	0.67	0.26
Garulia	4.82	2.61	54.15	3.82	79.25	25.10	0.27	0.37	0.1
Gayeshpur	23.35	1.94	8.31	5.01	21.46	13.15	0.02	0.29	0.27
Gram panchayat	967.11	167.95	17.37	244.11	25.24	7.88	0.06	0.33	0.27
Halisahar	9.73	4.44	45.63	7.06	72.56	26.93	0.19	0.39	0.2
Hooghly-Chinsurah	12.6	8.02	63.65	10.54	83.65	20.00	0.59	0.82	0.23
Howrah	46.55	31.55	67.78	38.9	83.57	15.79	0.59	0.51	−0.08
Kalyani	28.25	7.82	27.68	11.32	40.07	12.39	0.46	0.75	0.29
Kamarhati	11.86	8.82	74.37	9.68	81.62	7.25	0.38	0.67	0.29

(continued)

Table 12.3 (continued)

Name of the municipality	Total area (sq. km.)	Built-up 1990 (sq. km.)	Built-up density 1990	Built-up 2015 (sq. km.)	Built-up density 2015	Change in built-up density 1990–2015	Entropy 1990	Entropy 2015	Change in entropy value
Kanchrapara	9.44	4.02	42.58	5	52.97	10.38	0.23	0.52	0.29
Khardah	5.77	2.41	41.77	3.31	57.37	15.60	0.54	0.59	0.05
Kolkata	200.71	123.76	61.66	163.34	81.38	19.72	0.58	0.44	−0.14
Konnagar	5.17	3.21	62.09	4.24	82.01	19.92	0.49	0.59	0.1
Madhyamgram	25.92	5.72	22.07	6.85	26.43	4.36	0.35	0.69	0.34
Maheshtala	45	8.53	18.96	21.29	47.31	28.36	0.35	0.82	0.47
Naihati	8.7	4.78	54.94	6.17	70.92	15.98	0.35	0.31	−0.04
New Barrackpore	13.4	0.59	4.40	3.13	23.36	18.96	0.38	0.85	0.47
North Barrackpore	12.72	2.86	22.48	6.49	51.02	28.54	0.41	0.92	0.51
North Dumdum	17.4	5.08	29.20	13.16	75.63	46.44	0.34	0.88	0.54
Panihati	20.68	11.43	55.27	17.24	83.37	28.09	0.28	0.47	0.19
Pujali	8.31	1.42	17.09	2.2	26.47	9.39	0.13	0.65	0.52
Rajarhat	28.64	9.14	31.91	20.11	70.22	38.30	0.36	0.93	0.57
Rishra	3.48	2.09	60.06	2.85	81.90	21.84	0.34	0.57	0.23
Sonarpur/ Rajpur	55.89	5.68	10.16	15.09	27.00	16.84	0.19	0.63	0.44
South Dumdum	14.46	10.54	72.89	13.87	95.92	23.03	0.35	0.78	0.43
Srirampur	9.67	4.96	51.29	6.08	62.87	11.58	0.37	0.64	0.27
Titagarh	4.36	2.52	57.80	2.97	68.12	10.32	0.42	0.36	−0.06
Uluberia	39.53	6.51	16.47	7.59	19.20	2.73	0.38	0.51	0.13
Uttarpara	12.14	6.3	51.89	7.37	60.71	8.81	0.63	0.69	0.06

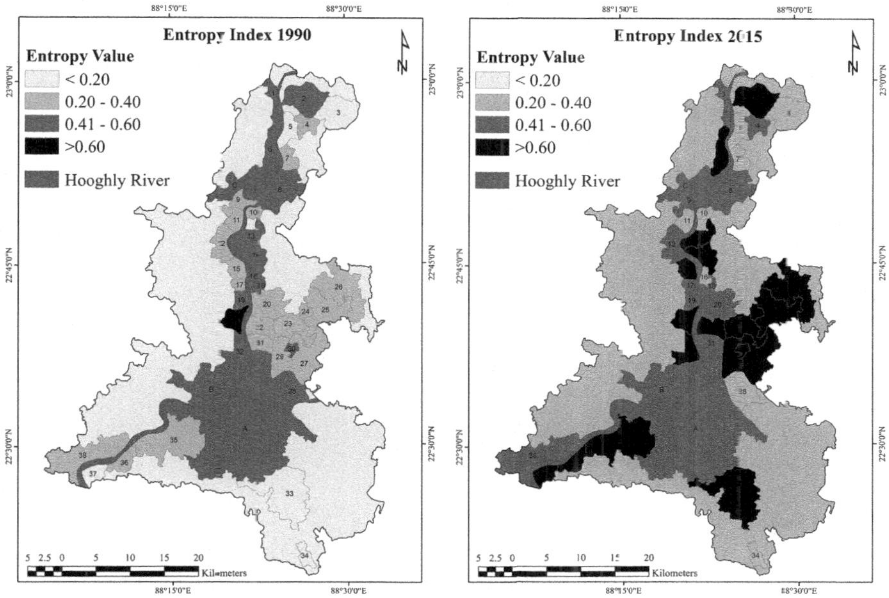

Fig. 12.4 Entropy index of 1990 and 2015 of Kolkata urban agglomeration

12.5 Conclusion

Kolkata is the world's 13th mostly inhabited and 8th biggest urban agglomeration in the world, having a population of approximately over 14 million. Urban organization indicates the diverse alignment of earthly objects and different land use characteristics which broadly include green space, built-up, water bodies and wetlands as well as open spaces in a region. So the perception and evaluation of urban dynamics in spatiotemporal context involve significant land transitions. Present attempt demonstrates the quantification of land use dynamics in the most popular urban agglomeration in Eastern India emphasizing on the spatiotemporal domain of urban sprawl through multi-temporal satellite data using Shannon's entropy model. Land use and land cover analysis for Kolkata urban agglomeration reveals that the area for agricultural land significantly has declined from 35% (1990) to 23% in 2015. Simultaneously agricultural fallow and built-up area increased from 6.65 (1990) to 7.2% (2015) and 30% (1990) to 43.5% (2015), respectively, during the 25 years. Comparing to this situation, built-up density of different constituent municipalities also increased significantly. Among them municipalities of north-eastern adjoin regions of KMC are noteworthy (i.e. North Dumdum, Rajarhat). Shannon's entropy values draw attention to the tendency of urban sprawl that urge for apt policy interventions to afford basic facilities as well as services in the regions. Spatially urban sprawl during the 25-year time span covered almost major four directions

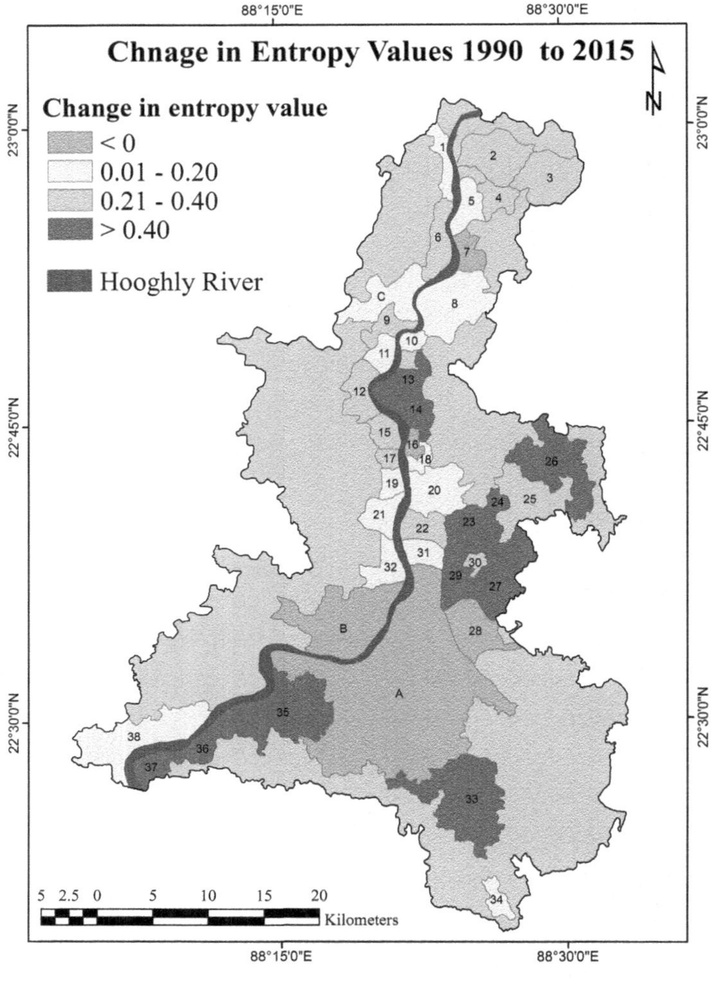

Fig. 12.5 Change in entropy value (1990–2015) in Kolkata urban agglomeration

concentrating in the main urban core of Kolkata MC starting from Barrackpore region in the north to Dumdum region in the north-eastern to Sonarpur and Budge Budge region in the south-eastern and south-western directions, respectively. In the present study, multi-temporal satellite data (1990 and 2015) under Shannon's entropy index has provided considerable versatile information in every order of spatial units (e.g. municipality and municipal corporation) for unfolding the spatial design of urban growth dynamics to maintain choice-based planning in Kolkata UA as a composite urban system.

References

Abbas R (2016) Internal migration and citizenship in India. J Ethn Migr Stud 42(1):150–168 https://doi.org/10.1080/1369183X.2015.1100067

Ahmed S, Bramley G (2015) How will Dhaka grow spatially in future?-modelling its urban growth with a near-future planning scenario perspective. Int J Sustain Built Environ 4(2):359–377

Al-shalabi M, Billa L, Pradhan B, Mansor S, Abubakr A, Al-Sharif A (2012) Modelling urban growth evolution and land-use changes using GIS based cellular automata and SLEUTH models: the case of Sana'a metropolitan city, Yemen. Environ Earth Sci 70:425–437 https://doi.org/10.1007/s12665-012-2137-6

Al-shalabi M, Billa L, Pradhan B, Mansor S, Al-Sharif AA (2013) Modelling urban growth evolution and land-use changes using GIS based cellular automata and SLEUTH models: the case of Sana'a metropolitan city, Yemen. Environ Earth Sci 70(1):425–437

Al-sharif AA, Pradhan B (2015) A novel approach for predicting the spatial patterns of urban expansion by combining the chi-squared automatic integration detection decision tree, Markov chain and cellular automata models in GIS. Geocarto Int 30(8):858–881

Angel S, Parent J, Civco D (2007) Urban sprawl metrics: an analysis of global urban expansion using GIS. Proceedings of ASPRS 2007 annual conference, Tampa, Florida May 7–11. http://clear.uconn.edu/publications/research/tech_papers/Angel_et_al_ASPRS2007.pdf

Awasthi A, Chauhan SS, Goyal SK (2011) A multi-criteria decision making approach for location planning for urban distribution centers under uncertainty. Math Comput Model 53(1–2):98–109

Bhagat RB, Mohanty S (2009) Emerging pattern of urbanization and the contribution of migration in urban growth in India. Asian Popul Stud 5(1):5–20

Bhatta B (2009) Analysis of urban growth pattern using remote sensing and GIS: a case study of Kolkata, India. Int J Remote Sens 30(18):4733–4746

Bhatta B (2010) Analysis of urban growth and sprawl from remote sensing data. Springer, Berlin

Brenner N, Schmid C (2014) The 'urban age'in question. Int J Urban Reg Res 38(3):731–755

Castle CJ, Crooks AT (2006) Principles and concepts of agent-based modelling for developing geospatial simulations, Centre for Advanced Spatial Analysis University College London, 1-19 Torrington Place, London, WC1E 6BT, UK

Census of India (1991) Primary census abstract, census of India. The government of India, Registrar General and Census Commissioner of India, Ministry of Home Affairs, New Delhi, India

Census of India (2011) Primary census abstract, census of India. The government of India, Registrar General and Census Commissioner of India, Ministry of Home Affairs, New Delhi, India

Cohen J (1968) Weighted kappa: nominal scale agreement provision for scaled disagreement or partial credit. Psychol Bull 70(4):213

Cohen B (2004) Urban growth in developing countries: a review of current trends and a caution regarding existing forecasts. World Dev 32(1):23–51. https://doi.org/10.1016/j.worlddev.2003.04.008

Cohen B (2006) Urbanization in developing countries: Current trends, future projections, and key challenges for sustainability. Technol Soc 28(1-2):63–80

Dasgupta S, Asvani K, Gosain K, Rao S, Roy S, Sarraf M (2013) A megacity in a changing climate: the case of Kolkata. Clim Chang 116:747–766 https://doi.org/10.1007/s10584-012-0516-3

Dewan AM, Yamaguchi Y (2009) Land use and land cover change in Greater Dhaka, Bangladesh: Using remote sensing to promote sustainable urbanization. Appl Geogr 29(3):390–401

Dhar Chakrabarti PG (2001) Urban crisis in India: New initiatives for sustainable cities. Dev Pract 11(2-3):260–272

Duany A, Plater-Zyberk E, Speck J (2001) Suburban nation: the rise of sprawl and the decline of the American dream. Macmillan, New York

Dutta S, Sahana M, Guchhait SK (2017) Assessing anthropogenic disturbance on forest health based on fragment grading in Durgapur Forest Range, West Bengal, India. Spat Inf Res 25 (3):501–512

Ghosh D, Sen S (1987) Ecological history of Calcutta's wetland conversion. Environ Conserv 14 (3):219–226

Grimm NB, Faeth SH, Golubiewski NE, Redman CL, Wu J, Bai X, Briggs JM (2008) Global change and the ecology of cities. Science 319(5864):756–760

Haack BN, Rafter A (2006) Urban growth analysis and modeling in the Kathmandu Valley, Nepal. Habitat Int 30(4):1056–1065

Hashem N, Balakrishnan P (2015) Change analysis of land use/land cover and modelling urban growth in greater Doha, Qatar. Ann GIS 21(3):233–247

Jat MK, Garg PK, Khare D (2008) Monitoring and modelling of urban sprawl using remote sensing and GIS techniques. Int J Appl Earth Obs Geoinf 10(1):26–43

Jamil M, Sahana M, Sajjad H (2018) Crop suitability analysis in the Bijnor District, UP, using geospatial tools and fuzzy analytical hierarchy process. Agricultural research, Springer 1-17. https://doi.org/10.1007/s40003-018-0335-5

Jiang F, Liu S, Yuan H, Zhang Q (2007) Measuring urban sprawl in Beijing with geo-spatial indices. J Geogr Sci 17:469–478

Jokar J, Helbich M, De-Noronha E (2013) Spatio-temporal simulation of urban growth patterns using agent-based modeling: the case of Tehran. Cities 32:33–42. https://doi.org/10.1016/j.cities.2013.01.005

KMC (2015) Basic statistics of Kolkata. https://www.kmcgov.in/KMCPortal/jsp/KolkataStatistics.jsp. Accessed May 2016

KMDA (2011) Kolkata metropolitan development authority. http://www.kmdaonline.org/html/about-us. Accessed May 2016

Li X, Yeh AGO (2004) Analyzing spatial restructuring of land use patterns in a fast growing region remote sensing and GIS. Landsc Urban Plan 69:335–354

Luo L, Mountrakis G (2010) Integrating intermediate inputs from partially classified images within a hybrid classification framework: an impervious surface estimation example. Remote Sens Environ 114(6):1220–1229

Mondal B, Das DN, Dolui G (2015) Modeling spatial variation of explanatory factors of urban expansion of Kolkata: a geographically weighted regression approach. Model Earth Syst Environ 1:29. https://doi.org/10.1007/s40808-015-0026-1

Mondal B, Das DN, Bhatta B (2016) Integrating cellular automata and Markov techniques to generate urban development potential surface: a study on Kolkata agglomeration. Geocarto Int. https://doi.org/10.1080/10106049.2016.1155656

Mukherjee M (2012) Urban growth and spatial transformation of Kolkata metropolis: a continuation of colonial legacy. ARPN J Sci Technol 2:365–380

Montgomery MR (2008) The urban transformation of the developing world. Sci 319 (5864):761–764

Nuissl H, Haase D, Lanzendorf M, Wittmer H (2009) Environmental impact assessment of urban land use transitions—a context-sensitive approach. Land Use Policy 26(2):414–424

Population Division (2014) World urbanization prospects: The 2014 revision, highlights (ST/ESA/SER.A/352)

Prenzel B (2004) Remote sensing-based quantification of land-cover and land-use change for planning. Prog Plan 61(4):281–299

Puyravaud JP (2003) Standardizing the calculation of the annual rate of deforestation. For Ecol Manag 177(1–3):593–596

Ramachandra TV, Aithal BH, Sowmyashree MV (2014) Urban structure in Kolkata: metrics and modelling through geo-informatics. Appl Geomat 6:229–244. https://doi.org/10.1007/s12518-014-0135-y

Roy D, Lees M, Palavalli B, Pfeffer K, Sloot M (2014) The emergence of slums: a contemporary view on simulation models. Environ Model Softw 59:76–90

Sahana M, Sajjad H (2017) Evaluating effectiveness of frequency ratio, fuzzy logic and logistic regression models in assessing landslide susceptibility: a case from Rudraprayag district, India. J Mt Sci 14(11):2150–2167

Sahana M, Sajjad H (2018) Assessing influence of erosion and accretion on landscape diversity in sundarban biosphere reserve, lower ganga basin: a geospatial approach in quaternary geomorphology in india. Springer, Cham, pp 191–203 2019
Sahana M, Hong H, Sajjad H (2018) Analyzing urban spatial patterns and trend of urban growth using urban sprawl matrix: A study on Kolkata urban agglomeration, India. Sci Total Environ 628–629:1557–1566
Sajjad H, Iqbal M (2012) Impact of urbanization on land use/land cover of Dudhganga watershed of Kashmir Valley, India. Int J Urban Sci 16(3):321–339
Sajjad H (2014) Living standards and health problems of lesser fortunate slum dwellers: evidence from an Indian City. Int J Environ Protect Policy 2(2):54–63. https://doi.org/10.11648/j.ijepp.20140202.13
Sang L, Zhang C, Yang J, Zhu D, Yun W (2011) Simulation of land use spatial pattern of towns and villages based on CA–Markov model. Math Comput Model 54(3–4):938–943
Shannon CE (1948) A mathematical theory of communication. Bell Syst Tech J 27(3):379–423
Shaw A, Satish MK (2007) Metropolitan restructuring in post-liberalized India: separating the global and the local. Cities 24(2):148–163
Su S, Xiao R, Jiang Z, Zhang Y (2012) Characterizing landscape pattern and ecosystem service value changes for urbanization impacts at an eco-regional scale. Appl Geogr 34:295–305
Sugiyama M (2008) The study on climate impact adaptation and mitigation in Asian coastal mega cities of integrated research system for sustainability science. University of Tokyo. Final Report to JICA
Taubenböck H, Wegmann M, Roth A, Mehl H, Dech S (2009) Urbanization in India–spatiotemporal analysis using remote sensing data. Comput Environ Urban Syst 33(3):179–188
Torrens PM, Alberti M (2000) Measuring sprawl. Centre for Advanced Spatial Analysis, London
UN (2007) World urbanization prospects: database. http://www.un.org/esa/population/publications/WUP2005/2005WUP_DataTables11.pdf
UN (2011) World urbanization prospects: the 2011 revision. United Nations Department of Economic and Social Affairs/Population Division, New York
United Nations Population Fund (2007) The state of world population 2007: unleashing the potential of urban growth, United Nations publications, and chapter 1
United Nations (2014) World urbanization prospects. The 2014 revision department of economic and social affairs population division New York
Velmurugan A, Sajjad H (2009) The study of land transformation and land degradation in Dehradun District, Uttrakhand. Deccan Geographer 48. (ISSN-0011-7269)
World Bank Group (2013) World development indicators 2013. World Bank Publications

Chapter 13
Automated Extraction of Urban Impervious Area from Spectral-Based Digital Image Processing Techniques

Suman Sinha

Abstract Urban area comprises a complex mix of diverse land cover types and materials; it is often difficult to separate these classes due to their heterogenic nature. Studying and monitoring urban areas and its environment are closely associated with the study of impervious surfaces, which are anthropogenic features through which water cannot infiltrate into the soil. In the present study, spectral indices were developed using spectral information from satellite remote sensing sensor. Several spectral indices like vegetation index, soil-adjusted vegetation index (SAVI) and normalized difference vegetation index (NDVI); water index, modified normalized difference water index (MNDWI); and urban indices, normalized difference built-up index (NDBI), built-up index (BUI) and index-based built-up index (IBI), were implemented in the study. The combination of various spectral indices can be used, and finally using NDBI, BUI and IBI, principal component analysis (PCA) was performed, the first component of which was classified through unsupervised classification through K-means algorithm to extract urban built-up impervious features. The methodology has the potential to identify and automatically extract urban impervious features from other land use-land cover classes and is established over the city of Kolkata (India). Maps showing effective classification of urban areas were developed. The approach is further successfully operated over a forested area in order to extract settlements within the forest patch that proves the transferability of the method and can be universally accepted.

Keywords Urban · Resourcesat LISS-III · Spectral indices · PCA · K-means · Classification

---

S. Sinha (✉)
Department of Civil Engineering, Haldia Institute of Technology, Haldia, India

P. Kumar et al. (eds.), *Applications and Challenges of Geospatial Technology*,
https://doi.org/10.1007/978-3-319-99882-4_13

13.1 Introduction

The Earth is experiencing serious alteration in its processes and environment primarily due to the anthropogenic activities, the consequences of which are severe, irresistible and irreversible. Urbanization, along with its allied detrimental consequences like industrialization, urban sprawl, population explosion, heat island effect etc., is a solemn issue associated with this. Urban represents a complex system with intermixing of several land use and land cover (LULC) types, including water, vegetation or plantation, various soil types and structures, impervious structures, etc. The quantitative and qualitative extent of impervious surface in a landscape is an indicator of environmental quality (Arnold and Gibbons 1996). Impervious surfaces are surfaces through which water cannot infiltrate and are associated with transportation and building rooftops (Bauer et al. 2004). Hence, imperviousness becomes an inevitable factor in runoff, non-point pollution source, water quality, land use-land cover change, land and forest degradation, energy balance, urban heat island, habitat degradation, fragmentation and other ecological services. As the rural environment changes to urban and suburban environment, the proportion of the impervious areas also enhances, and this change of land use has serious effect over the entire ecosystem, primarily owing to decrease in water and forest cover. Urban system has an intricate intermingling of diverse land cover types and materials. Hence, an efficient monitoring of urban impervious areas for timely and accurate calculation at high spatial and temporal resolution has become a serious challenge that can be confronted by the use of satellite remote sensing-based techniques.

Remote sensing technology offers novel prospects to establish significant contributions in urban ecosystems through objective and verifiable characterization of urban composition (Ridd 1995). It provides an easy cost-effective solution to handle this cumbersome problem to efficiently map the impervious surfaces due to the synoptic coverage of any area. Huge load of archived remotely sensed data is easily available freely in a raster-based digital form that allows immense computer-assisted processing with precise and accurate interpretation, as compared to the tedious, time-consuming and erroneous manual interpretations.

Extraction of the impervious features and differentiating from the other LULC types, specifically with multispectral optical remote sensing, are of utmost challenge. The use of these optical satellite sensors for several aspects of the urban environment has been demonstrated by Patino and Duque (2013). The development of the concept started from the vegetation-impervious surface-soil (VIS) model of the urban set-up designed by Ridd (1995) that became the backbone of spectral-based LULC classification which marked the basic strategy to differentiate between vegetation, impervious and soil agglomerates for urban ecosystems. However, the water bodies were not addressed in this model that led to inconveniences in forthcoming urban-related studies using this model, and hence, this became the most serious drawback of the stated model. Xu (2008) targeted this shortcoming by developing IBI that included NDBI (Zha et al. 2003), soil-adjusted vegetation index (SAVI) (Huete 1988) and modified normalized difference water index (MNDWI) (Xu 2006) simultaneously, for extracting built-up areas with an overall accuracy of 96.7%. The fact that spectral response of built-up lands has higher reflectance in the middle-infrared (MIR) or short-wave infrared (SWIR) wavelength range than in the near-infrared (NIR) wavelength range led to the conceptualization of NDBI however, the drier vegetation can show higher reflectance in the MIR or SWIR wavelength range than in the NIR range, which led to built-up land classification often mixed with plant noise (Xu 2008). This eventually led to the development of Index-based Built-up Index (IBI). Urban index (UI), with structurally and functionally similar equation to that of NDBI, developed by Kawamura et al. (1996) behaves similarly as NDBI and IBI; however, UI reveals more mixing of built-up lands with the bare lands than the other two indices (Li et al. 2017) and shows no better classification accuracies while inheriting similar drawbacks of that of NDBI. UI involves SWIR2 band information, while the other indices use information of SWIR from SWIR1 band. BUI is not any new concept that has already been considered during the analysis of NDBI by Zha et al. (2003), however, without producing comparable advancement over NDBI and IBI, due to noises from other classes that lead to misclassifications in built-up classification. UI, NDBI, BUI and IBI result in speedy mapping of built-up areas or bare land areas but are unable to differentiate between these two classes efficiently, while enhanced built-up and bareness index (EBBI) can successfully distinguish among these two classes (As-syakur et al. 2012). Thermal remote sensing using thermal infrared (TIR) band information has the potential to improve LULC classification, as every LULC parcel has a unique distinctive emissivity (As-syakur et al. 2012; Sinha et al. 2014, 2015b; Wang et al. 2015; Li et al. 2017). Relationship exists between land surface temperature (LST) variation derived from MODIS thermal data products and urban built-up density that helps in better understanding the complex dynamics of urban microclimate and urban heat island (UHI) effect (Morabito et al. 2016).

Spectral-based indices have proved beneficial than the original spectral bands in such feature identification. In a study using thermal information from thermal remote sensing sensors, it was observed that thermal-based vegetation indices produced an overall accuracy of more than 90% in comparison to the overall accuracy of 85% for LULC classification using original multispectral band information (Sinha et al. 2015b). Simultaneously, an overall classification accuracy of 85% was obtained by Masek et al. (2000) using unsupervised classification on NDVI image products for detecting built-up features. Vegetation and impervious features are negatively correlated, and this concept of studying vegetation distribution for mapping impervious features was demonstrated by Gillies et al. (2003) and Bauer et al. (2004). However, this perception has serious limitations arising due to seasonal variations and vegetation dynamism. Zha et al. (2003) developed NDBI to delineate urban areas with 92.6% of classification accuracy but involved NDVI corrections owing to the interference of vegetation noise in the classification. The approach of using NDBI was improved by He et al. (2010) where the proposed automatic segmentation method resulted in classification accuracy of 20% higher than the original method for extracting the built-up areas. Improvement in NDBI was also targeted by Varshney (2013) via an automated kernel-based thresholding algorithm. Xu (2007) applied SAVI, MNDWI and NDBI concurrently to classify urban areas with accuracy lying between 91.5 and 98.5%. The same indices were simultaneously used to develop IBI to effectively map out the built-up areas by Xu (2008) with 96.7% as the overall classification accuracy. Likewise, Lee et al. (2010) also developed BUI for mapping built-up areas using inputs from NDBI and NDVI. Thermal data were used by As-Syakur et al. (2012) for designing EBBI for mapping built-up and even separated the bare land from the built-up with high accuracy. Moreover, Normalized Difference Bare Land Index (NBLI) formulated by Li et al. (2017) performed even better than EBBI in distinguishing bare lands from built-up and other lands with an overall classification accuracy of 92%. Normalized Difference Impervious Surface Index (NDISI) developed by Xu (2010) also helped classify impervious surface but produced poor results. Liu et al. (2013) transformed this index to Modified Normalized Difference Impervious Surface Index (MNDISI); however, it too suffered due to its intrinsic complexities. Simultaneously, Wang et al. (2015) also proposed another index for mapping impervious surface, namely, Normalized Difference Impervious Index (NDII), which was developed using Landsat TM thermal data and attained an overall classification accuracy of 91.4%. However, the major challenge is to use visible multispectral bands for mapping built-up impervious surface, and till date, IBI has the best potential for mapping built-up impervious areas.

Mapping of land features started with the use of multispectral satellite images, however, with satisfactory level of classification accuracy owing to spectral mixture of heterogeneous land features (Xu 2008). The methods for image classification using satellite images involve unsupervised classification (Masek et al. 2000), supervised classification (Kumar et al. 2013), interpretation keys (Sinha et al. 2013), principal component analysis (Masek et al. 2000), knowledge-based expert classification using decision tree algorithm (Sharma et al. 2013), texture measures (Weeks et al. 2007), object-based image analysis (Blaschke 2010), spectral index-based classification (Sinha et al. 2015b), artificial neural networks (Hu and Weng 2009) and then different integrated methods and algorithms for image classification (Xu 2002; Xian and Crane 2005). Apart from optical sensors, SAR has also been used for built-up studies (Abdikan et al. 2016; Lv et al. 2015; Aghababaee et al. 2013; Chen et al. 2012); however, it is limited. Synergy of optical and SAR has potential for innovative openings in this context (Sinha et al. 2016; Corbane et al. 2008; Shao et al. 2016; Johnson et al. 2017).

The available literature implies that major approaches dealing with the remote sensing-based estimation methods for urban detection over the last decade comprise of pixel-based, subpixel-based, object-oriented algorithms and artificial neural networks (Weng 2012). Pixel-based techniques include image classification, regression, etc., while subpixel-based methods include linear spectral unmixing, imperviousness as the complement of vegetation fraction, etc. Techniques, such as data or image fusion, textural measures, expert systems, data mining and contextual classification methods, have also been investigated. Various indices have been developed for automated mapping of built-up areas. The majority of research efforts have been made for mapping urban landscapes at various scales from global to local and regional and on a wide range of spatial resolution requirements from coarse to very high resolution for such mapping. Numerous image processing techniques are also being investigated. The current study integrates spectral-based built-up indices and certain digital image processing techniques to automatically extract out the built-up areas.

This study targets the extraction of urban impervious areas with a unique approach that uses principal component analysis (PCA) technique applied to a pseudo-false colour composite (FCC) image formed from three built-up indices, namely, NDBI, BUI and IBI, which are generated from spectral band information of the satellite imagery, with a view of extracting the urban impervious built-up features from the information provided by the spectral-based built-up indices through PCA.

13.2 Materials and Methods

13.2.1 Study Area and Data Sets

The megacity of Kolkata, capital of the state of West Bengal, India, and its surroundings, located at 22.57°N and 88.36°E, is taken as the study site. This megacity mainly encompasses residential and commercial areas with a backdrop of huge proportions of anthropogenic structures and high population density. The site represents a purely urban scenario with heterogeneous features distributed throughout with an average altitude of 17 feet from mean sea level. Kolkata, previously known as Calcutta, is the capital city of the state of West Bengal (India). Kolkata is the most important commercial, cultural and educational centre of Eastern India that is located on the eastern bank of River Hooghly, a distributary of River Ganges, well within the Bengal basin. The annual mean temperature is 26.8 °C; monthly mean temperatures lie in between 19 °C and 30 °C. Maximum temperatures exceed over 40 °C during summer in the months of May and June. Winters occur during December to early February when the minimum temperatures generally lie between 12 °C and 14 °C. The highest recorded temperature is 43 °C and the lowest is 5 °C. Kolkata has a tropical wet-and-dry climate, with summer monsoons dominated by strong southwesterly monsoon winds. Annual average rainfall is 1582 mm. The city generally experiences spells of thunderstorm and heavy rains during early summer. August marks the maximum precipitation, while the monsoon generally occurs during June–September. Temperature inversions occasionally lead to mists during early mornings and smog during evenings. The wind and cyclone zones of Kolkata possess 'very high damage risk'. Also the city lies within the seismic zone III on a scale ranging from I to V in order of increasing susceptibility to earthquakes.

Resourcesat LISS-III imagery acquired on November 23, 2009, is used. It has a spatial resolution of 23.5 m and a swath width of 141 km. The four spectral bands are, namely, the band 1 corresponding to green (0.52–0.59 μm), band 2 corresponding to red (0.62–0.68 μm), band 3 for near infrared or NIR (0.76–0.86 μm) and band 4 for short-wave infrared or SWIR (1.55–1.70 μm). The data was obtained from the National Remote Sensing Centre (NRSC, ISRO, India) (Fig. 13.1).

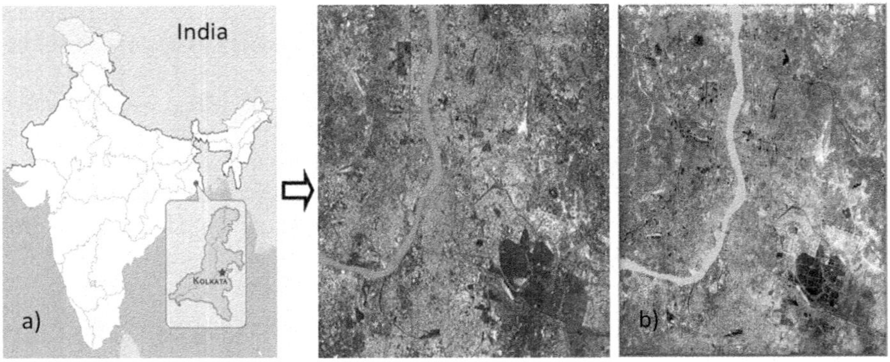

Fig. 13.1 (**a**) Location map of the study site with (**b**) standard FCC of LISS-III satellite imagery

13.2.2 Methodology

The entire methodological flow diagram is sketched out in Fig. 13.2. The first step was the proper rectification and preprocessing of the satellite data. Secondly the spectral-based indices were calculated. Next, PCA was performed using the built-up indices as input parameters. Lastly, the PCA image was classified based on the indices, and classification accuracy was assessed with finally extracting the built-up areas. NEST and Erdas Imagine software were used for this procedure.

Satellite images of LISS-III were co-registered and geometrically rectified in reference to the mosaicked Survey of India (SOI) toposheet for analogous distinct identifiable objects on the toposheets, ground (GT points via Global Positioning System or GPS) and preregistered images of the same region. The satellite image was geocoded with UTM projection, datum WGS-84 and Zone 45 North comprising of green (G), red (R), near-infrared (NIR) and short-wave infrared (SWIR) bands on a 1:50,000 scales (Sinha et al. 2013).

The following spectral-based indices considered in the study are enlisted in Table 13.1. SAVI (with an additional soil factor with NDVI), MNDWI and NDBI were used to extract vegetation, water and built-up feature classes, respectively. These three indices, being mutually negatively correlated, were combined together resulting to IBI (Table 13.1). Simultaneously BUI was also calculated. NDBI, BUI and IBI images were combined to be used as the input parameter for carrying out PCA. As the first principal component has the greatest variance, it produced the highest data variability. So the first principal component (PCA1) was considered for classification through k-means cluster analysis, thus resulting to a binary classified map that delineated the urban impervious areas from the non-impervious urban areas. Accuracy assessment for the classification was performed generating 300 ground control points (GCPs). Based on the accuracy, the method was established and validated over two different regions.

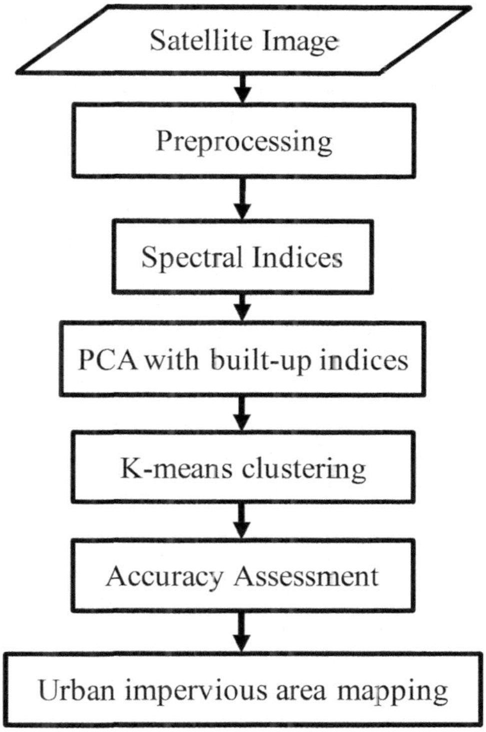

Fig. 13.2 Methodological workflow

Table 13.1 Formulae used in the study

Eq.	Formula	Source	Remarks
1	Normalized difference vegetation index $NDVI = \frac{(NIR-R)}{(NIR+R)}$	Rouse et al. (1974)	Surface reflectance in near-infrared (NIR) and red (R) spectral bands
2	Soil-adjusted vegetation index $SAVI = \frac{(NIR-R)\times(1+L)}{(NIR+R+L)}$	Huete (1988)	Surface reflectance in near-infrared (NIR) and red (R) spectral bands, L is a constant whose value depends on the soil properties
3	Normalized difference built-up index $NDBI = \frac{(SWIR-NIR)}{(SWIR+NIR)}$	Zha et al. (2003)	Surface reflectance in short-wave infrared (SWIR) or middle-infrared (MIR) and near-infrared (NIR) spectral bands
4	Modified normalized difference water index $MNDWI = \frac{(G-SWIR)}{(G+SWIR)}$	Xu (2006)	Surface reflectance in short-wave infrared (SWIR) or middle-infrared (MIR) and green (G) spectral bands
5	Built-up index $BUI = NDBI - NDVI$	Lee et al. (2010)	NDBI and NDVI from Eqs. 3 and 1, respectively
6	Index-based built-up index $IBI = \frac{[NDBI-(SAVI+MNDWI)/2]}{[NDBI+(SAVI+MNDWI)/2]}$	Xu (2008)	NDBI, SAVI and MNDWI from Eqs. 3, 2 and 4, respectively; NDVI from Eq. 1 can also be used instead of SAVI

13.3 Results and Discussion

The nature of the DN response for the feature class shows that vegetation can be well demarcated from the difference in NIR and red bands. Likewise, water can be well identified from the difference in green and NIR or SWIR. Built-up discrimination requires inputs from SWIR; however, any particular spectral information is inappropriate for discriminating built-up/impervious from other feature classes. Figure 13.3 depicts the mean DN values of the four spectral bands in the study area. The maximum variability was observed from the B3, red, and B4: NIR bands. The graph of the relationship between the red and NIR shows the vegetation along the NIR (y-axis), water along red (x-axis) and built-up along the diagonals (Fig. 13.4). Vegetation pixel values in SWIR are observed to be well below NIR values; so dry vegetation that can pose noise in built-up classification is absent.

SAVI and MNDWI maps in Fig. 13.5a, b show an efficient demarcation of vegetation and water bodies, respectively. Moreover, the three built-up indices, namely, NDBI, BUI and IBI maps in Fig. 13.5c, d and e, respectively, could also extract out the built-up areas. Figure 13.6 depicts the scattergrams of spectral feature space of the index images derived from LISS-III bands showing separability of three main land cover classes, namely, water (marked in ‘blue’), vegetation (marked in ‘green’) and urban impervious (marked in ‘red’). Here, the relationship between the built-up indices is demarcated as (a) BUI (x-axis)-IBI (y-axis), (b) NDBI (x-axis)-IBI (y-axis) and (c) NDBI (x-axis)-BUI (y-axis). Figure 13.6 shows positive correlation between the indices, which represents clear separation of the built-up and urban impervious feature class from the remaining two of the major classes, namely, water and vegetation. On the other hand, these two classes are inseparable as depicted in the figure.

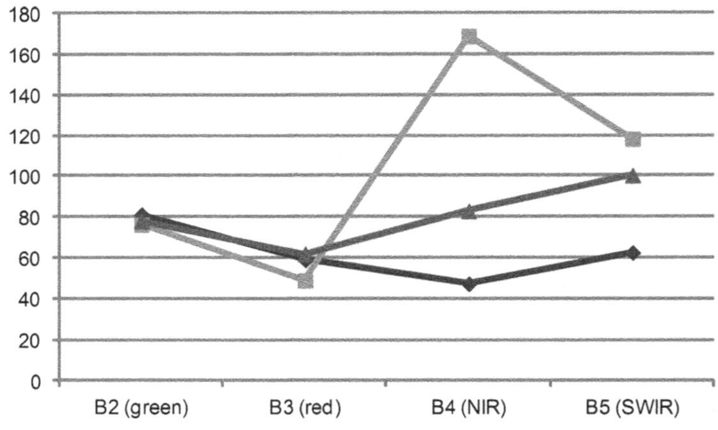

Fig. 13.3 Mean DN values (y-axis) of the four spectral bands (B2, green; B3, red; B4, NIR; B5, SWIR) for water (blue), vegetation (green) and built-up/impervious (red) feature classes for the study area

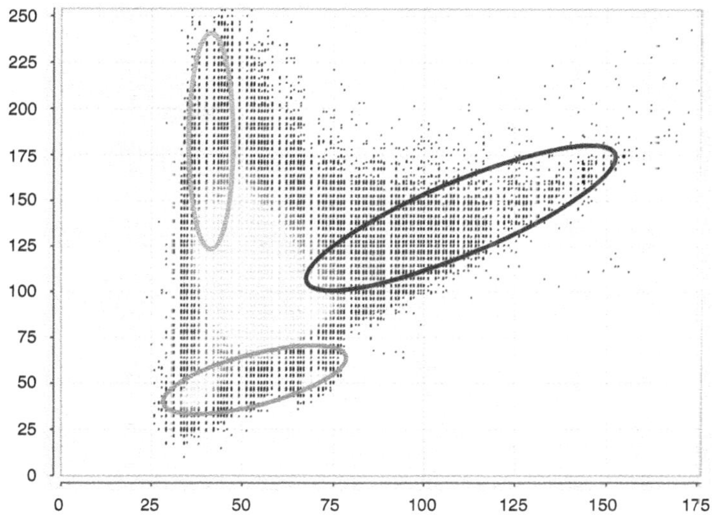

Fig. 13.4 Scatter plot of the spectral clusters in red (x-axis) and NIR (y-axis) bands for water (blue), vegetation (green) and built-up/impervious (red) for the study area

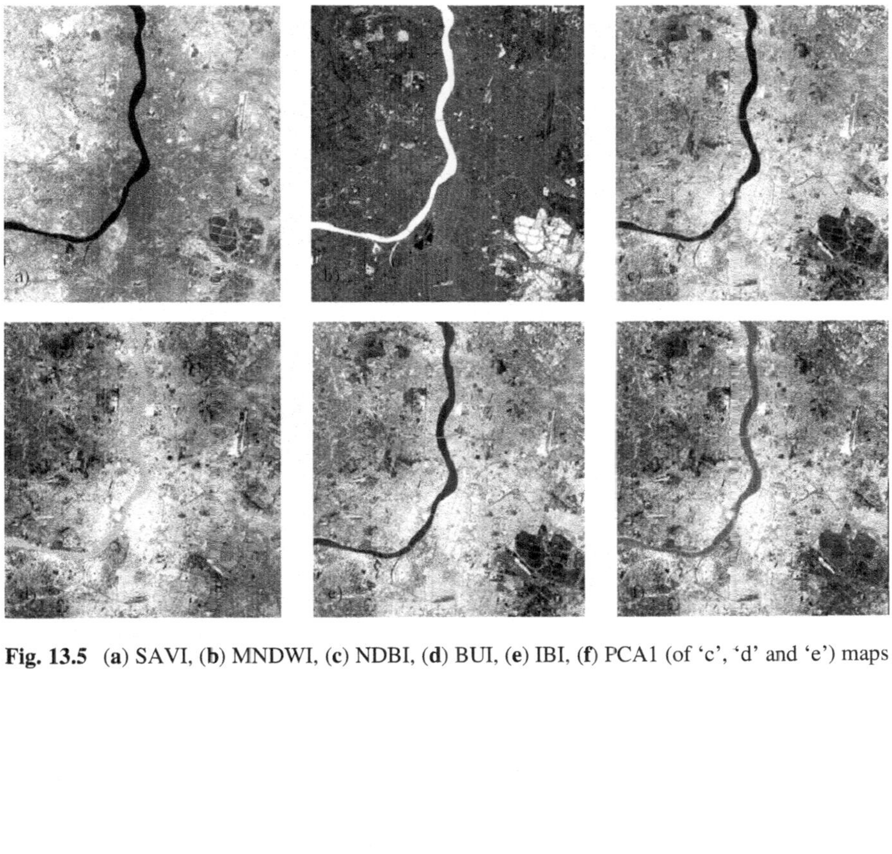

Fig. 13.5 (**a**) SAVI, (**b**) MNDWI, (**c**) NDBI, (**d**) BUI, (**e**) IBI, (**f**) PCA1 (of 'c', 'd' and 'e') maps

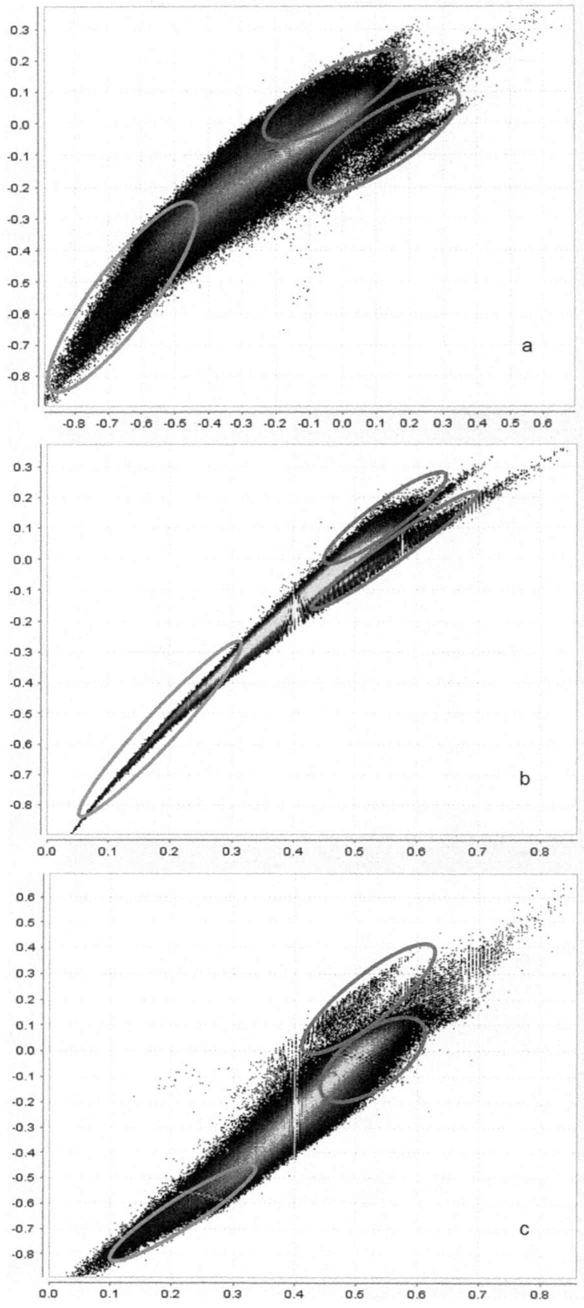

Fig. 13.6 Scattergrams of spectral feature space of the index images derived from LISS-III bands showing separability of classes water (blue), vegetation (green) and urban impervious (red): (**a**) BUI (x-axis)-IBI (y-axis), (**b**) NDBI (x-axis)-IBI (y-axis), (**c**) NDBI (x-axis)-BUI (y-axis)

Table 13.2 Accuracy assessment report

BUI						***IBI***					
	IS	**NIS**		**OA**	**k**		**IS**	**NIS**		**OA**	**k**
IS	183	14	197	*86*	*0.68*	**IS**	186	11	197	*90*	*0.77*
NIS	28	75	103			**NIS**	19	84	103		
	211	89	300				205	95	300		
NDBI						***PCA***					
	IS	**NIS**		**OA**	**k**		**IS**	**NIS**		**OA**	**k**
IS	185	12	197	*89*	*0.75*	**IS**	190	7	197	*96*	*0.91*
NIS	21	82	103			**NIS**	5	98	103		
	206	94	300				195	105	300		

Both the NDBI and IBI maps seem quite similar. On properly classifying the NDBI and IBI maps into impervious (IS) and non-impervious (NIS) areas, it is evident that IBI maps produced more accurate result, as revealed from the accuracy assessment reports (Table 13.2) for the classification. BUI map showed the least accuracy for classifying the urban impervious areas. All these three indices, namely, NDBI, BUI and IBI, are for identifying built-up areas via spectral information, but each of them suffers from certain limitations. PCA over these indices extract out the data variability, hence, ending up with the maximum information from all the three indices. Henceforth, the binary classified map derived from the PCA presented in Fig. 13.7d gave the best results for extracting the impervious built-up areas.

The scatterplots between the NIR and PCA1 in Fig. 13.7d reveal the zone for built-up/impervious area discrimination. The 'yellow' portion in the graph encircled in 'white' colour concentrates the built-up feature class. This figure describes the efficiency of PCA generated from built-up indices (NDBI, BUI and IBI), which resulted in the well-defined zone of concentration of the built-up areas when plotted against the NIR band. The zone of built-up area concentration is not observed as such when the individual built-up indices are plotted against the NIR band (Fig. 13.7a, b and c). Figure 13.7a (i.e. NDBI and NIR) shows complete absence of any dominant zone, while in Fig. 13.7b, c, respectively, for BUI versus NIR and IBI versus NIR, a not so well-defined but a zone can be identified for built-up area demarcation. Among BUI and IBI, the zone is more prominent for IBI as marked in Fig. 13.7c. However, the best of all is shown in Fig. 13.7d, where a clear well-defined intense zone is highlighted for the built-up area. This clearly proves the effectiveness of PCA in the analysis over the individual indices.

Figure 13.8 demarcates the impervious built-up areas in 'red' colour, while the white portions are non-built-up regions. Table 13.2 states the overall accuracy (OA) and kappa statistics (k) for classification. According to the table, based on the classification accuracy, the performance can be arranged as follows: BUI (OA = 86%, k = 0.68) < NDBI (OA = 89%, k = 0.75) < IBI (OA = 90%, k = 0.77) < PCA (OA = 96%, k = 0.91).

The same approach is applied to another area in order to establish the method and reproduce the results, so as to test for the transferability and validity of the approach.

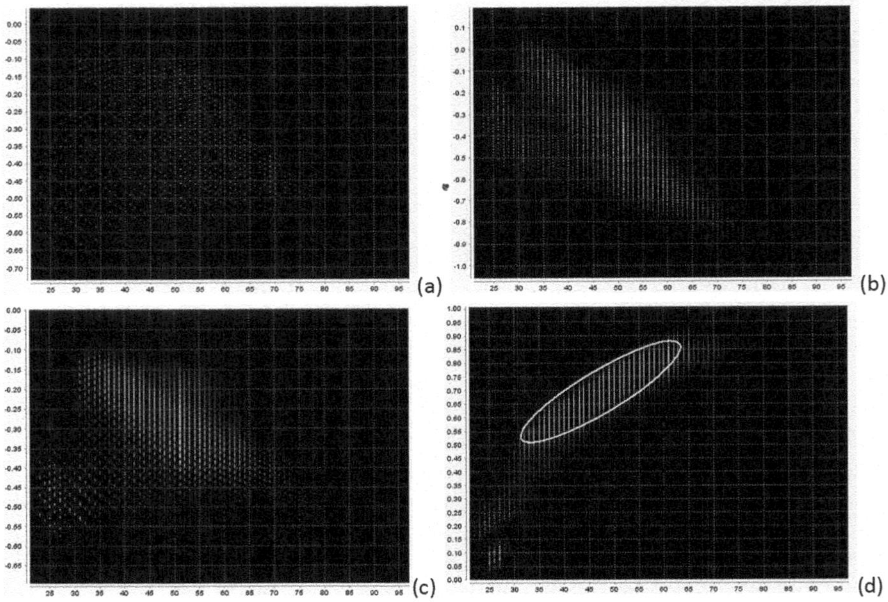

Fig. 13.7 Scatterplots for NIR in the x-axis and (**a**) NDBI, (**b**) BUI, (**c**) IBI and (**d**) PCA1 in y-axis highlighting the built-up encircled in 'white'

Fig. 13.8 Binary classified map derived from PCA1 (red colour = IS, white colour = NIS)

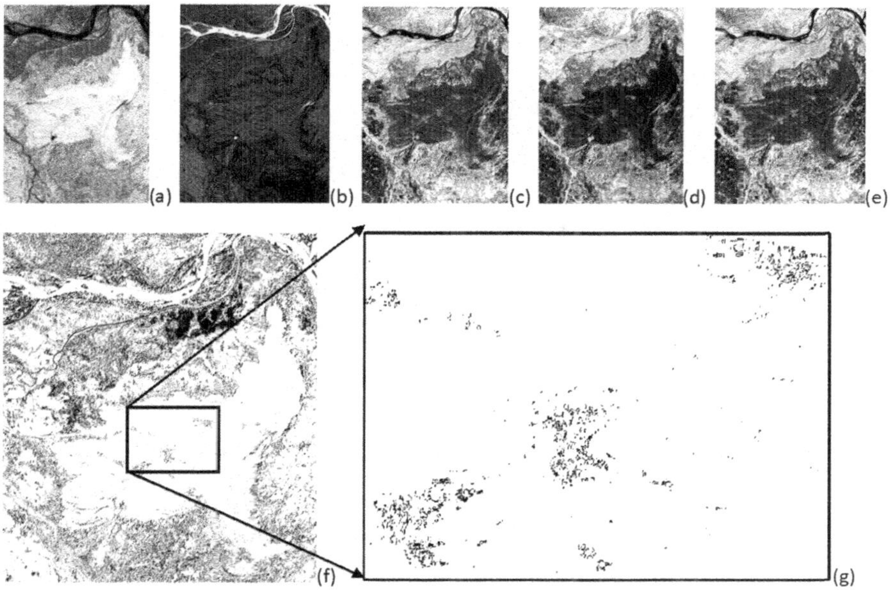

Fig. 13.9 (**a**) SAVI, (**b**) MNDWI, (**c**) NDBI, (**d**) BUI, (**e**) IBI, (**f**) binary classified PCA1 (of 'c', 'd' and 'e') and (**g**) zoomed built-up areas from classified maps (black colour = IS, white colour = NIS)

The second area under investigation is the tropical deciduous forested area of Munger in Bihar, India, with geographic coordinates of 25°19′30″N–24°56′50″N latitudes and 86°33′33″E–86°11′51″E longitudes. Details of the study site have been discussed by Sinha and Sharma (2013). The method is applied to automatically extract out the built-up areas within the forests (Fig. 13.9). Figure 13.9 illustrates the index maps, namely, the SAVI, MNDWI, NDBI, BUI and IBI maps, and the binary classified map of the derived PCA1 product. The method is observed to provide similar significant results with high level of accuracy for classification of in-forest built-up areas (OA = 92.86%, k =0.82). The result showed considerable improvements in the accuracy when IBI was used alone to extract the built-up areas within forests, where overall accuracy value was 89% with k value of 0.78 (Sinha et al. 2018).

13.4 Conclusion

The study targets in efficient extraction of built-up impervious areas within an urban landscape that cannot be easily demarcated due to unavailability of pure pixels from low or moderately low spatial resolution satellite images as the urban set-up is well intermixed with vegetation, plantation, water bodies and other non-impervious features. This becomes more challenging in a forest landscape, where settlement areas are scattered heterogeneously within the forest patch. Spectral indices provide better information in comparison to original spectral bands of any remote sensing satellite sensor. The study reveals the use of combined built-up spectral indices fed into PCA for extracting of built-up impervious areas from other non-impervious built-up features with high accuracy. PCA serves as a powerful and effective image processing technique that can extort information from the original data. Hence, PCA proved to be an extremely useful tool for feature extraction. The unique approach adopted in the study is not region-specific and can be successfully implemented over other regions as well. Hence, this approach has the global acceptability and can be reproduced over any region of interest, ranging over from small to large scales.

Deforestation, forest degradation and illegal cutting, logging and felling of trees can also be well demarcated within any forest patch by this approach, which can prove advantageous in the framework of REDD. Sensors, other than optical remote sensing sensors, need to be further investigated. In this context, the future scope is to use SAR (synthetic aperture radar) or radar remote sensing that can provide new openings, as the target-wave interaction depends on the scattering properties of the features on the ground, based on which backscatter values are generated (Sinha et al. 2015a). Index based on multi-frequency and multi-polarized SAR data is a prospective thrust area that needs development, as limited research has been done in this context. Future scopes comprising of PCA and similar such statistical and image processing techniques applied to SAR multi-polarized and hyperspectral data need to be properly investigated for automated detection of impervious areas. Mapping of impervious surface area can be useful inputs to planning and management activities arising due to land use shift from non-impervious to impervious surfaces that are closely correlated with several detrimental ecological and environmental impacts.

Acknowledgements The author acknowledges Science and Engineering Research Board (SERB), Department of Science and Technology (DST), India, for providing funds under SERB National Post-Doctoral Fellowship scheme (File No.: PDF/2015/000043).

References

Abdikan S, Sanli FB, Ustuner M, Calò F (2016) Land cover mapping using sentinel-1 SAR data. The international archives of the photogrammetry. Remote Sens Spat Inf Sci 41:757–761

Aghababaee H, Niazmardi S, Amini J (2013) Urban area extraction in SAR data. International Archives of the Photogrammetry, Remote Sensing and Spatial Information Sciences 40. SMPR 2013, Tehran, Iran

Arnold CL, Gibbons CJ (1996) Impervious surface coverage: the emergence of a key environmental indicator. J Am Plan Assoc 62(2):243–258. https://doi.org/10.1080/01944369608975688

As-syakur AR, Adnyana IWS, Arthana IW, Nuarsa IW (2012) Enhanced built-up and bareness index (EBBI) for mapping built-up and bare land in an urban area. Remote Sens 4 (10):2957–2970

Bauer ME, Heiner, NJ, Doyle JK, Yuan F (2004) Impervious surface mapping and change monitoring using Landsat remote sensing. Paper presented at the ASPRS Annual Conference Proceedings, Denver, Colorado

Blaschke T (2010) Object based image analysis for remote sensing. ISPRS J Photogramm Remote Sens 65:2–16

Chen Z, Zhang Y, Guindon B, Esch T, Roth A, Shang J (2012) Urban land use mapping using high resolution SAR data based on density analysis and contextual information. Can J Remote Sens 38:738–749

Corbane C, Faure J-F, Baghdadi N, Villeneuve N, Petit M (2008) Rapid urban mapping using SAR/optical imagery synergy. Sensors 8:7125–7143. https://doi.org/10.3390/s8117125

Gillies RR, Box JB, Symanzik J, Rodemaker EJ (2003) Effects of urbanization on the aquatic fauna of the Line Creek watershed, Atlanta- a satellite perspective. Remote Sens Environ 86:411–422

He C, Shi P, Xie D, Zhao Y (2010) Improving the normalized difference built-up index to map urban built-up areas using a semiautomatic segmentation approach. Remote Sensing Letters 1:213–221

Hu X, Weng Q (2009) Estimating impervious surfaces from medium spatial resolution imagery using the self-organizing map and multi-layer perceptron neural networks. Remote Sens Environ 113(10):2089–2102. https://doi.org/10.1016/j.rse.2009.05.014

Huete AR (1988) A soil-adjusted vegetation index (SAVI). Remote Sens Environ 25(3):295–309. https://doi.org/10.1016/0034-4257(88)90106-X

Johnson BA, Iizuka K, Bragais MA, Endo I, Magcale-Macandog DB (2017) Employing crowdsourced geographic data and multi-temporal/multi-sensor satellite imagery to monitor land cover change: a case study in an urbanizing region of the Philippines. Comput Environ Urban Syst 64:184–193

Kawamura M, Jayamana S, Tsujiko Y (1996) Relation between social and environmental conditions in Colombo Sri Lanka and the urban index estimated by satellite remote sensing data. Int Arch Photogramm Remote Sens 31:321–326

Kumar P, Sharma LK, Pandey PC, Sinha S, Nathawat MS (2013) Geospatial strategy for tropical forest-wildlife reserve biomass estimation. IEEE J Sel Top Appl Earth Observ Remote Sens 6:917–923

Lee J, Lee SS, Chi KH (2010) Development of an urban classification method using a built-up index. Paper presented at the Selected Topics in Power Systems and Remote Sensing, Sixth WSEAS International Conference on Remote Sensing, Iwate Prefectural University, Japan

Li H, Wang C, Zhong C, Su A, Xiong C, Wang J, Liu J (2017) Mapping urban bare land automatically from Landsat imagery with a simple index. Remote Sens 9(3):249. https://doi.org/10.3390/rs9030249

Liu C, Shao Z, Chen M, Luo H (2013) MNDISI: a multi-source composition index for impervious surface area estimation at the individual city scale. Remote Sens Lett 4:803–812

Lv Q, Dou Y, Niu X, Xu J, Xu J, Xia F (2015) Urban land use and land cover classification using remotely sensed SAR data through deep belief networks. J Sens 2015:1–10

Masek JG, Lindsay FE, Goward SN (2000) Dynamics of urban growth in the Washington DC metropolitan area, 1973-1996, from Landsat observations. Int J Remote Sens 21 (18):3473–3486. https://doi.org/10.1080/014311600750037507

Morabito M, Crisci A, Messeri A, Orlandini S, Raschi A, Maracchi G, Munafò M (2016) The impact of built-up surfaces on land surface temperatures in Italian urban areas. Sci Total Environ 551-552:317–326

Patino JE, Duque JC (2013) A review of regional science applications of satellite remote sensing in urban settings. Comput Environ Urban Syst 37:1–17

Ridd MK (1995) Exploring a V-I-S (vegetation-impervious surface-soil) model for urban ecosystem analysis through remote sensing: comparative anatomy for cities. Int J Remote Sens 16:2165–2185

Rouse JW, Haas RH, Schell JA, Deering DW (1974) Monitoring vegetation systems in the Great Plains with ERTS. Paper presented at the Proceedings of the Third ERTS Symposium, Washington, DC

Shao Z, Fu H, Fu P, Yin L (2016) Mapping urban impervious surface by fusing optical and SAR data at the decision level. Remote Sens 8:1–21

Sharma LK, Nathawat MS, Sinha S (2013) Top-down and bottom-up inventory approach for above ground forest biomass and carbon monitoring in REDD framework using multi-resolution satellite data. Environ Monit Assess 185(10):8621–8637. https://doi.org/10.1007/s10661-013-3199-y

Sinha S, Sharma LK (2013) Investigations on potential relationship between biomass and surface temperature using thermal remote sensing over tropical deciduous forests. Res Rev J Space Sci Technol 2(3):13–18

Sinha S, Sharma LK, Nathawat MS (2013) Integrated geospatial techniques for land-use/land-cover and forest mapping of deciduous Munger forests (India). Univers J Environ Res Technol 3:190–198

Sinha S, Pandey PC, Sharma LK, Nathawat MS, Kumar P, Kanga S (2014) Remote estimation of land surface temperature for different LULC features of a moist deciduous tropical forest region. In: Srivastava PK, Mukherjee S, Gupta M, Islam T (eds) Remote sensing applications in environmental research. Springer International Publishing, Cham, pp 57–68. https://doi.org/10.1007/978-3-319-05906-8_4

Sinha S, Jeganathan C, Sharma LK, Nathawat MS (2015a) A review of radar remote sensing for biomass estimation. Int J Environ Sci Technol 12(5):1779–1792. https://doi.org/10.1007/s13762-015-0750-0

Sinha S, Sharma LK, Nathawat MS (2015b) Improved land-use/land-cover classification of semi-arid deciduous forest landscape using thermal remote sensing. Egypt J Remote Sens Space Sci 18(2):217–233. https://doi.org/10.1016/j.ejrs.2015.09.005

Sinha S, Jeganathan C, Sharma LK, Nathawat MS, Das AK, Mohan S (2016) Developing synergy regression models with space-borne ALOS PALSAR and Landsat TM sensors for retrieving tropical forest biomass. J Earth Syst Sci 125:725–735

Sinha S, Santra A, Mitra SS (2018) Automated extraction of built-up areas within forests using remote sensing. In: Santra A, Yadav NK (eds) Proceedings of national conference on advancement in civil engineering practice and research. Excel India Publishers, New Delhi, pp 96–99

Varshney A (2013) Improved NDBI differencing algorithm for built-up regions change detection from remote-sensing data: an automated approach. Remote Sensing Letters 4:504–512

Wang Z, Gang C, Li X, Chen Y, Li J (2015) Application of a normalized difference impervious index (NDII) to extract urban impervious surface features based on Landsat TM images. Int J Remote Sens 36(4):1055–1069. https://doi.org/10.1080/01431161.2015.1007250

Weeks JR, Hill A, Stow D, Getis A, Fugate D (2007) Can we spot a neighborhood from the air? Defining neighborhood structure in Accra, Ghana. GeoJournal 69(1–2):9–22. https://doi.org/10.1007/s10708-007-9098-4

Weng Q (2012) Remote sensing of impervious surfaces in the urban areas: requirements, methods, and trends. Remote Sens Environ 117:34–49

Xian G, Crane M (2005) Assessments of urban growth in the Tampa Bay watershed using remote sensing data. Remote Sens Environ 97(2):203–215. https://doi.org/10.1016/j.rse.2005.04.017

Xu H (2002) Spatial expansion of urban/town in Fuqing and its driving force analysis. Remote Sens Technol Appl 17:86–92

Xu H (2006) Modification of normalised difference water index (NDWI) to enhance open water features in remotely sensed imagery. Int J Remote Sens 27(14):3025–3033. https://doi.org/10.1080/01431160600589179

Xu H (2007) Extraction of urban built-up land features from Landsat imagery using a thematic-oriented index combination technique. Photogramm Eng Remote Sens 73:1381–1391

Xu H (2008) A new index for delineating built-up land features in satellite imagery. Int J Remote Sens 29(14):4269–4276. https://doi.org/10.1080/01431160802039957

Xu H (2010) Analysis of impervious surface and its impact on urban heat environment using the normalized difference impervious surface index (NDISI). Photogramm Eng Remote Sens 76:557–565

Zha Y, Gao J, Ni S (2003) Use of normalized difference built-up index in automatically mapping urban areas from TM imagery. Int J Remote Sens 24(3):583–594. https://doi.org/10.1080/01431160304987

Part V
Future Challenges of Natural Resource Management and their Potential over Space and Time

Chapter 14
Hybrid Polarimetric Synthetic Aperture Radar for the Detection of Waterlogged Rice Fields

Bambang H. Trisasongko

Abstract Rice has been the most important agricultural commodity in Asia. Various monitoring schemes based on remotely sensed data have been dedicated for rice research purposes, and responsible agencies currently seek an efficient, operational paddy field monitoring. Multispectral datasets serve as the backbone to the application; nonetheless, their successful implementation in tropical regions is somewhat fluctuating due to persistent cloud cover. Options in the use of synthetic-aperture radar (SAR) data are currently available, from X-, C-, or L-band spaceborne systems. The latter is preferable as long wavelength is less susceptible to the attenuation of high precipitation often seen in tropical regions. In this chapter, hybrid polarization as one of the emerging SAR techniques is investigated to retrieve waterlogged rice fields as a proxy for the commencement of a new planting season. Two popular hybrid polarimetric representations, i.e., modulus of covariance matrix and polarimetric features of Raney decompositions, are discussed. Information extraction was done using 11 supervised learners. The findings indicated that modulus of covariance matrix generally performed inferior than Raney decomposition datasets. The latter amplified the overall accuracy to around 95%, with about 20% discrepancy to the covariance matrix. Although modern data mining methods including random forests and support vector machines were preferable than conventional methods such as single tree approach, this research indicated that some variants of random forests and support vector machines may yield overall accuracy below the expectation. The research also discovered that Raney decomposition features outweighed fully polarimetric backscatter coefficients, although the difference is considerably low (about 5%). Hence, it could be summarized that hybrid polarimetry may provide an efficacious solution to large-scale monitoring of active rice fields.

B. H. Trisasongko (✉)
School of Physical, Environmental and Mathematical Sciences, UNSW Canberra at Australian Defence Force Academy, Campbell ACT, Australia

Department of Soil Science and Land Resources, Bogor Agricultural University, Bogor, Indonesia
e-mail: trisasongko@apps.ipb.ac.id

P. Kumar et al. (eds.), *Applications and Challenges of Geospatial Technology*,
https://doi.org/10.1007/978-3-319-99882-4_14

Keywords Coherence matrix · Data mining · Hybrid polarimetry · Raney decomposition · Rice · Synthetic-aperture radar

14.1 Introduction

Rice remains the staple food for most Asian countries including Indonesia. Lowland areas with abundant water resources have been the most favorable environment to this commodity. Supported by suitable soil types, Java, Sumatra, and Sulawesi islands are top rice-producing areas in Indonesia, largely with two or three crop rotations per year. In addition, upland paddy fields are also common in major Indonesian highlands, generally with only one planting cycle each year during the rainy season. To some communities situated in East Sumatran wetlands, several rice varieties capable to handle frequent waterlogged condition have also been cultivated. Although its importance has been acknowledged in decades, monitoring scheme of rice fields is understudied. Especially in Indonesia, complexity of the problem remains large, partially due to small-scale farming and inhomogeneity of planting epoch. Rice planting season starts with waterlogging fields; hence, with the nature of the condition, it is possible by remote sensors to identify the starting phase of the season.

Multispectral sensors have been the primary source of data to this need. Landsat, for instance, has been one of the most popular datasets for agricultural mapping, including rice monitoring, essentially due to its long temporal data availability and free access. Kontgis et al. (2015) demonstrated its utility for mapping rice extent in a Vietnamese site. A lesson learned from the research includes that time stack of remote sensing data could be useful not only to map the extent of rice production but also to provide essential information on annual cropping frequency. Estimation of yield has also been conducted, mostly with the aid of vegetation indices. Normalized Difference Vegetation Index (NDVI) at the peak vegetation period was notably important to the estimation of rice yield (Siyal et al. 2015). Nonetheless, rigid conclusion is yet to be developed as site dependency appears having high influence and transferability of models has been far understudied. For tropical countries like Indonesia, this type of sensor has significant drawback because it has a greater sensitivity to atmospheric disturbance such as cloud, haze, and smoke. This dependency then limits multispectral dataset implementation for operational basis.

Synthetic-aperture radar (SAR) systems have a greater insensitiveness to atmospheric condition; hence, they have inherent advantages for tropical agriculture monitoring. Single-polarization radar was initially used (Panigrahy et al. 1999); however, with the advance of SAR technology, dual-polarization backscatter coefficients have been the primary dataset for rice field monitoring. Pei et al. (2011), for instance, examined dual-polarized X-band TerraSAR for rice field identification in a Chinese site. With the limited number of predictors in SAR data, fusion with multispectral data is commonly implemented to improve the outcome. This approach was successfully demonstrated in India using a combined RADARSAT and the Advanced Wide Field Sensor (Haldar and Patnaik 2010). SAR came into a new era of spaceborne SAR capable to generate fully polarimetric (FP) data since 2006 when Phased Array L-Band SAR (PALSAR) on board Advanced Land Observing Satellite (ALOS) became operational. Improvement of the outcomes, compared to dual- or single-polarization data, was studied elsewhere (Lee et al. 2001; Cloude 2009). Investigation of FP data with an application for rice monitoring was done to separate basmati and non-basmati rice yielding an accuracy around 85% (Kumar et al. 2016). This level of accuracy is somewhat similar to the earlier outcome of Wu et al. (2011), using the same sensor configuration. While the use of fully polarimetric data is strongly envisaged, the issue of data availability remains problematic. FP imaging mode requires higher amount of energy, which makes it complicated when the sensor is hosted in a spaceborne platform. Fully polarimetric mode, therefore, prohibits frequent data capture, a situation where is unfavorable for operational rice field monitoring.

Compact (or hybrid) polarity of SAR data, therein CP, is one of the emerging solutions to power and swath constraints in spaceborne SAR. The idea was firstly proposed by Souyris et al. (2005) through measuring reoriented waves. Conventional FP measurements include horizontal and vertical orientations, both in transmission and reception. In CP mode, only single transmission is used, either in circular or in 45° orientation, while normal horizontal and vertical positions are employed in the receiver. With this approach, about a half of the power is therefore conserved and allows full-swath scanning. While long debates are progressing on the utility of the technique, several SAR vendors such as Japan, Canada, and India have announced their readiness to provide CP data acquisition. Experimentation on its advantages and drawbacks, however, needs to be done in various aspects of remote sensing data utilization for improving previous knowledge gained in land cover monitoring (Turkar et al. 2013), forestry (Lardeux et al. 2010; Trisasongko 2015), and agriculture (McNairn et al. 2017).

To alleviate the understanding on the behavior and performance of CP, the central goal of this chapter is to observe the feasibility of CP SAR in comparison to FP dataset. In particular, the main focus is on the utilization of varieties of features derived from CP, which have been lightly studied. With the availability of numerous data mining methods, this chapter also discusses the implementation of popular classification methods such as random forests (RFs) and support vector machines (SVMs) and their variants to retrieve waterlogged condition of rice fields.

14.2 Materials and Methods

14.2.1 Test Site and SAR Data

The research was situated in Jember Regency, East Java Province, Indonesia. This area is considered as one of the largest rice cultivation areas in the province. This is

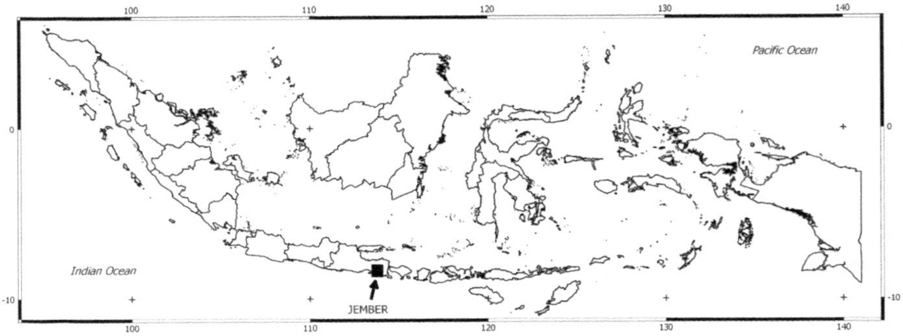

Fig. 14.1 Jember (Indonesia) test site

supported by low-lying terrain; only a few locations of the regency are occupied by hills and mountains. In hilly terrain, majority of land use includes forest, timber, coffee, and rubber plantations. In the rest of the area, irrigated fields dominate with paddy and sugarcane that serve as the primary commodities. Irrigations were vastly developed taking advantage of the humid climate with sufficiently high rainfall throughout the years. Abundance of spring water from Ijen-Raung volcanic complex and Mt. Argopuro in the northern part of the research site leads to adequately large amount of rivers and creeks. Rice agriculture is also supported by alluvium soil types influenced by two aforementioned volcanoes. Figure 14.1 presents the site location.

The main dataset used in this research was FP Phased Array-type L-band SAR (PALSAR-2), acquired in 31 March 2015 by the Japan Aerospace Exploration Agency (JAXA). Preliminary calibration was performed by the vendor before delivery. Nonetheless, since L-band data are susceptible to ionospheric condition, prior to analysis, Faraday rotation was estimated using an algorithm developed by Bickel and Bates (1965). Preliminary analysis showed that the rotation was 3.38°, which is lower than 5°, the threshold suggested by Wright et al. (2003). Complex scattering matrix S_2 data were subsequently calibrated using Ainsworth et al. (2006) technique. From this data, three forms of dataset were then derived.

14.2.2 Data Analysis

The first dataset was backscatter coefficients, which are the most popular form of SAR data, derived by taking the amplitude for each S_2 main components, i.e., S_{HH}, S_{HV}, and S_{VV}. Notations H and V, respectively, denote linear scattering element of transmitted and received wave; hence, S_{HV} indicates cross polarization of horizontal and vertical wave propagation between transmission and reception. Since receiver and transmitter modules are placed in the same SAR antenna, the reciprocity theorem applies, that is, $S_{HV} = S_{VH}$. Hence, only S_{HV} is considered in this analysis.

Due to unavailability of actual CP data, in this research, they were simulated from FP dataset following the procedure conducted in earlier publications (Chen et al. 2014; White et al. 2014). The π/4 method (Souyris et al. 2005), also known as slant linear, was adopted to simulate covariance matrix C_2 data. From this matrix, three amplitudes (i.e., C_{11}, C_{12}, and C_{22}) were extracted, serving as the second dataset. The last dataset in this research was polarimetric features derived from Raney et al. (2012) decomposition, also known as m-χ decomposition. The decomposition takes input from Stokes parameters calculated from C_2, resulting in six main features. The first three components, i.e., m, χ, and δ, respectively, denote degree of polarization, Poincare ellipticity, and relative phase between horizontal and linear polarization. Remaining components mimic features derived from model-based FP decomposition comprising contributions of double-bounce, volume, and surface scattering in the area of illumination. The latter group of three features is easier to interpret and may be linked to physical processes, perhaps, the main reason that they are more popular in the development of polarimetric decomposition theorems. Double-bounce scattering, for instance, is a key parameter to identify man-made structure. In the case of vegetative cover, it relates to sparse stand density which allows sequential wave interaction process with ground surface and the stem before reception by SAR antenna. Volume, also known as random, scattering is beneficial for vegetation studies, while surface scatterers are often identified in smooth surface like open water.

PALSAR-2 data were preprocessed using PolSARPro 5.0 software, freely available from European Space Agency (ESA, https://earth.esa.int/web/polsarpro/home). Further geometrical processing to match baseline map was done in SNAP (Sentinel Application Platform, http://step.esa.int/main/toolboxes/snap/) version 5.0, also a free toolkit from ESA. Since SAR data are strongly influenced by topography, this research employed SAR Simulation Terrain Correction module in SNAP to compensate the effect. Final data have 15 m spatial resolution with World Geodetic System (WGS) 1984-based geographic coordinate system. All datasets were then fed into R statistical software version 3.4.3 (https://cran.r-project.org/) for modeling and prediction procedures.

Information extraction targeting waterlogged and non-waterlogged rice fields was done by employing both conventional and contemporary classification methods applied to three aforementioned datasets. Summary of classification methods is presented in Table 14.1. In this research, CART technique was used as the benchmark considering its widely utilization in remote sensing. Variants of techniques, whenever available, were also evaluated to improve understanding of their implementations and extent for rice monitoring. Prior to classification, samples were taken throughout the image guided by the baseline map and field survey conducted in December 2015. Gap between satellite data acquisition and field survey was minimized by collecting information about normal pattern of planting schedule from local farmers and by obtaining planting seasons from local government bodies. Samples were subsequently divided into training and testing data with proportion of 3:1. Tenfold cross-validation with three repeats was adopted to minimize bias and to ensure that the outcome would statistically be reliable. Finally, the performance of statistical models was measured using overall accuracy.

Table 14.1 Employed classification models

Model name	Code	Reference
Classification and regression trees	CART (for original algorithm)	Breiman et al. (1984)
	CARTBag (for bagged version)	
Multivariate adaptive regression splines	MARS (for original algorithm)	Friedman (1991)
	MARSBag (for bagged version)	
Random forest	RF	Breiman (2001)
Rotation forest	RFrot	Rodriguez and Kuncheva (2006)
Weighted subspace RF	RFws	Xu et al. (2012)
Gradient boosting	GB	Friedman (2001)
Extreme gradient boosting	GBx	Chen and Guestrin (2016)
Support vector machines	SVr	Vapnik (2000)
Least squares SVM	SVls	Suykens and Vandewalle (1999)

14.3 Results and Discussion

14.3.1 The Importance of Dataset Selection

With large choices of remotely sensed data available today, selecting "the best option" would not be an easy task. In the case of optical data, hyperspectral data are perhaps the ideal one. Nonetheless, complexity of processing has been a long issue in data processing, and there is no guarantee that the outcome may be optimum. This also applies to FP dataset, where users can opt to retrieve conventional backscattering coefficients along with numerous features derived from polarimetric decomposition theorems. While FP data have been advocated (consult Charbonneau et al. (2010) for recent review), paucity of proofs in tropical rice monitoring is evident.

Figure 14.2 depicts variations in overall accuracy calculated using proposed models. This clearly summarizes that C_2 data were unable to fully distinguish two designated classes: waterlogged and non-waterlogged rice fields. The average accuracy of this data using available models was about 74.7%, with the highest provided by least squares SVM (76.0%). In contrast, backscatter coefficients of FP data and Raney features provide promising results. The figure indicates that the outcomes of backscatter coefficients were interestingly consistent irrespective of the complexity of selecting a good classifier. With the average of about 89.1%, backscatter coefficient may serve as the primary option when preference of classification methods is restricted. In this test case, classification using Raney features generally delivered a higher accuracy. They gave average accuracy about 95.9%.

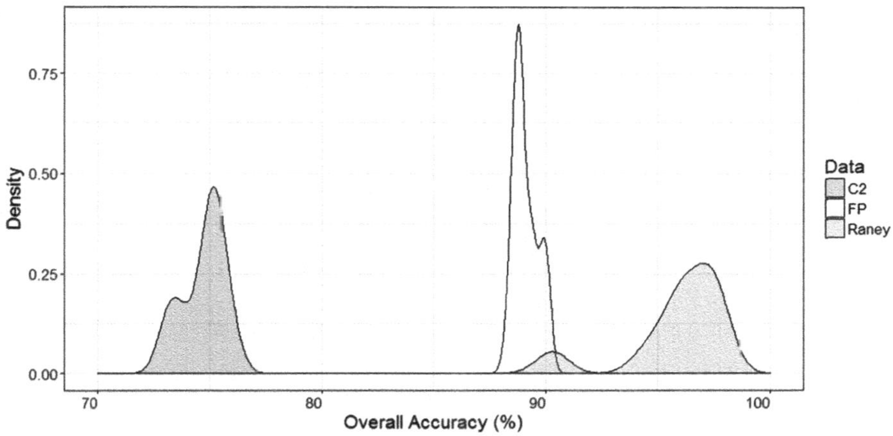

Fig. 14.2 Probability to obtain suitable accuracy among datasets

Previous reports found that accuracy obtained from CP was generally inferior than the one yielded from FP. Singh et al. (2014), for instance, discovered that FP mode, instead of CP, was suggested for mapping five different classes in a glacier area of India. The m-χ parameters coupled with model-based decomposition features used in that research achieved lower accuracy, suggesting a limitation of CP mode in a specific mapping purpose. For wetland vegetation monitoring, assessment using simulated RADARSAT Constellation Mission (RCM) was reported by White et al. (2017). The research found that acceptable performance of CP could only be achieved through a fusion with optical and terrain data. The issue may be linked with the nature of C-band used in RCM. Shorter wavelength generally interacts with top canopy layer of the woody vegetation; hence, actual tree structure as one of the key identifiers of arboreous stands is less detectable. This research, however, discovered that in a very specific application such as the detection of waterlogged rice fields, the outcomes of CP mode would be as high as the ones demonstrated by FP data. It was suspected that data selection fed into statistical models was one of important keys in successful class separation. While covariance matrix of FP mode was shown useful in a previous report (Amani et al. 2017), this research indicated that it was not the case in CP SAR application for rice field observation. Due to paucity of further information, the reason of this situation remains unknown and warrants further investigation. Although the variance is shown relatively high compared to FP backscattering coefficients, the span of CP's overall accuracy is less than 10%. The main contributor of lower accuracy was CART and MARSBag. This indicates that single tree models are less adaptable to this situation. Modern data mining models, in contrast, provided a much consistent outcome. Leaving CART and MARSBag out, CP yielded overall accuracy between 94.5% and 97.6%, which is nearly as consistent as FP. Similar conclusion was drawn from a research in Spanish test site (Lopez-Sanchez et al. 2014), showing competitiveness of CP as an potential alternative of FP. Likewise, Charbonneau et al. (2010) findings suggested that some CP forms could be able to provide an improved accuracy than the one given by fully polarimetric decomposition features. This research therefore suggests that the accuracy produced by CP remains high irrespective of the classification model being employed.

14.3.2 Performance of Prediction Models

In actual applications such as rice monitoring, complexity becomes bigger with the abundance of classification models available from the literature. Although modern classification techniques such as RFs or SVMs have become "standards" in data analysis, the extent of obtaining an acceptable outcome is generally understudied, partially because of limited number of publications assessing group of models. In this research, only ten models were further evaluated, and the overall accuracy is presented in Fig. 14.3. MARSBag model was found less robust, with the average of overall accuracy of about 20%. Hence, for the sake of brevity, this approach is excluded from discussion.

In general, all models yielded sufficiently high results as the average spans around 90%. As expected, CART as the benchmark model in this study provided least accurate results. SVM and its variant tended to produce a slightly higher overall accuracy than the remaining models. A previous study favored SVM over other classification models such as RF (Park and Im 2016), although abundant research also showed that similarity of performance exists between SVM and RF (Chan et al. 2012; Attarchi and Gloaguen 2014) or neural networks (Yousefi et al. 2015). Competing algorithms such as tree-based models (RF and GB) have a slightly lower average. Nonetheless, the difference appears insignificant; therefore one may expect comparable capability among modern data mining methods.

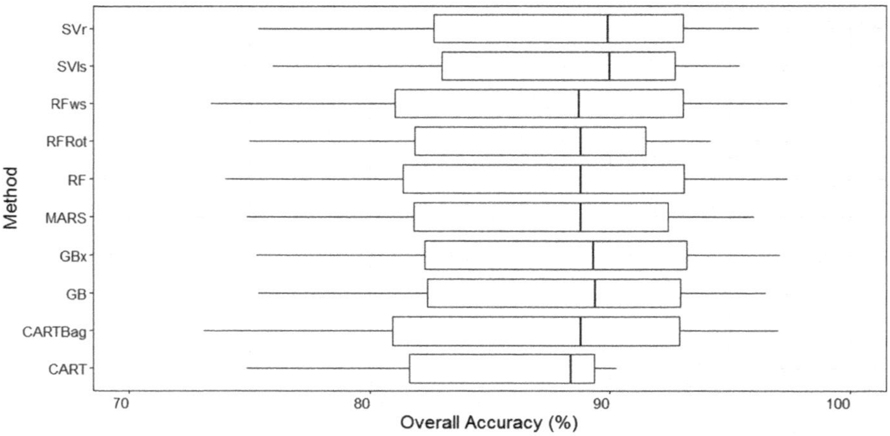

Fig. 14.3 Model performance

Interestingly, frequently neglected models including MARS could well compete with contemporary classification methods. This is probably due to the nature of simpler discrimination of only two classes with sufficient numbers of predictors, which makes the separation straightforward.

14.3.3 Best Performing Data and Models

As previously discussed, Raney decomposition features provided the most promising outcomes. Capability of separating classes by Raney components is presented in Fig. 14.4. As shown, some Raney components such as m and χ were less robust for the discrimination, which suggests a partial disagreement to previous research (Lopez-Sanchez et al. 2014). In contrast, the remaining features provided a better proficiency to distinguish waterlogged rice fields. Odd-bounce component, also known as surface scattering, contributed the most. This is understandable since radar has interesting ability to detect smooth Earth surfaces like open water. Their interaction with incoming waves would result a weaker returning signal as the surface would deflect most SAR waves away from the receiver. Lopez-Sanchez et al. (2014) reported that surface scattering would remain dominant until the age about 20 days after sowing. Nonetheless, this situation could vary depending on many factors, including SAR system parameters (wavelength and incidence angle) and object properties (genetic varieties, etc.).

Along with odd-bounce, double-bounce scattering also contributed to the discrimination. This is possibly due to wave augmentation after sequential interaction with ground surface and then rice stalks before returning to SAR sensor. The phenomenon was previously reported in Japanese rice fields with regular planting spaces using an L-band sensor (Ouchi et al. 2006). Investigation of FP data in Spain also indicates similar situation where double-bounce scattering would be in its existence until about 30 days after sowing (Lopez-Sanchez et al. 2014). Detection of double-bounce component in waterlogged rice field, therefore, indicates that sowing has already commenced.

Random (or volume) scattering is a distinctive feature of vegetative cover. Hence, it is explicable that in-production rice fields have a stronger random scattering component from vegetative cover. In this specific case, bimodal structure was observed in waterlogged rice fields. In addition to weak random scattering (nearly 0.1) which suggests the existence of water-dominated pixels, secondary peak at about 0.35 was observed. Possible reason is that the remaining rice stumps with small fraction of leaves exist and act as random scatterers. In general, stumps are not entirely removed from rice fields by local farmers; therefore, they acted as additional random scatterers in waterlogged fields. Another reason is the inclusion of seedling plots which are often situated within the field. Spatial resolution of the SAR image was about 15 m which was quite excessive for small fields and often occupied two adjacent terraced fields in the northern part of the site. Therefore, a mixed condition

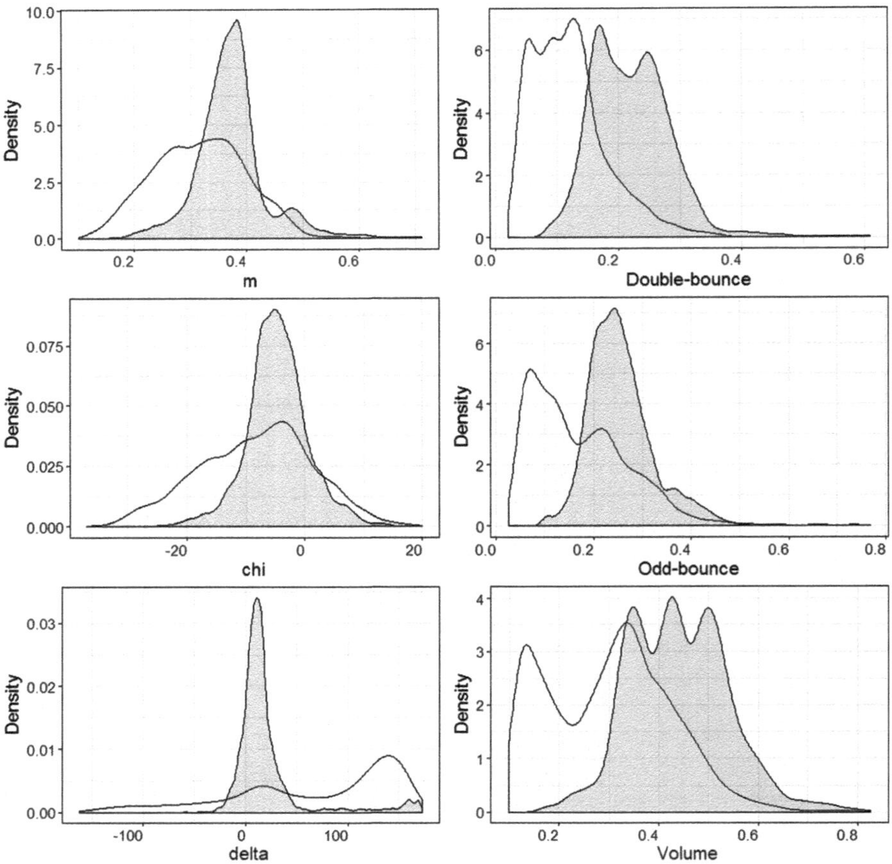

Fig. 14.4 Histogram indicating class separability of each component. Responses from waterlogged paddy fields are presented in light yellow

often occurs within single pixel. This suggests an interesting information that CP could provide a possibility to identify pure and mixed waterlogged paddy fields.

Although SVM has a greater chance in yielding preferable results (Fig. 14.3), individual measurement of model performance shows that RF and its variants consistently achieved the best (Table 14.2). The best performing SVM with radial basis function (RBF) kernel produced 96.1% accuracy, while least squares SVM yielded a half percent lower. Previous record showed that least squares SVM was in favor due to its capability to handle noises and to offer an improved accuracy (Shi et al. 2009); however, in the case of rice monitoring, significance of SVM variant was considerably low. In this research, RF and its variant yielded equally; nonetheless, in terms of computation time during training phase, weighted subspace algorithm learned much slower than original RF. Hence, preference would be given to original RF technique for the prediction of the whole scene.

Table 14.2 Top five performing models, all from Raney decomposition features

Model name	Overall accuracy (%)
RF	97.6
RFws	97.6
CARTBag	97.5
GBx	97.2
GB	96.6

CART's performance was considerably at moderate level due to its monolithic decision scheme, although its bagging variant yielded a much appreciable outcome. The reason to significant contribution includes mechanisms for recursive partitioning allowing more evaluation during the construction of decision-making. Although developed in the same era with RF, popularity of GB has considerably been much lower. This may be linked with slower learning process than RF, which in turn requires more computing time and may not be the perfect choice when it comes into implementation of real applications. Performance of GB in comparison to RF, SVM, and ANN for rural land cover mapping was discussed in Trisasongko et al. (2017). The research found that GB may compete with RF and SVM in complex classification problems, without the necessity to tune its parameters. While modern data mining techniques progressively deliver improvement to the outcome, many algorithms require detailed parameter settings to achieve the expected accuracy. It appears that GB and RF algorithms did not well respond to parameter setting; hence, this may be the main benefit of using both techniques to seek suitable outcomes despite their slow training procedures. Experiments on tuning data mining parameters, as discussed above, remain an important subject for research and should be continuously expanded using various datasets and problem definitions.

Figure 14.5 shows the outcome of classification using RF model, segmented using irrigation blocks. Non-rice field land cover classes such as built-up and waterbodies are masked (colored white) to emphasize the classification result. It is shown that most central irrigation regions have been in a new planting cycle, supported by primary irrigation channels. Almost all northern irrigation districts also commenced new plantation cycle, except Jubung village, which is situated slightly far from the main irrigation channels. More rice fields in southern irrigation districts, especially in Ajung and Tegal Besar villages, stayed behind the schedule.

14.4 Conclusion

Capability of monitoring rice fields in very large spatial extent is one of the significant contributions of remote sensing. With greater options on data selection recently available, the application needs some insight whether specific sensor is suitable to the purpose. For many cases in tropical regions, FP SAR data are favorable due to their richness in information, similar to modern multispectral

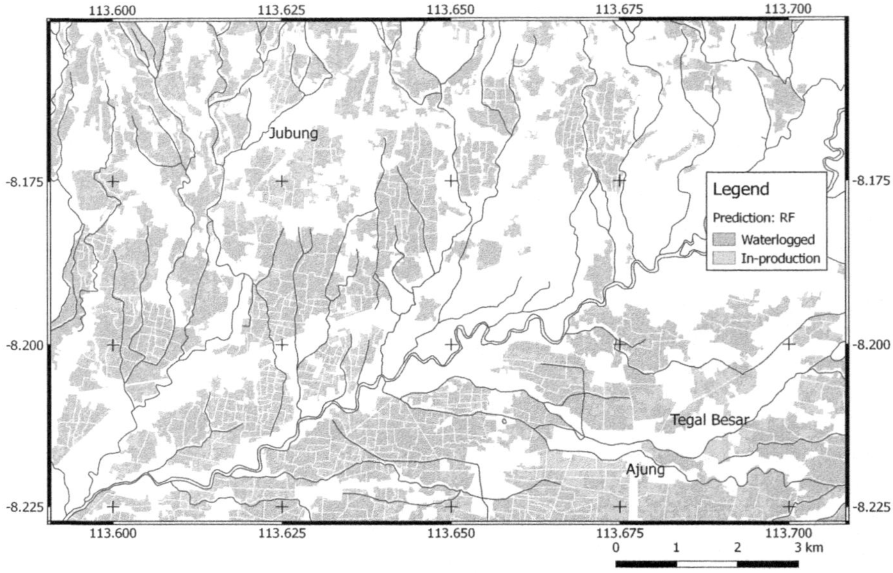

Fig. 14.5 Prediction map overlaid by irrigation compartments

imaging systems. Nonetheless, they have severe consequences in terms of data generation and processing, leading to current development of CP SAR. To date, applications of CP are unfortunately understudied.

Similar to FP mode, several forms of CP data are available for research and implementation. It was found that features derived from Raney decomposition theorem provided an equally accurate result to conventional FP backscatter coefficients. Considering lower power requirement than FP, the outcome of this research suggests that CP could be more efficient in the detection of new planting season in tropical region. This indicates that spaceborne CP SAR systems such as Japanese PALSAR-2, Canadian RCM, and Indian RISAT-1 satellites may offer global-scale information on rice production seasons with reasonable accurateness. The research also showed that RF, among other modern data methods, was capable to provide highly accurate estimation without additional tuning experiment. It yielded overall accuracy about 97%, with negligible difference to its weighted subspace variant. RF models were in favor for mapping waterlogged fields since they have fewer parameter settings, compared to other contemporary approaches. Nonetheless, slow learning process during ensemble tree modeling as adopted in RF encourages more studies of learning strategies in the future. Conventional models such as CART may be used only if it was further improved through bagging mechanisms, yielding a much improved outcome with insignificant discrepancy to the accuracy provided by RF.

Acknowledgment The author would like to thank Japan Aerospace Exploration Agency (JAXA) for data provision through RA6-3004 project. Partial funding was obtained from UNSW Australia through UIPA scholarship. Fieldworks were assisted by Dyah R. Panuju, Annisa P. Trisasongko, and Rizqi I. Sholihah; their contributions to this work, therefore, are greatly acknowledged. Finally, the author expresses his gratitude to Dr. David Paull of UNSW Canberra for his encouragement and support.

References

Ainsworth TL, Ferro-Famil L, Lee J-S (2006) Orientation angle preserving a posteriori polarimetric SAR calibration. IEEE Trans Geosci Remote Sens 44(4):994–1003. https://doi.org/10.1109/TGRS.2005.862508

Amani M, Salehi B, Mahdavi S, Granger J, Brisco B (2017) Wetland classification in Newfoundland and Labrador using multi-source SAR and optical data integration. GISci Remote Sens 54 (6):779–796. https://doi.org/10.1080/15481603.2017.1331510

Attarchi S, Gloaguen R (2014) Classifying complex mountainous forests with L-band SAR and landsat data integration: a comparison among different machine learning methods in the Hyrcanian forest. Remote Sens 6(5):3624–3647. https://doi.org/10.3390/rs6053624

Bickel SH, Bates RHT (1965) Effects of magneto-ionic propagation on the polarization scattering matrix. Proc IEEE 53(8):1089–1091. https://doi.org/10.1109/PROC.1965.4097

Breiman L (2001) Random forests. Mach Learn 45(1):5–32. https://doi.org/10.1023/A:1010933404324

Breiman L, Friedman JH, Olshen RA, Stone CJ (1984) Classification and regression trees. Chapman and Hall/CRC, Boca Raton

Chan JCW, Beckers P, Spanhove T, Borre JV (2012) An evaluation of ensemble classifiers for mapping Natura 2000 heathland in Belgium using spaceborne angular hyperspectral (CHRIS/Proba) imagery. Int J Appl Earth Obs Geoinf 18(1):13–22. https://doi.org/10.1016/j.jag.2012.01.002

Charbonneau FJ, Brisco B, Raney RK, McNairn H, Liu C, Vachon PW, Shang J, De Abreu R, Champagne C, Merzouki A, Geldsetzer T (2010) Compact polarimetry overview and applications assessment. Can J Remote Sens 36(SUPPL. 2):S298–S315

Chen T, Guestrin C (2016) XGBoost: a scalable tree boosting system. Paper presented at the Proceedings of the 22nd ACM SIGKDD international conference on knowledge discovery and data mining, San Francisco, California, USA

Chen H, Goodenough DG, Cloude SR (2014) Mapping forest fire scars with simulated RCM compact-pol data:1572–1575. https://doi.org/10.1109/IGARSS.2014.6946740

Cloude SR (2009) Dual versus quadpol: a new test statistic for radar polarimetry. In: Polinsar 2009, Frascati, Italy, 26–30 Jan 2009

Friedman JH (1991) Multivariate adaptive regression splines. Ann Stat 19(1):1–141

Friedman JH (2001) Greedy function approximation: a gradient boosting machine. Ann Stat 29 (5):1189–1232. https://doi.org/10.2307/2699986

Haldar D, Patnaik C (2010) Synergistic use of multi-temporal Radarsat SAR and AWiFS data for Rabi rice identification. J Indian Soc Remote Sens 38(1):153–160. https://doi.org/10.1007/s12524-010-0006-x

Kontgis C, Schneider A, Ozdogan M (2015) Mapping rice paddy extent and intensification in the Vietnamese Mekong River Delta with dense time stacks of Landsat data. Remote Sens Environ 169:255–269. https://doi.org/10.1016/j.rse.2015.08.004

Kumar V, Kumari M, Saha SK (2016) Discrimination of basmati and non-basmati rice types using polarimetric target decomposition of temporal SAR data. Curr Sci 110(11):2166–2169. https://doi.org/10.18520/cs/v110/i11/2166-2169

Lardeux C, Niamen D, Routier JB, Giraud A, Frison PL, Pottier E, Rudant JP (2010) Use of PalSAR polarimetric data for tropical forest stratification and comparison of simulated dual and compact polarimetric modes. Paper presented at the international geoscience and remote sensing symposium (IGARSS)
Lee JS, Grunes MR, Pottier E (2001) Quantitative comparison of classification capability: fully polarimetric versus dual and single-polarization SAR. IEEE Trans Geosci Remote Sens 39 (11):2343–2351. https://doi.org/10.1109/36.964970
Lopez-Sanchez JM, Vicente-Guijalba F, Ballester-Berman JD, Cloude SR (2014) Polarimetric response of rice fields at C-band: analysis and phenology retrieval. IEEE Trans Geosci Remote Sens 52(5):2977–2993. https://doi.org/10.1109/TGRS.2013.2268319
McNairn H, Homayouni S, Hosseini M, Powers J, Beckett K, Parkinson W (2017) Compact polarimetric synthetic aperture radar for monitoring crop condition. Paper presented at the international geoscience and remote sensing symposium (IGARSS)
Ouchi K, Wang H, Ishitsuka N, Saito G, Mohri K (2006) On the Bragg scattering observed in L-band synthetic aperture radar images of flooded rice fields. IEICE Trans Commun E89-B (8):2218–2225. https://doi.org/10.1093/ietcom/e89-b.8.2218
Panigrahy S, Manjunath KR, Chakraborty M, Kundu N, Parihar JS (1999) Evaluation of RADARSAT standard beam data for identification of potato and rice crops in India. ISPRS J Photogramm Remote Sens 54(4):254–262. https://doi.org/10.1016/S0924-2716(99)00020-9
Park S, Im J (2016) Classification of croplands through fusion of optical and sar time series data. Paper presented at the international archives of the photogrammetry, remote sensing and spatial information sciences – ISPRS archives
Pei Z, Zhang S, Guo L, Mc Nairn H, Shang J, Jiao X (2011) Rice identification and change detection using TerraSAR-X data. Can J Remote Sens 37(1):151–156. https://doi.org/10.5589/m11-025
Raney RK, Cahill JTS, Patterson GW, Bussey DBJ (2012) The m-chi decomposition of hybrid dual-polarimetric radar data with application to lunar craters. Journal of Geophysical Research: Planets 117(E12):n/a–n/a. https://doi.org/10.1029/2011JE003986
Rodriguez JJ, Kuncheva LI (2006) Rotation forest: a new classifier ensemble method. IEEE Trans Pattern Anal Mach Intell 28:1619–1630. https://doi.org/10.1109/TPAMI.2006.211
Shi W, Zheng S, Tian Y (2009) Adaptive mapped least squares SVM-based smooth fitting method for DSM generation of LIDAR data. Int J Remote Sens 30(21):5669–5683. https://doi.org/10.1080/01431160802709237
Singh G, Yamaguchi Y, Park SE, Boerner WM, Cui Y, Venkataraman G (2014) Categorization of the glaciated terrain of Indian Himalaya using CP and FP mode SAR. IEEE J Sel Top Appl Earth Obs Remote Sens 7(3):846–854. https://doi.org/10.1109/JSTARS.2013.2266354
Siyal AA, Dempewolf J, Becker-Reshef I (2015) Rice yield estimation using Landsat ETM+ data. J Appl Remote Sens 9(1). https://doi.org/10.1117/1.JRS.9.095986
Souyris JC, Imbo P, Fjørtoft R, Mingot S, Lee JS (2005) Compact polarimetry based on symmetry properties of geophysical media: the π/4 mode. IEEE Trans Geosci Remote Sens 43 (3):634–645. https://doi.org/10.1109/TGRS.2004.842486
Suykens JAK, Vandewalle J (1999) Least squares support vector machine classifiers. Neural Process Lett 9(3):293–300. https://doi.org/10.1023/A:1018628609742
Trisasongko BH (2015) Potential use of hybrid synthetic aperture radar polarimetry in earth surface monitoring. AIP Conf Proc 1677:060013. https://doi.org/10.1063/1.4930693
Trisasongko BH, Panuju DR, Paull DJ, Jia X, Griffin AL (2017) Comparing six pixel-wise classifiers for tropical rural land cover mapping using four forms of fully polarimetric SAR data. Int J Remote Sens 38(11):3274–3293. https://doi.org/10.1080/01431161.2017.1292072
Turkar V, De S, Rao YS, Shitole S, Bhattacharya A, Das A (2013) Comparative analysis of classification accuracy for RISAT-1 compact polarimetric data for various land-covers. Paper presented at the international geoscience and remote sensing symposium (IGARSS)
Vapnik VN (2000) The nature of statistical learning theory, 2nd edn. Springer Verlag, New York

White L, Landon A, Dabboor M, Pratt A, Brisco B (2014) Mapping and monitoring flooded vegetation and soil moisture using simulated compact polarimetry. 1568–1571. https://doi.org/10.1109/IGARSS.2014.6946739

White L, Millard K, Banks S, Richardson M, Pasher J, Duffe J (2017) Moving to the RADARSAT constellation mission: comparing synthesized compact polarimetry and dual polarimetry data with fully polarimetric RADARSAT-2 data for image classification of peatlands. Remote Sens 9 (6). https://doi.org/10.3390/rs9060573

Wright PA, Quegan S, Wheadon NS, Hall CD (2003) Faraday rotation effects on L-band spaceborne SAR data. IEEE Trans Geosci Remote Sens 41(12 PART I):2735–2744. https://doi.org/10.1109/TGRS.2003.815399

Wu F, Wang C, Zhang H, Zhang B, Tang Y (2011) Rice crop monitoring in South China with RADARSAT-2 quad-polarization SAR data. IEEE Geosci Remote Sens Lett 8(2):196–200. https://doi.org/10.1109/LGRS.2010.2055830

Xu B, Huang JZ, Williams G, Wang Q, Ye Y (2012) Classifying very high-dimensional data with random forests built from small subspaces. Int J Data Warehousing Min 8(2):44–63. https://doi.org/10.4018/jdwm.2012040103

Yousefi S, Khatami R, Mountrakis G, Mirzaee S, Pourghasemi HR, Tazeh M (2015) Accuracy assessment of land cover/land use classifiers in dry and humid areas of Iran. Environ Monit Assess 187(10). https://doi.org/10.1007/s10661-015-4847-1

Chapter 15
Assessment of Forest Species Diversity in Sariska Tiger Reserve, Rajasthan, India

Pavan Kumar, Haroon Sajjad, Sufia Rehman, and Purva Jain

Abstract This study makes an attempt to assess tree species diversity in Sariska Tiger Reserve (STR), Rajasthan, India, using Sentinel-2A data. We collected tree samples from ten plots in STR through random variable probability selection method. A total of 62 different species and 584 individual trees were selected from the plots using a principal coordinates of neighborhood matrices (PCNM). Four ecological indicator indices, namely, Margalef index (*SR*), Simpson's diversity (*D*) index, Shannon-Wiener index (H'), and Pielou's index (J), were utilized for measuring species diversity. Results revealed that Simpson's diversity (*D*) index was more suited for determining species diversity, while Shannon-Wiener index (H') was found to be the best index for assessing species richness. The methodology used in this study can help forest managers, environmentalist, and conservationist for formulating policies for management of forest ecosystem at various scales. This approach will be instructive in examining varied tree species and their richness with Simpson's diversity (*D*) index and Shannon-Wiener index (H').

Keywords Species diversity · Species richness · Simpson's diversity index · Shannon-Wiener index · Sariska Tiger Reserve

15.1 Introduction

Tropical forest as an important repository of terrestrial biological diversity is the most diverse ecosystem on the earth. Tropical forests act as benevolence for the life forms living in the tropics and provide natural resources and habitat conditions (Tomar et al. 2013; Thomson et al. 2010; Legendre et al. 2009). These forests play a crucial role in lessening the effects of global climate change by sequestering carbon from the atmosphere than any other terrestrial biological community (Fearnside and Laurance

P. Kumar · H. Sajjad (✉) · S. Rehman · P. Jain
Department of Geography, Jamia Millia Islamia, New Delhi, India
e-mail: pavan.jamia@gmail.com

P. Kumar et al. (eds.), *Applications and Challenges of Geospatial Technology*,
https://doi.org/10.1007/978-3-319-99882-4_15

2004). However, these forests are under degradation due to increasing anthropogenic impacts and natural adversity (Nagendra 2001; Chiarucci and Bonini 2005; Quéré et al., Le Quéré et al. 2013). These losses perturb the natural functioning of this terrestrial biome and strongly influence the carbon emission and diversification of species (Fahrig 2003; Gibbs et al. 2007). Hence, appropriate policies are required to be made for maintaining diversification in species and carbon stocks of tropical forests. Geospatial techniques can help in proper planning and management of species and carbon stock (Turner et al. 2003). Developing flexible methods from remotely sensed data allows critical mapping of the elements influencing genre diversity and their biomass (Hernández-Stefanoni et al. 2011; Gillespie et al. 2008; Balvanera et al. 2002). Forest stand can be scrutinized on the basis of diversity in position, dimension, and species (Ozdemir and Karnieli 2011; Pommerening 2002; O'Hara 2014; St-Louis et al. 2009; D'Alessandro and Fattorini 2002). These three aspects of structural diversity can be utilized for forest management planning. Bettinger and Tang (2015) codified and optimized tree species diversity amalgamation for the management of forest on structural basis (Lexerød and Eid 2006; Bettinger and Tang 2015).

The conventional method for estimating biodiversity relied on the evaluation of diversity at the local level (alpha diversity), species dissimilarities between sites (beta diversity), and the aggregation of these two measures (Foody and Cutler 2006; Whittaker 1972; Gallardo- Cruz et al. 2009). Understanding and identification of specified information such as rarity, diversity of species, composition, and threats are critical for the efficient conservation of biodiversity efforts (Pérez-García et al. 2009; Morlon et al. 2008; Palmer 1995, 2005; Laurance 1991). Examining and analyzing the spatial distribution of species have long been attracted the attention of ecologists. However, traditional methods of such assessments are cost-effective and time-consuming. Geospatial technology, on the other hand, provides timely and accurate analysis of spatial distribution of vegetation species (He et al. 2009; St-Louis et al. 2009; Rocchini et al. 2009; Oldeland et al. 2010a, b; Koellner et al. 2004). The chemical and physical attributes of each delineated species can be assigned with unique spectral signature using remotely sensed data (Palmer et al. 2008; Ferretti and Chiarucci 2003; Kalkhan et al. 2007; Skidmore et al. 2003).

Ground surveys and field inventories partially help in the evaluation of spatial distribution of plant species. Remote sensing serves as an efficacious source for obtaining such collateral data at regular interval of time (Dale and Fortin 2002; Chust et al. 2006; Rocchini et al. 2011). Traditional methods generally involve assessment of influencing the factors of biodiversity like habitat extent and terrain pattern. Advancement in resolutions of remote sensing data has enabled the researchers to delineate plant and tree canopies and identification of idiosyncratic species of trees (Oindo and Skidmore 2002; Fairbanks and McGwire 2004; Rocchini 2007; Feilhauer and Schmidtlein 2009; St-Louis et al. 2009). The parameters derived through remote sensing data can help in the assessment of distribution; pool of species and their richness at microlevel is feasible through various remotely sensed derived data parameters (Bock et al. 2005). Structural and compositional complexity of habitat can be analyzed either by species indices (NDVI) variability or by establishing spectral variability using multiple bands (Kirby and Thomas 2000; Rocchini et al. 2005; Fairbanks and McGwire 2004).

Spectral signature of species heterogenetic in nature can be determined by analyzing the remotely sensed data (Gallardo-Cruz et al. 2010; Palmer et al. 2002; Haralick et al. 1973; Wilson 2000). The influence of environment and spatial elements of biodiversity is currently being given in ecological studies (Oindo and Skidmore 2010; Mejía-Domínguez et al. 2012). The information derived through these studies can be helpful in assessing diversity based on characteristics of tree species (Jones et al. 2008). Thus, this study intends to assess diversity of tree species and their richness using ecological indicator indices. Principal coordinates of neighborhood matrices (PCNM) was used for analyzing spatial variability of species diversity at local level (alpha diversity) and species dissimilarities between sites (beta diversity).

15.2 Materials and Methods

15.2.1 Study Area

Sariska Tiger Reserve is located in the Alwar district of Rajasthan. It stretches between 27°13′ N and 27°31′ N latitudes and 76°15′ E and 76°33′ E longitudes (Fig. 15.1). Its elevation ranges between 200 and 250 meters with the maximum

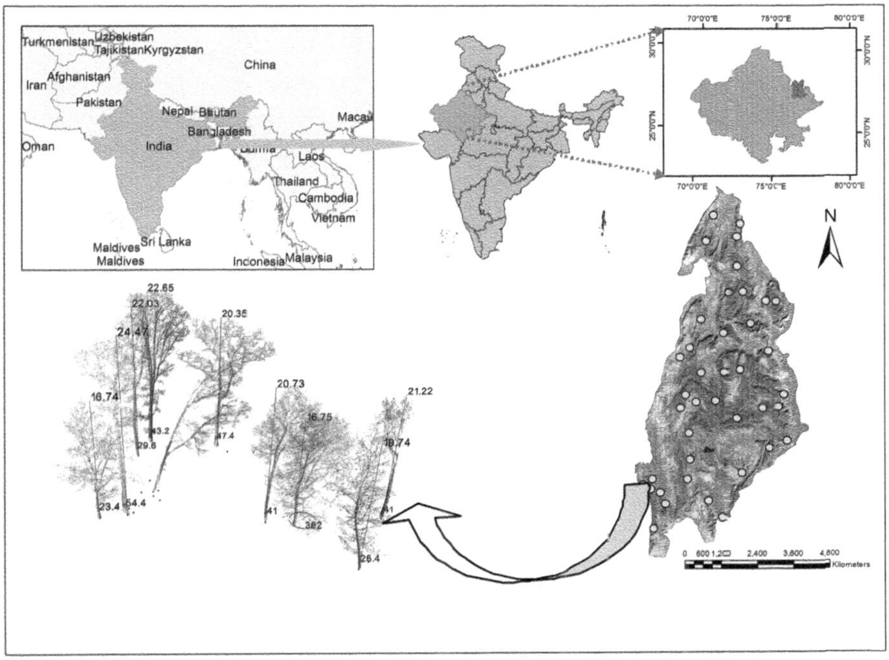

Fig. 15.1 Location map of the study area

elevation of 230 m. The average rainfall in the study area amounts to be 621 mm. The Sariska Tiger Reserve is dotted with diversified landscape including denudational valleys, rocks, grasslands, and scrub-thorn arid forest (Kumar et al. 2015; Jain and Sajjad 2016b). The vegetation of the reserve can be categorized into two major classes, i.e., tropical thorn and tropical dry deciduous forest (Jain and Sajjad 2016a). However, many types of forest can be observed on the basis of edaphic variations.

The dominant species of the reserve are *Anogeissus pendula* Edgew., *Acacia catechu*, and *Boswellia serrata* Planch. *Acacia catechu* and *Anogeissus pendula* species are found along the buffer zone of STR (Jain et al. 2016). Other species identified in the study area are *Lannea coromandelica* Houtt. and *Boswellia serrata* Planch., found along the Aravalli hills in Rajputana to Bundelkhand. These species are utilized for fuel purposes.

15.2.2 Data Used

Multispectral instrument (MSI) of Sentinel-2A satellite data with 12 spectral bands, 13 spectral channels, and 10 m spatial resolution was procured on September 2017 for developing NDVI indices of the sampled tree species. The effectiveness of Sentinel-2A data in monitoring land use at optimal ground resolution helps in monitoring land use/land cover changes due to wildfire, forest change, drought, urbanization, climate change, etc (Kumar et al. 2013b). Atmospheric and geometric corrections of satellite image were carried out using image processing software ERDAS IMAGINE (v. 2014). False color composite (FCC) was developed for interpretation of different land use/cover categories using element image interpretation (Kumar et al. 2012, 2013a). Delineated tree species on satellite image were verified with geo-tagged surface species through ground truth.

15.2.3 Determination of Sample Size and Species Sampling Distribution

The in situ data were collected from ten different transect plots (Fig. 15.2) of the STR in both reserved and non-reserved forest areas (Table 15.1). The total number of all sampled from n random variables is denoted by X_1, X_2,. . . ..X_n. The distribution for which all these sample plots were taken is determined by the function $Y{=}Y(X_1, X_2, \ldots, X_n)$. Sample size of transect plots was determined by using the equation:

$$P[Y = y] = \sum P[X_1 = x_1, X_2 = x_2, \ldots\ldots, X_\mathrm{n} = x_\mathrm{n}]$$

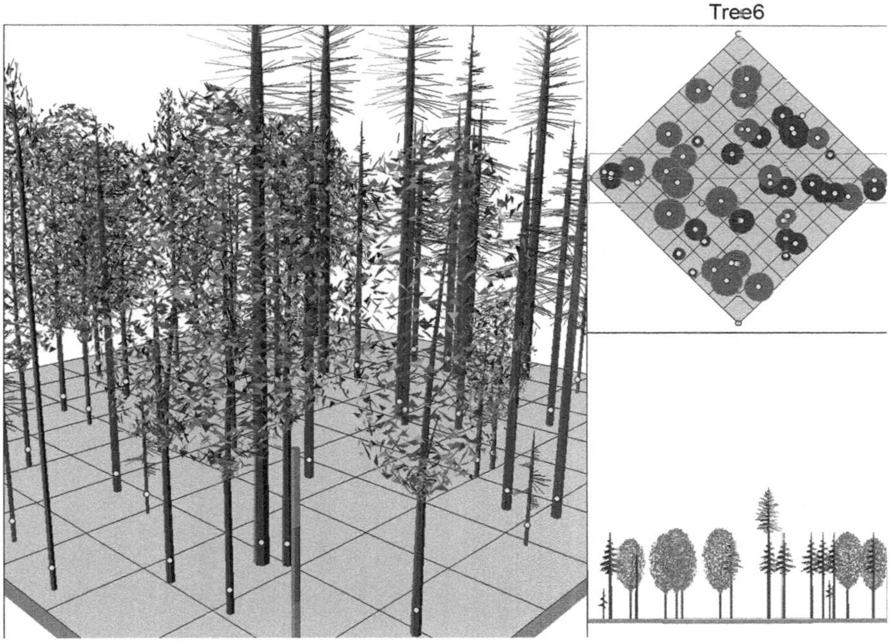

Fig. 15.2 Plot layout design of study area

15.2.4 Data Assembly and Basal Area

The basal area/genera and species/hectare were analyzed for each plot. Diversity indices of all the sampled species in STR were calculated, and their significance was determined using the analysis of variance (ANOVA). Basal area (BA) was calculated as:

$$A = \pi r^2 \text{ (i)}$$

where

BA = basal area of the tree (m^2)
r = radius (cm)

Tree species were collected from each sampled transect plot. For each species total number of individual species were determined. Natural log of species was calculated using total number of individual species. Mean number of genera/ha was calculated from total no. of species.

Table 15.1 Stand structure analysis of sampled strata

Variables	T_1	T_2	T_3	T_4	T_5	T_6	T_7	T_8	T_9	T_{10}
Total no. of species	8	11	4	2	7	9	11	5	2	3
Total no. of individual (N)	47	69	42	58	44	75	65	77	64	43
Natural log of species	2.079	2.398	1.386	0.693	1.946	2.197	2.398	1.386	0.693	1.099
Mean no. of genera ha^{-1}	3.850	4.234	3.738	4.078	3.784	4.317	1.174	4.174	4.159	3.761

15.2.5 Empirical Analysis of Species Richness Diversity and Evenness Indices

The most quantifying matrices to measure the species richness are α diversity and β diversity. Remote sensing has been widely used for estimating species richness. Species richness can be determined using very high temporal resolution data (Kark et al. 2008). Shannon-Wiener index (H/), Margalef index of species richness (SR), and Simpson's diversity (D) index were used for determining species diversity, and Pielou's index (J) was used for determining species evenness.

15.2.5.1 Shannon-Wiener Index (H')

Shannon's index (H′) was utilized to estimate species diversity (Shannon and Weiner 1949):

$$H' = \sum_{i=1}^{S} p_i \ln p_i$$

where

H'= Shannon-Wiener index
p_i= proportion of individual belonging to species i
ln= natural log

15.2.5.2 Margalef Index of Species Richness (SR)

Margalef index was used for evaluating species richness (Margalef 1958; R. E. Ulanowicz 2001):

$$SR = \frac{S - 1}{\ln (N)}$$

SR = Margalef index of species richness
S = number of species
N = total number of individuals

15.2.5.3 Simpson's Diversity (D) Index

Plant diversity increases by species richness and their regularity (Simpson 1949). Simpson's diversity index was utilized for determining species diversity:

$$D = 1 - \frac{\sum n(n-1)}{N(N-1)}$$

n= total number of organisms of a particular species
N=total number of organisms of all species

15.2.5.4 Pielou's Index (J)

Species evenness was determined following Pielou's index (Bray and Curtis 1957):

$$J = \frac{H^{/}}{\ln(S)}$$

J= Pielou's measure of species evenness
$H^{/}$=Shannon-Wiener index
S= total number of species/sample

15.3 Results and Discussion

15.3.1 Land Use/Land Cover and Forest Stand Structure Analysis

The LULC classification was prepared using spectral analysis of remote sensing data and ground truthing. Nine main classes of land use/cover were identified (Fig. 15.3). Of the total forest area (51%), open forest occupies the largest area. Barren land has been encroached in settlement and agriculture. Area under different LULC classes revealed that the largest area was found under open forest (39.4%) followed by fallow land (24.3%), cropland (16.3%), barren land (9.0%), degraded forest (8.2%), dense forest (1.1%), water body (0.7%), settlement (0.3%), and forest blank (0.1%).

15.3.2 Forest Type Classification

Six major types of forest species were identified in Sariska Tiger Reserve, namely, *Anogeissus pendula* Edgew., *Boswellia serrata* Planch., *Acacia-Ziziphus*, *Butea monosperma*, mixed forest, and riverine forest (Fig. 15.4). The forest of the study

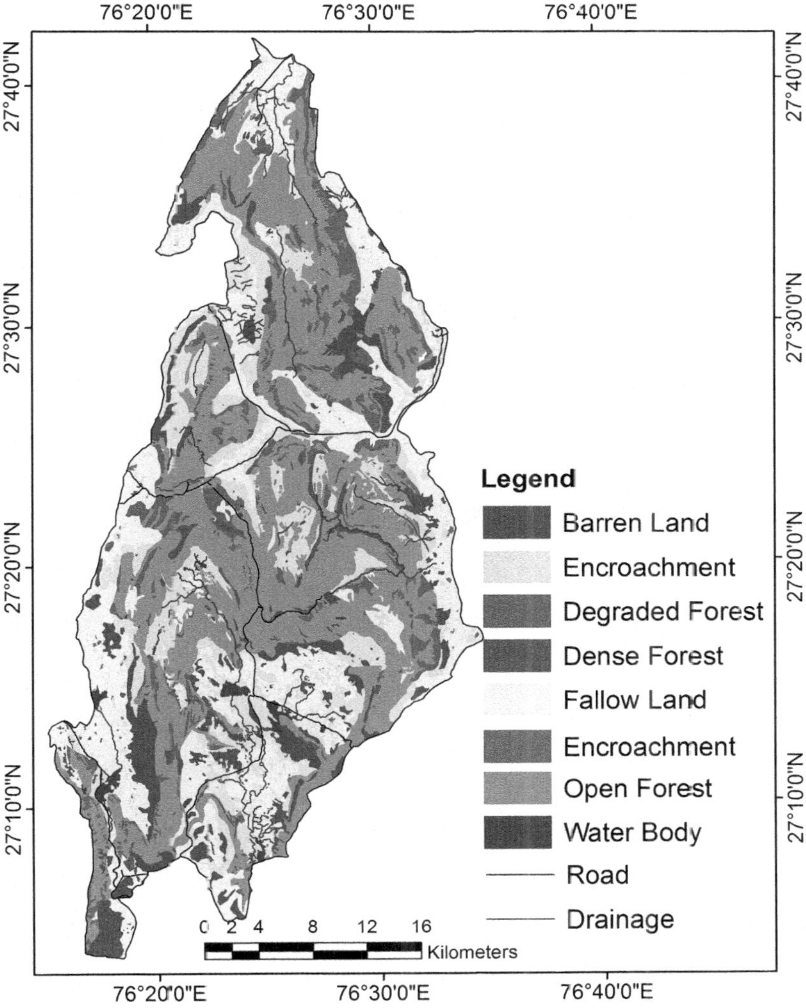

Fig. 15.3 LULC classes

area was divided into four density classes, viz., >70%, 70–40%, 40–10%, and < 10%. Of the total forest area of the reserve (48.78%), about 22.75% area was under the forest having 40–70% density followed by 16.81% area having 10–40%, 7.83% area having <10%, and 1.13% area having >70% density.

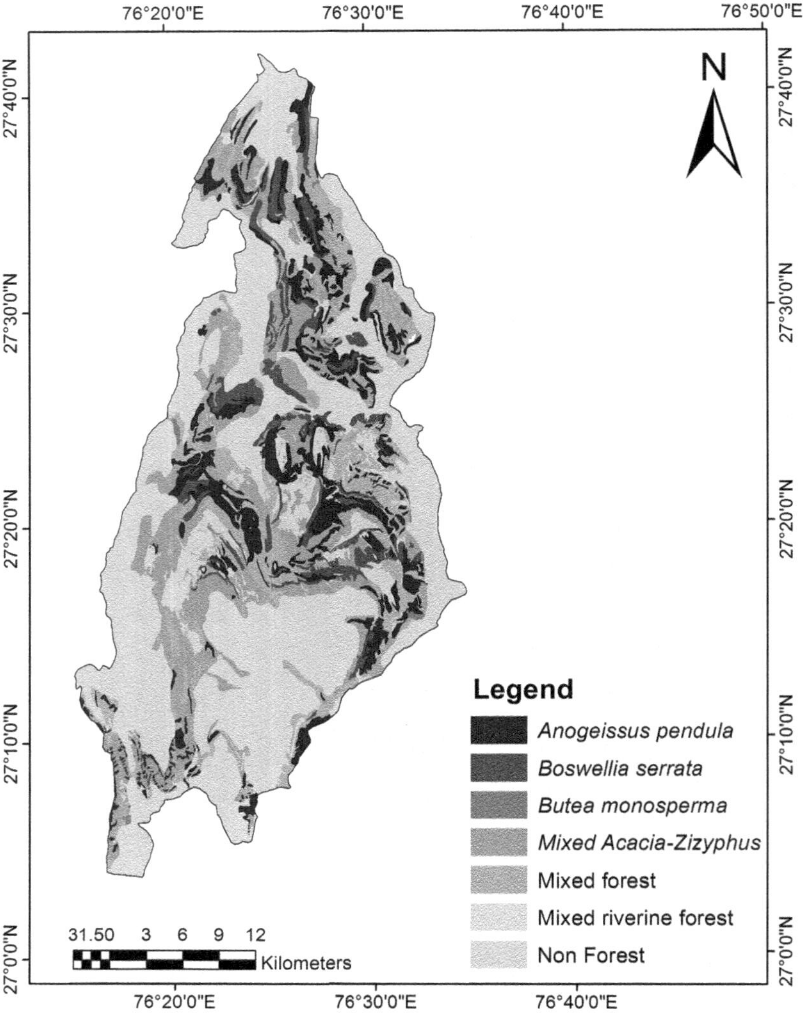

Fig. 15.4 Forest cover type

15.4 Species Diversity

A total of 62 species and 584 individual trees of ≥10 cm DBH belonging to 43 genera and 23 families were distinguished in 10 transect plots (Table 15.2). The Shannon-Wiener index (H′) was used to compare species diversity between transects. The index (H′) for T_1–T_{10} was determined as 1.915, 2.146, 1.04, 0.086, 1.643, 1.819, 2.301, 1.304, 0.139, and 1.017, respectively. Species diversity was found complex in transect T7, while T4 was simple in composition of species. The

Table 15.2 Tree species diversity in Sariska Tiger Reserve

Variables	T_1	T_2	T_3	T_4	T_5	T_6	T_7	T_8	T_9	T_{10}
Margalef index of species richness (*SR*)	1.818	2.362	0.803	0.245	1.586	1.853	2.396	0.719	0.240	0.532
Simpson's diversity (*D*) index	6.052	7.203	2.443	1.034	4.420	5.00	9.165	3.460	1.064	2.579
Shannon-Wiener index (*H'*)	1.915	2.146	1.040	0.086	1.643	1.819	2.301	1.304	0.139	1.017
Pielou's index (J)	0.921	0.895	0.750	0.124	0.844	0.828	0.960	0.941	0.201	0926

Shannon-Wiener index (H′) (ranged between 0.086 and 2.301) in our study was found to be lower than other tropical forests. However, species diversity cannot be compared in view of differences in geographical location and different structural characteristics of species (selection of plot size and diameter). Closeness of the species to environment evenness is measured by Pielou's index (J). The maximum evenness was measured by plotting the individual species with each transect (Bray and Curtis 1957, Kumar et al. 2014). The highest and lowest values of Pielou's index were observed at T10 and T4 transects, respectively.

The biological diversity through the Simpson's indices is a quantified value to measure the species richness and evenness (Hernandez-Stefanoni and Ponce-Hernandez 2004). Simpson's diversity (*D*) index was utilized to measure the species richness in terms of population size of each of the species. Species diversity could not be assessed effectively by Simpson's diversity index as T_4 and T_9 transects had similar values. Simpson's diversity (*D*) index was found to be higher compared to other indices as the density of the indigenous species was more pronounced than the other species indices. The more is the species index value of the Simpson's diversity (*D*) index, the more is the value of species individual population present. Simpson's diversity (*D*) index was found to be the best diversity index for determining species diversity. Similar patterns were observed for species richness (SR). Habitat heterogeneity within the transects was assessed using Margalef index of species richness (SR). Maximum heterogeneity (2.396) was found in transect T7 and lowest (0.245) in T4. Further analysis revealed more dissimilarities in species between sites.

15.5 Determination of Diversity Through Clustering

The species clustering of transect plot is associated with two methods. One is the agglomerative method, where the species have their own separate cluster. The optimum cluster number of any chosen distribution is taken into account in this clustering method. The second one is the divisive method where probable distribution of all the individual species is clustered and kept in one sample. The agglomerative method is more often usable than divisive methods, because of the usefulness of Euclidean distance.

In general, if the p variables are measured as $X_1, X_2, \ldots, X_p$ on a species of n plots, the detected data for transect can be denoted by $x_{i1}, x_{i2}, \ldots, x_{jp}$ and the non-detected data for other transects by $x_{j1}, x_{j2}, \ldots, x_{jp}$.

The Euclidean distance between these two plots is calculated as:

$$d_{ij} = \sqrt{\left(x_{i1} - x_{j1}\right)^2 + \left(x_{i2} - x_{j12}\right)^2 + \cdots\cdots\cdots + \left(x_{1p} - x_{jp}\right)^2}$$

The maximum, minimum, and mean values of clustering were determined for species richness of ten transect plots. The highest clustering species value richness was assigned by plotting the graphs in the form of the total plot with species richness.

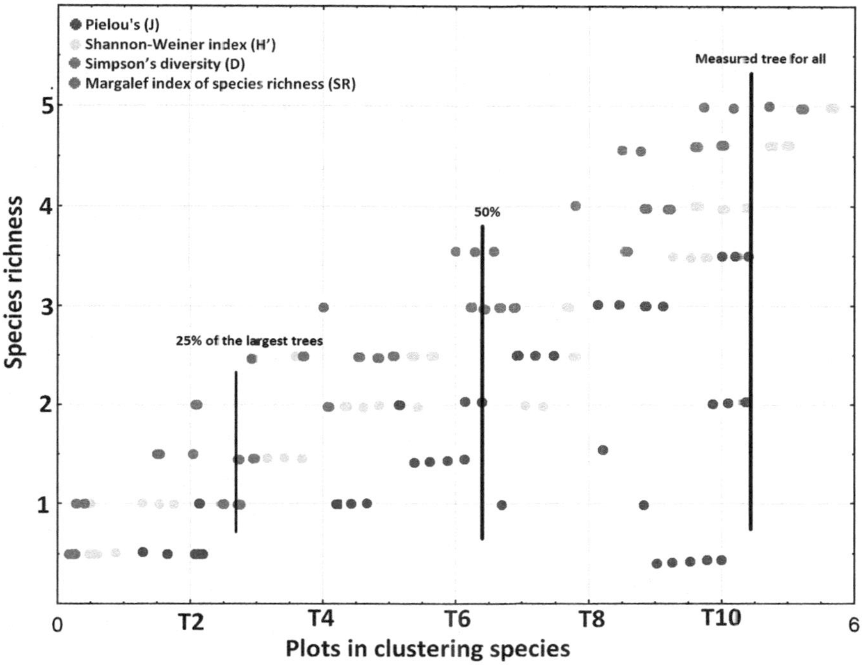

Fig. 15.5 Species richness for clustering with k values

Clustering pattern was analyzed through plotting of species richness and diversity indices (Fig. 15.5). The calculated Shannon-Wiener index ($H^{/}$) from the clustering showed significant association between the values derived from field species. Margalef index, Simpson's diversity (D) index, and Pielou's index (J) were found to have lower values of species richness. The increase of clusters (k) does not seem to have affected the result. The clustering of Shannon-Wiener index increased with k value.

The clustering value was found to be closer to the species richness value obtained through filed data. Thus, it is clear from analysis revealed that Simpson's diversity (D) index was found to be the best method for determining diversity of species, while Shannon-Wiener index ($H^{/}$) was found to be the best method for species richness.

15.6 Conclusion

This paper examined tree species diversity and their richness in STR, India. Land use/land cover and trees species types were mapped using Sentinel-2A data. Six major types of forest species were identified in Sariska Tiger Reserve, namely,

Anogeissus pendula Edgew., *Boswellia serrata* Planch., *Acacia-Ziziphus*, *Butea monosperma*, mixed forest, and riverine forest. Ten transect plots of 0.1 ha were delineated using Chacko's formula, and from all the selected plots, 62 species and 584 individuals were selected. Height, volume, and diameter at breast height (DBH) and circumference at breast height (CBH) were measured for each individual species within 0.1 ha plot. We evaluated the effectiveness of Shannon-Wiener index ($H^{\prime}$), Margalef index, Simpson's diversity (D) index, and Pielou's index (J) for assessing trees species diversity. Tree species richness was determined by agglomerative and divisive cluster methods. The findings revealed that Simpson's diversity (*D*) index was more suited for determining species diversity, while Shannon-Wiener index ($H^{\prime}$) was found to be the best index for assessing species richness. The methodology adopted in this study will be instructive for future researches in various forest ecosystems at spatial scales.

References

Balvanera P, Lott E, Segura G, Siebe C, Islas A (2002) Patterns of β-diversity in a Mexican tropical dry forest. J Veg Sci 13:145–158

Bettinger P, Tang M (2015) Tree-level harvest optimization for structure-based forest management based on the species mingling index. Forests 6:1121–1144

Bock M, Rossner G, Wissen M, Remm K, Langanke T, Lang S, Klug H, Blaschke T, Vrscaj B (2005) Spatial indicators for nature conservation from European to local scale. Ecol Indic 5:322–338

Bray JR, Curtis JT (1957) An ordination of the upland forest communities of southern Wisconsin. Ecol Monogr 27(4):325–349

Chiarucci A, Bonini I (2005) Quantitative floristics as a tool for the assessment of plant diversity in Tuscan forests. For Ecol Manag 212:160–170

Chust G, Chave J, Condit R, Aguilar S, Lao S, Pérez R (2006) Determinants and spatial modeling of tree beta-diversity in a tropical forest landscape in Panama. J Veg Sci 17:83–92

Dale MRT, Fortin MJ (2002) Spatial autocorrelation and statistical tests in ecology. Ecoscience 9:162–167

D'Alessandro L, Fattorini L (2002) Resampling estimators of species richness from presence–absence data: why they don't work. Metro 61:5–19

Fahrig L (2003) Effects of habitat fragmentation on biodiversity. Annu Rev Ecol Evol Syst 34:487–515

Fairbanks HK, McGwire KC (2004) Patterns of floristic richness in vegetation communities of California: regional scale analysis with multi-temporal NDVI. Glob Ecol Biogeogr 13:221–235

Ferretti M, Chiarucci A (2003) Design concepts adopted in long-term forest monitoring programs in Europe-problems for the future? Sci Total Environ 310:171–178

Fearnside PM, Laurance WF (2004) Tropical deforestation and greenhouse gas emissions. Ecol Appl 14:982–986

Feilhauer H, Schmidtlein S (2009) Mapping continuous fields of forest alpha and beta diversity. Appl Veg Sci 12:429–439

Foody GM, Cutler MEJ (2006) Mapping the species richness and composition of tropical forests from remotely sensed data with neural networks. Ecol Model 195:37–42

Gallardo-Cruz JA, Pérez-García EA, Meave JA (2009) β-Diversity and vegetation structure as influenced by slope aspect and altitude in a seasonally dry tropical landscape. Landsc Ecol 24:473–482

Gallardo-Cruz JA, Meave JA, Pérez-García EA, Hernández-Stefanoni JL (2010) Spatial structure of plant communities in a complex tropical landscape: implications for alpha and beta diversity. Commun Ecol 11:202–210

Gibbs HK, Brown S, Niles JO, Foley JA (2007) Monitoring and estimating tropical forest carbon stocks: making REDD a reality. Environ Res Lett 2(4):045023

Gillespie TW, Foody GM, Rocchini D, Giorgi AP, Saatchi S (2008) Measuring and modelling biodiversity from space. Prog Phys Geogr 32:203–221

Haralick RM, Shanmugam K, Dinstein I (1973) Textural features for image classification. IEEE Trans Syst Man Cybern SMC-3:610–621

He KS, Zhang J, Zhang Q (2009) Linking variability in species composition and MODIS NDVI based on beta diversity measurements. Acta Oecol 35:14–21

Hernandez-Stefanoni JL, Ponce-Hernandez R (2004) Mapping the spatial distribution of plant diversity indices in a tropical forest using multi-spectral satellite image classification and field measurements. Biodivers Conserv 13(14):2599–2621

Hernández-Stefanoni JL, Gallardo-Cruz JA, Meave JA, Dupuy JM (2011) Combining geostatistical models and remotely sensed data to improve tropical plant richness mapping. Ecol Indic 11:1046–1056

Jain P, Ahmed R, Sajjad H (2016) Assessing and monitoring forest health using a forest fragmentation approach in Sariska Tiger Reserve, India. J Geogr 70(5):306–315

Jain P, Sajjad H (2016a) Analysis of willingness for relocation of the local communities living in the critical Tiger habitat of the Sariska Tiger Reserve, India. Local Environ 21(11):1409–1419

Jain P, Sajjad H (2016b) Household dependency on forest resources in the Sariska Tiger Reserve (STR), India: implications for management. J Sustain For 35(1):60–74

Jones MM, Tuomisto H, Borcard D, Legendre P, Clark DB, Olivas PC (2008) Explaining variation in tropical plant community composition: influence of environmental and spatial data quality. Oecologia 155:593–604

Kalkhan MA, Stafford EJ, Stohlgren TJ (2007) Rapid plant diversity assessment using a pixel nested plot design: a case study in beaver meadows, Rocky Mountain National Park, Colorado, USA. Divers Distrib 13(4):379–388

Kark S, Levin N, Phinn S (2008) Global environmental priorities: making sense of remote sensing: reply to TREE letter: satellites miss environmental priorities by Loarie et al. (2007). Trends Ecol Evol 23:181–182

Kirby KJ, Thomas RC (2000) Changes in the ground flora in Wytham woods, southern England from 1974 to 1991: implications for nature conservation. J Veg Sci 11:871–880

Koellner T, Hersperger AM, Wohlgemuth T (2004) Rarefaction method for assessing plant species diversity on a regional scale. Ecography 27:532–544

Kumar P, Kumar D, Mandal VP, Pandey PC, Rani M, Tomar V (2012) Settlement risk zone recognition using high resolution satellite data in Jharia coal field, Dhanbad, India. Life Sci J 9 (1s):1–6

Kumar P, Sharma LK, Pandey PC, Sinha S, Nathawat MS (2013a) Geospatial strategy for tropical forest-wildlife reserve biomass estimation. 6(2):917–923

Kumar P, Singh BK, Rani M (2013b) An efficient hybrid classification approach for land use/land cover analysis in a semi-desert area using ETM+ and LISS-III sensor. IEEE Sensors J 13 (6):2161–2165

Kumar P, Pandey PC, Kumar V, Singh BK, Tomar V, Rani M (2014) Efficient recognition of forest species biodiversity by inventory based geospatial approach using LISS IV. IEEE Sensors J 13 (6):2161–2165

Kumar P, Pandey PC, Kumar V, Singh BK, Tomar V, Rani M (2015) Efficient recognition of forest species biodiversity by inventory-based geospatial approach using LISS IV sensor. IEEE Sensors J 15(3):1884

Laurance WF (1991) Edge effects in tropical forest fragments: application of a model for the design of nature reserves. Biol Conserv 57:205–219

Le Quéré C, Peters GP, Andres RJ, Andrew RM, Boden T, Ciais P, Friedlingstein P, Houghton RA, Marland G, Moriarty R et al (2013) Global carbon budget 2013. Earth Syst Sci Data Discuss 6:689–760
Legendre P, Mi X, Ren H, Ma K, Yu M, Sun I-F, He F (2009) Partitioning beta diversity in a subtropical broad-leaved forest of China. Ecology 90:663–674
Lexerød NL, Eid T (2006) An evaluation of different diameter diversity indices based on criteria related to forest management planning. For Ecol Manag 222:17–28
Margalef FR (1958) Information theory in ecology. Int J Gen Syst 3:36–71
Mejía-Domínguez NR, Meave JA, Díaz-Ávalos C (2012) Spatial structure of the abiotic environment and its association with sapling community structure and dynamics in a cloud forest. Int J Biometeorol 56(2):305–318
Morlon H, Chuyong G, Condit R, Hubbell S, Kenfack D, Thomas D, Valencia R, Green JL (2008) A general framework for the distance-decay of similarity in ecological communities. Ecol Lett 11:904–917
Nagendra H (2001) Using remote sensing to assess biodiversity. Int J Remote Sens 22:2377–2400
O'Hara K (2014) Multiaged silviculture: managing for complex forest stand structures. Oxford University Press, Oxford, p 17
Oindo BO, Skidmore AK (2002) Interannual variability of NDVI and species richness in Kenya. Int J Remote Sens 23:285–298
Oindo BO, Skidmore AK (2010) Interannual variability of NDVI and species richness in Kenya, pp 285–298
Oldeland J, Wesuls D, Rocchini D, Schmidt M, Jürgens N (2010a) Does using species abundance data improve estimates of species diversity from remotely sensed spectral heterogeneity? Ecol Indic 10:390–396
Oldeland J, Wesuls D, Rocchini D, Schmidt M, Jürgens N (2010b) Does using species abundance data improve estimates of species diversity from remotely sensed spectral heterogeneity? Ecol Indic 10:390–396
Ozdemir I, Karnieli A (2011) Predicting forest structural parameters using the image texture derived from worldview-2 multispectral imagery in a dryland forest, Israel. Int J Appl Earth Obs Geoinf 13:701–710
Palmer MW (1995) How should one count species? Nat Areas J 15:124–135
Palmer MW (2005) Distance decay in an old-growth neotropical forest. J Veg Sci 16:161–166
Palmer MW, Earls P, Hoagland BW, White PS, Wohlgemuth T (2002) Quantitative tools for perfecting species lists. Environmetrics 13:121–137
Palmer MW, McGlinn DJ, Birrer S (2008) Artifacts and artifictions in biodiversity research. Folia Geobot 43:245–257
Pérez-García EA, Sevilha AC, Meave JA, Scariot A (2009) Floristic differentiation in limestone outcrops of southern Mexico and Central Brazil: a beta diversity approach. Bol Soc Bot Méx 84:45–58
Pommerening A (2002) Approaches to quantifying forest structures. Forestry 75:305–324
Rocchini D (2007) Effects of spatial and spectral resolution in estimating ecosystem α-diversity by satellite imagery Remote Sens Environ 111:423–434
Rocchini D, Andreini Butini S, Chiarucci A (2005) Maximizing plant species inventory efficiency by means of remotely sensed spectral distances. Glob Ecol Biogeogr 14:431–437
Rocchini D, Nagendra H, Ghate R, Cade BS (2009) Spectral distance decay: assessing species beta-diversity by quantile regression. Photogramm Eng Remote Sens 75:1225–1230
Rocchini D, Hortal J, Lengyel S, Lobo JM, Jiménez-Valverde A, Ricotta C, Bacaro G, Chiarucci A (2011) Accounting for uncertainty when mapping species distributions: the need for maps of ignorance. Prog Phys Geogr 35:211–226
Shannon CE, Weaver W (1949) The mathematical theory of communication. The University of Illinois Press, Urbana, pp 1–117
Simpson EH (1949) Measurement of diversity. Nature 163(4148):688–688

Skidmore AK, Oindo BO, Said MY (2003) Biodiversity assessment by remote sensing. In: Proceedings of the 30th international symposium on remote sensing of the environment: information for risk management and sustainable development, November 10–14, 2003, Honolulu, Hawaii, 4p

St-Louis V, Pidgeon AM, Clayton MK, Locke BA, Bash D, Radeloff VC (2009) Satellite image texture and a vegetation index predict avian biodiversity in the Chihuahuan Desert of New Mexico. Ecography 32:468–480

Thomson A, Calvin K, Chini L, Hurtt G, Edmonds J, Ben Bond-Lamberty B, Frolking S, Wise MA, Janetos AC (2010) Climate mitigation and the future of tropical landscapes. Proc Natl Acad Sci U S A 107:19633–19638

Tomar V, Kumar P, Rani M, Gupta G, Singh J (2013) A satellite-based biodiversity dynamics capability in tropical forest. Electron J Geotech Eng 18(Bund. F):1171–1180

Turner W, Spector S, Gardiner N, Fladeland M, Sterling E, Steiniger M (2003) Remote sensing for biodiversity science and conservation. Trends Ecol Evol 18:306–314

Ulanowicz RE (2001) Information theory in ecology. Comput Chem 25(4):393–399

Whittaker RH (1972) Evolution and measurement of species diversity. Taxon 21:213–251

Wilson SD (2000) Heterogeneity, diversity and scale in plant communities. In: Hutchings MJ, John EA, Stewart AJA (eds) The ecological consequences of environmental heterogeneity. Blackwell Science, Oxford, pp 53–69

Chapter 16
Future Challenges and Perspective of Remote Sensing Technology

Haroon Sajjad and Pavan Kumar

Abstract During the last few decades, the world has experienced a remarkable and rapid advancement in the field of remote sensing, acquisition of geospatial information and mapping. The technology has gained momentum in its utilization and application in different fields. With increased demand of geospatial data, various space-borne, airborne and ground-based remote sensing technologies have been instigated and put into venture. Consequently, various research institutes, government agencies as well as private sectors are excessively utilizing this technology for getting timely data. This has resulted into easy accessibility and utilization of remote sensing data for finding solutions to more complicated problems. However, it experienced many challenges. Handling large volume of data and complex data formats with complex processing is a major issue to be tackled. Hence, real-time data processing and open GIS conversing data formats will go a long way for better utilization of spatial data. The technology has wide potential in analysing land, vegetation and water resources. Landsat, SPOT, Sentinel, IRS, IKONOS, QuickBird, etc. high-resolution data are in great use for making inventory and analysing various aspects of human interaction with land, water and vegetation. In this backdrop, the book presents an overview of the potential challenges and the perspective of the fast-growing geospatial technology.

Keywords Remote sensing · Potential · Application · Challenges

Earth-observing satellites are increasing in space and remote sensing data are undergoing an explosive growth. These satellites will produce large volume of remote sensing data and will give rise to complexity and diversity in high-dimensional characteristics of data. However, it will be extremely difficult for various end users and advance user communities to efficiently use such traditional remote sensing data sets. At the same time, data processing and distribution will be a new challenge. It will also become difficult for user community to maintain the increasing volume of data

H. Sajjad · P. Kumar (✉)
Department of Geography, Jamia Millia Islamia, New Delhi, India

P. Kumar et al. (eds.), *Applications and Challenges of Geospatial Technology*,
https://doi.org/10.1007/978-3-319-99882-4_16

and to build and launch hardware into space. Large volume of data, complex data formats, different map projections, applications of GIS, communication capability and processing time are crucial issues for the wide use of remote sensing data. Therefore, real-time remote sensing data in different formats will help users in analysing future earth-observing satellite systems. Preprocessing, data format conversing, open GIS applications and real-time data processing will help in better utilization of future earth-observing satellite systems.

Remote sensing technology has a wide range of scope from agriculture to urban and geosciences to disaster management. The study of plant characteristics in the agricultural field through the crop reflectance at various stages of growth is assessed to estimate the crop acreage and yield forecasting. It also studies about the identification of crops and landholding size through high-resolution data. Conventional visual interpretation method of land use and land cover classification has given way to high-tech machine learning tools. Remote sensing technology is widely used for planning and management of forest resources. The technology has proved useful in assessing forest biomass, deforestation, forest fragmentation, spatial distribution of tree species, classification of forest, etc. Landsat, SPOT, Sentinel and IRS data are extensively used in forest resource assessment and management. Very high spatial resolution data are used for classification and identification of built-up area in urban studies. Satellite sensors like IKONOS and QuickBird with high spatial resolution (<1 m) are used for cadastral mapping and urban planning. However, the growth of urban built-up, urban agglomeration and urban sprawl can be assessed using moderate resolution data acquired through Landsat, Sentinel and IRS sensors.

Hyperspectral images are widely used by geologists for identification of rocks and minerals. However, multispectral images are also used for detection and identification of geological structures and rock formation with the help of image interpretation keys. Geostationary satellite data is preferred for studying extreme weather events and cyclone due to its constant monitoring. MODIS and Landsat 8 sensors having coastal and aerosol spectral band are used for aerosol optical depth. Remote sensing data along with geographical information system (GIS) has also proved significant tools for disaster management. Digital elevation model (DEM) is extensively used to analyse the slope instability and occurrence of landslide. Remote sensing data are also used for mapping the past and current occurrence of landslide and analysing its trend and patterns. Weather and climate variability analysis, climate change analysis, natural hazards mapping and vulnerability assessment are being efficiently carried out with the help of remote sensing and GIS techniques. Earthquake mapping and flood mapping are also carried out by remote sensing data. Multispectral, hyperspectral and DEM are utilized for micro-zonation mapping of earthquake-prone areas, while microwave data along with DEM are extensively used for assessment of flood inundation.

Remote sensing technology is experiencing shift from optical and multispectral to hyperspectral remote sensing mainly due to limited spectral bands. Hyperspectral remote sensing or imaging spectroscopy has invited the attention of researchers since the last decade or so. First sensor of imaging spectrometer (Hyperion) was launched into orbit in 2000. Subsequently, access to such data became easier due to availability of various imaging spectrometer sensors, namely, HyMap, DAIS and AVIRIS.

These sensors record spectral signature with hundreds or thousands of elements from a pixel in a targeted area. Hyperspectral remote sensing improves the capability of object identification in extremely harsh conditions due to the availability of large number of spectral bands. However, unmanned aerial vehicles (UAVs) are used nowadays for monitoring and collecting of data for small areas. Flexibility to sensor selection in UAVs has made it easy to collect data as per requirement.

The main challenges for remote sensing technology are preprocessing and validation. Various remote sensing data products are available for site-specific researches, but their resolutions varied from each other. For example, the assessment of forest fire needs high spatial and temporal resolution data. This is not possible for one sensor to provide both high resolutions, for example, if MODIS data is utilized for assessing forest fire, it will provide high temporal resolution of 1 day but will yield poor spatial resolution. Similarly, real-time flood assessment needs high temporal and spectral resolution data for bypassing cloud cover and detecting water spread. Hyperspectral remote sensing data with high spectral resolution will hold promising key to the solution to the future challenges of remote sensing technology. This will reduce the time of image processing for various spectral bands. However, hyperspectral remote sensing data will also need to be specifically enhanced for spatial resolution for providing better results. Hyperspectral image needs large storage and heavy processing configuration which will be a hurdle for digital image processing. Finally, remote sensing model validation is a complex task for researchers. There are large numbers of models based on remote sensing data applied by the researchers in various fields. All mathematical models don't have compatibility with all sensors, and these also don't work in all geographical locations. Therefore, validation of model is essential with in situ data. The on-ground data can also be utilized for model improvement through its integration with remote sensing data. This will improve the accuracy of results carried out by remote sensing data.

Printed by Printforce, the Netherlands